Advanced Flight Dynamics with Elements of Flight Control

Advanced Flight Dynamics with Elements of Flight Control

Nandan K. Sinha and N. Ananthkrishnan

CRC Press
Taylor & Francis Group
Boca Raton London New York

CRC Press is an imprint of the
Taylor & Francis Group, an **informa** business

CRC Press
Taylor & Francis Group
6000 Broken Sound Parkway NW, Suite 300
Boca Raton, FL 33487-2742

© 2017 by Taylor & Francis Group, LLC
CRC Press is an imprint of Taylor & Francis Group, an Informa business

No claim to original U.S. Government works

Printed on acid-free paper

International Standard Book Number-13: 978-1-4987-4604-5 (Hardback)

Library of Congress Cataloging-in-Publication Data

Names: Sinha, Nandan K., author. | Ananthkrishnan, N., author.
Title: Advanced flight dynamics with elements of flight control / Nandan K. Sinha and N. Ananthkrishnan.
Description: Boca Raton : Taylor & Francis, CRC Press, 2017. | Includes bibliographical references and index.
Identifiers: LCCN 2016056393| ISBN 9781498746045 (hardback : alk. paper) | ISBN 9781138746039 (pbk. : alk. paper) | ISBN 9781498746069 (ebook)
Subjects: LCSH: Aerodynamics. | Flight control. | Bifurcation theory. | Continuation methods.
Classification: LCC TL570 .S529 2017 | DDC 629.132/3--dc23
LC record available at https://lccn.loc.gov/2016056393

Visit the Taylor & Francis Web site at
http://www.taylorandfrancis.com

and the CRC Press Web site at
http://www.crcpress.com

Contents

Preface

Most aerospace curricula around the world have distinct levels of courses dealing with the subject of aircraft performance, stability, and control. The very first course usually deals with the performance of the aircraft in relation to its design parameters using appropriate equations of equilibrium under different flight conditions. In a second course on aircraft flight dynamics, the emphasis is on studying the effect of various design parameters on the stability of the aircraft; an important exercise here being the derivation of the equations of six degrees of freedom (6 DOF) motion of a rigid airplane, followed by the modal analysis of aircraft behavior, usually in low-angle-of-attack cruise flight using a linearized model and aero-control derivatives. To some extent, this course may also introduce the topic of flight control law design using linearized (state-space, transfer function) models and linear control theory. An advanced-level course on flight dynamics could deal with simulation aspects gradually developing into using advanced control theory to flight control design problems. Several excellent books covering aspects of flight modeling, simulation, and control based on a systems and control perspective are presently available. However, most books, but for some cursory material on nonlinear phenomena, tend to focus exclusively on linearized flight dynamics using linear systems and controls analysis tools. Another question, a very pertinent one, most often asked by students of flight dynamics is how good the simulations using the linear models are when compared to real-life flight maneuvers.

Exclusively focusing on modeling, simulation, and control aspects of aircraft flight dynamics, this book is an attempt to introduce readers to a systematic approach to "what to do with aircraft models?"—typically, coupled 6 DOF dynamic models with fairly extensive nonlinear aerodynamic models without making any simplification or sacrifice of rigor, which is usually the cause of most confusion in dealing with the subject, this book directly delves into the analysis and simulation of the full-order equations of aircraft motion. Three aircraft models available in the public domain, two of them being models for high- and low-angles-of-attack dynamics of F-18 and another one for studying the roll coupling behavior of airplanes, are used as illustrative examples. Beginning from setting up simulations of simple flight conditions, viz. level flight, landing, take-off, vertical loop in longitudinal plane, and horizontal turn, simulations are extended further to investigate the nonlinear behavior of the aircraft resulting from the onset of instability. Bifurcation analysis and continuation theory-based methodologies for investigation and prediction of aircraft nonlinear flight dynamics and nonlinear control techniques for devising the recovery strategy of airplanes from loss-of-control scenario are presented. The performance of aircraft models has been treated as an integral aspect in this presentation and computed simultaneously with trim and stability under different constrained flight conditions. The analysis of constrained maneuvers in this manner also reveals various control interconnect laws (schedules) required to fly the aircraft in specific state-constrained flights. The investigation of dynamics throws up avenues for problem-specific intuitive control law development, which fits well in this unified framework. Direct control strategies to avoid aircraft departures resulting from the onset of instability and control law prototyping based on nonlinear control techniques for different flight control problems adequately supplement the integrated treatment of the subject. These features are intended to add pedagogical value and distinguish the text from other works on this topic.

The expected reader group for this book would ideally be senior undergraduate and graduate students, practicing aerospace/flight simulation engineers/scientists from industry as well as researchers in various organizations. It is expected that readers would have had exposure in some form to basic aerodynamics, numerical simulation tools, and knowledge of calculus, linear algebra, etc. The computing environments used are MATLAB® for time simulations and the numerical continuation algorithms AUTO and MATCONT available in the public domain for steady-state analysis and control schedules. The chapters are distributed fairly uniformly, touching on different aspects of the subject.

Chapter 1 covers the modeling aspects of rigid-body aircraft 6 DOF motion. Aircraft motion and control variables are introduced and the equations of motions are derived in detail. For specific problems of interest, simpler sets of equations are extracted from the full 6 DOF equations, and equations with wind as external input are also presented.

Chapter 2 introduces the aerodynamic and thrust model with appropriate description in relation to aircraft design parameters. Three aircraft models with various nonlinearities available in the public domain are used to set up the equations for simulation.

Chapter 3 provides adequate introduction to dynamical systems theory and bifurcation and continuation-based tools for the analysis of nonlinear aircraft models.

Chapter 4 is devoted to the application of the theoretical and computational approach developed in Chapter 3 to study the performance and stability aspects of the aircraft in longitudinal flight.

Chapter 5 provides some background and a framework for a typical flight control system.

Carrying forward from Chapter 4, in Chapter 6, the analysis of coupled dynamics using the full 6 DOF aircraft model is carried out in continuation framework.

Chapter 7 covers the analysis of coupled dynamics of the aircraft in detail. The use of bifurcation methods applied to different aircraft models to capture various nonlinear phenomena both at low as well as at high angles of attack is shown. Aircraft recovery from loss-of-control scenario using bifurcation analysis results and nonlinear control algorithms, and aircraft maneuver design for real-time flight simulation of complex maneuvers using sliding mode control techniques constitute the latter part of the chapter.

Chapter 8 presents a case study showing the application of the nonlinear dynamic inversion technique introduced previously for control prototyping of a divert attitude thruster system.

Words cannot describe enough our gratitude toward the people who have directly or indirectly contributed to the successful completion of this book. We thank Amit, Anuj, Anshul, Rohith, and Ramesh, a fantastic set of students, for helping generate many of the results. Thanks are due to Gagandeep Singh, the commissioning editor, Mouli Sharma from the editorial team, and several others from CRC Press for keeping us on schedule. Special thanks are also due to the reviewers whose valuable inputs to our proposal helped improve the contents and presentation. We have benefited immensely from our interactions with experts in the field, both in academia and industry—Fred Culick, Mikhail Goman, T.G. Pai, M.S. Bhat, V. Ramanujachari, and K. Sudhakar, to name but a few. Last, but not the least, the role of our families as our constant source of energy in all our endeavors is humbly acknowledged, particularly our children who have been excited throughout the preparation of this manuscript—Anand, Purna, Aanchal—we are in awe of the strength that you provide us at your tender age, and your belief and trust in our endeavors in times when our much-needed attention to you is compromised, the source of these attributes can be none other but only our better halves!

Nandan K. Sinha
N. Ananthkrishnan

MATLAB® is a registered trademark of The MathWorks, Inc. For product information, please contact:

The MathWorks, Inc.
3 Apple Hill Drive
Natick, MA 01760-2098 USA
Tel: 508 647 7000
Fax: 508-647-7001
E-mail: info@mathworks.com
Web: www.mathworks.com

Authors

Dr. Nandan K. Sinha is on the faculty of Department of Aerospace Engineering at the Indian Institute of Technology (IIT) Madras, India, where he has been a professor since July 2014. Dr. Sinha has bachelor's, master's, and PhD degrees in aerospace engineering all from IITs followed by a post-doctoral visiting scholar position during 2003–2006 at the TU-Darmstadt, Germany, before taking up the faculty position at IIT Madras in 2006. Dr. Sinha has over a decade of experience teaching courses in vibrations, flight mechanics and controls, aircraft design, space technology, nonlinear dynamics, and so on. His research interests revolve around design, dynamics, control, and guidance of aerospace vehicles, working on several funded projects from the Indian aerospace industry. He is known for his popular video lecture series on flight dynamics available on YouTube and web-based lecture series on Introduction to Space Technology, both via NPTEL resources, an initiative of MHRD, Government of India. He has coauthored the book *Elementary Flight Dynamics with an Introduction to Bifurcation and Continuation Methods* with Dr. N. Ananthkrishnan, published by CRC Press, Taylor & Francis in 2014. His professional services as subject expert extend to many national committees for various assignments and as reviewer to many national and international journals and conferences.

Dr. N. Ananthkrishnan is an independent consultant presently based out of Mumbai, India. He has more than 22 years of experience in academia and industry in multidisciplinary research and development across

a wide spectrum, from combustion systems to airplane aerodynamics to flight control and guidance. Over the past decade, he has largely worked with businesses in the Mumbai–Pune area and Bangalore in India, and in Daejeon, South Korea, and with a few select academic institutes. His recent work has focused on the broad area of aerospace systems design and integration with emphasis on atmospheric flight mechanics and control and air-breathing propulsion systems. He has previously served on the faculty of aerospace engineering at the Indian Institute of Technology (IIT) Bombay at Mumbai, India, and as a visiting faculty member at the California Institute of Technology at Pasadena, California, USA and at KAIST, Daejeon, South Korea. He received the "Excellence in Teaching" award at IIT Bombay in the year 2000. He has authored a textbook (with Nandan K. Sinha), *Elementary Flight Dynamics with an Introduction to Bifurcation and Continuation Methods*, published by CRC Press, Taylor & Francis (2014). He received his education in aerospace engineering at the Indian Institutes of Technology, majoring in flight mechanics and control, aerodynamics, aircraft design, and nonlinear systems. He is an associate fellow, American Institute of Aeronautics & Astronautics (AIAA), and has served a term as a member of the AIAA Atmospheric Flight Mechanics Technical Committee.

CHAPTER 1

Six Degrees of Freedom Equations of Motion

W HAT IS COMMON TO the three different types of flying vehicles pictured below?

(From left to right) Space launch vehicle (Indian GSLV Mark III), electric-powered airplane (Solar Flight's Sunseeker), and the Zeppelin NT airship. (From Wikipedia.org.)

The answer—despite the many differences between them that we shall highlight later—is that all of them follow the same equations of motion in flight. In fact, any rigid body in atmospheric flight obeys the same set of equations, which can then be used to predict its trajectory and orientation during flight. These are popularly called the six degrees of freedom (6 DOF) equations of motion of a rigid body. The physical, analytical, and computational study of the 6 DOF equations of motion for a flight vehicle constitutes the subject of flight dynamics.

Flight dynamics is the linchpin of any aerospace vehicle design and analysis activity. Many of the requirements and mission objectives for a flight vehicle are naturally specified in terms of flight dynamic quantities. These are usually classified under two heads—performance, and stability and control, both of which we shall explore throughout this text.

As depicted in Figure 1.1, flight dynamics is central to the field of aeronautics and is the meeting point of other disciplines or components of the flight vehicle. While the nature of the engine, payload, and airframe, and the type of requirements demanded from and the performance and stability and control parameters obtained may vary from one kind of flight vehicle to another, the flight dynamics block with the 6 DOF equations of motion will invariably be a constant in every case.

Among the example flight vehicles showcased on page 1, the airframes are obviously visually different. The flight regimes and the aerodynamic forces acting on the vehicles also vary greatly—the space launch vehicle covers a range of speeds from low subsonic to hypersonic, whereas the other two vehicles are mostly confined to low-speed flight. While the space launch vehicle has hardly any lifting surface, the solar-powered airplane has a high-aspect-ratio wing producing aerodynamic lift, and the airship uses an aerostatic lifting mechanism. The engines in use also vary greatly between them—launch vehicles typically use solid or liquid rocket engines (and sometimes engines with cryogenic fuel), the solar-powered airplane uses an electric engine while the airship employs propellers powered by piston engines. Their missions are again quite different, from carrying a payload into space to flying for long range/endurance to just loitering at a

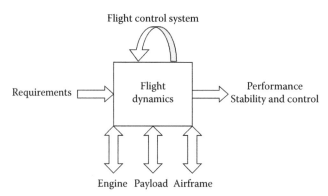

FIGURE 1.1 Flight dynamics as the linchpin of an aerospace vehicle design/ analysis exercise and the interactions with other disciplines/components of the vehicle.

designated station for a specified period. Obviously, their requirements in terms of speeds, flight times, launch and recovery, wind conditions tolerated, etc. are vastly different and so are their performance and stability and control parameters. Likewise, their flight control systems would be configured quite differently. Nevertheless, their flight can be described by the same set of 6 DOF equations of motion and the general principles of flight dynamics and control do apply quite consistently to flight vehicles across the spectrum.

Having said that, there are a few caveats that must be mentioned; for example, space launch vehicles carry huge amounts of fuel that are consumed in flight, so the change in mass, center of gravity (CG) location, and moments of inertia (MI) must be accounted for within the 6 DOF equations of motion. Also, as each stage is jettisoned, there are sudden changes in mass/MI and CG values. Additionally, the liquid fuel in the tanks tends to slosh, setting up additional dynamics that couples with the launch vehicle dynamics—in this case, additional equations for the extra slosh degrees of freedom would need to be appended to the vehicle 6 DOF equations. Most airplanes have movable surfaces such as elevators, rudders, ailerons, flaps, slats, spoilers, and foldable fins—depending on the circumstance, one needs to adjudge whether their deflection involves any change in CG/MI, but a change in the aerodynamic forces/moments with the deflection is almost inevitable; in fact it is sorely desirable. Inertial effects from rotating parts such as propellers or fans in the engines must also be accounted for.

Other vehicles may display a variety of effects that need to be considered either within or separately from the standard 6 DOF equations of rigid body dynamics. For instance, the Concorde is known to transfer fuel between its aft and fore tanks in flight—there is obviously an effect due to this on the CG and MI. Then, as depicted in Figure 1.2, the

FIGURE 1.2 (From left to right) The Concorde with nose droop, the Helios UAV with flexed wing, and the F-111 variable-sweep airplane. (From Wikipedia.org.)

Concorde features a drooping nose as it approaches to land—besides the visible change in the aerodynamic configuration, the MI is also likely to be affected. A more extreme case is that of the Helios UAV, a flying wing configuration, whose wing flexes to a considerable extent in flight, as pictured in Figure 1.2. Should the 6 DOF equations for the Helios be augmented by the equations for the flexural dynamics of the wing? One more example is that of the F-111, an airplane with variable-sweep wings, shown in Figure 1.2. Clearly, there are large changes in CG/MI as well as the aerodynamics in the different wing-sweep positions. Can these be handled within the 6 DOF equations of motion for a rigid body?

Some more examples are shown in Figure 1.3: the parafoil with an under-slung payload is actually two rigid bodies connected together. The motion of each rigid body is described by a set of 6 DOF equations but the connection implies that these two sets are not completely independent but are bound together by constraints. In the usual case, there are three constraint conditions that must be added to the equations of motion, so the net dynamics of the parafoil–payload system effectively has nine degrees of freedom. The other example pictured in Figure 1.3 is that of a tilt-rotor aircraft. How about the dynamics of the rotor (a) in the airplane mode, (b) in the rotorcraft mode, and (c) during the transition from one mode to the other? Should additional rotor dynamics equations be appended to the aircraft 6 DOF equations in this case? The final example is that of an ornithopter, an airplane with flapping wings, somewhat like a bird. What modifications or additions are necessary to the 6 DOF equations of motion to capture the flight dynamics of such a flying vehicle?

Thus, while the 6 DOF equations of motion for a flying vehicle are generally known, there are many twists and turns that must be carefully considered when modeling the flight dynamics of a particular vehicle

FIGURE 1.3   (Left to right) A parafoil with under-slung payload, the V-22 Osprey tilt-rotor aircraft, UTIAS ornithopter. (From Wikipedia.org.)

in a specific configuration and flight condition. Most prominent are the modeling of the aerodynamics, the thrust, the variations in mass and CG/MI, and additional dynamics due to flexibility, liquid slosh, rotating/ moving parts, under-slung loads, and so on. Every situation introduces novel physical effects that must be understood and properly modeled so that the relevant flight dynamics effect is correctly captured and analyzed.

For the most part in this text, we will consider the flight dynamics and control of conventional airplane geometry—wing–body–tail configuration with standard aerodynamic control surfaces, well approximated in flight as a rigid body with fixed mass, CG, and MI. Much of the modeling, analysis, and numerical procedures presented here will be applicable in large measure to other flying vehicles with additional features subject to the comment at the end of the previous paragraph. We shall demonstrate this for a nonstandard flight vehicle in the last chapter of this book.

Homework Exercise: Check out the Internet or your library for information about the flight dynamics and modeling peculiarities of some of these vehicles: (a) air-breathing hypersonic vehicles or *waveriders*, (b) quadcopters or more generally multicopters, and (c) hovercraft, also called ekranoplan.

1.1 DEFINITION OF AXIS SYSTEMS

There are three different axis systems that are important to the development of rigid aircraft 6 DOF equations of motion. The first is a body-fixed axis system whose precise definition is related to the geometry of the aircraft in question (see Figure 1.4a and b). The second is an inertial axis system that may be used as a reference to determine the instantaneous position and orientation of the aircraft; an Earth-fixed inertial axis system (shown in Figure 1.4c) is conventionally used in flight dynamics. The body- and Earth-fixed axes are common to other disciplines such as robotics, for instance. The third axis system, unique to atmospheric flight, is the wind-fixed axis system that defines the orientation and angular rate of the relative wind as seen by the aircraft. The aerodynamic forces on the airplane depend on the relative orientation and angular rate between the body- and wind-fixed axes. All three axis systems are conventionally defined as right-handed orthogonal axis systems.

1.1.1 Body-Fixed Axis System

The body-fixed axis system, as its name suggests, is fixed to the airplane (body) and translates and rotates with the airplane. Thus, when compared

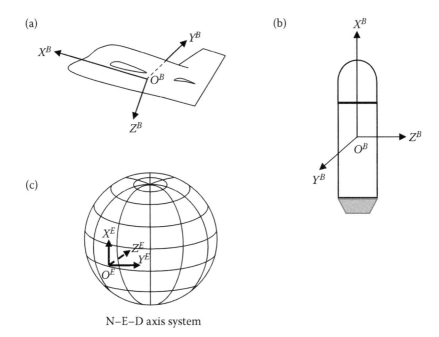

(a)

(b)

(c)

N–E–D axis system

FIGURE 1.4 Body-fixed axis system for an airplane and an Earth-fixed axis system. (a) Airplane-fixed axis system. (b) Launch vehicle-fixed axis system. (c) Earth-fixed inertial axis system.

with an Earth-fixed reference axis, the body-fixed axes can indicate changes in the airplane's position and orientation with respect to the Earth. Likewise, when compared with a wind-fixed axis system (also carried on the airplane as explained next), the body-fixed axes reveals the orientation and relative angular rate of the relative wind incident on the airplane.

Usually, the origin of the body-fixed axis system $X^B Y^B Z^B$ is placed at the CG of the aircraft. Refer Figure 1.4a and b. Following the usual practice for airplanes, the axis X^B is taken to point toward the nose of the airplane but there are various choices for how it may be located. For example, it can be taken along the fuselage reference line (FRL), a straight line in the plane of symmetry $X^B Z^B$ passing through the center of fuselage sections at various locations. Or, it can be selected such that the axes $X^B Y^B Z^B$ form the principal axes for the airplane, that is, the cross products of inertia are all zero. Other options are possible, but it is largely a matter of convenience; all of them are equally acceptable. Nothing changes much between one choice and the other except the moments and cross products of inertia and the reference condition for the aerodynamic angles. The Z^B axis is taken orthogonal to the X^B axis such that the plane $X^B Z^B$ defines the plane of symmetry for the airplane. Finally, the Y^B axis is taken along the starboard and completes the coordinate system according to the right-hand rule. The $X^B Z^B$ plane is also known as the longitudinal plane of the airplane and the X^B axis as the longitudinal axis.

The body-fixed axes are defined similarly for a different flight vehicle such as a space launch rocket; see Figure 1.4b. In this case, the nose usually points near vertically and so does the X^B axis. The examples from Figure 1.4 suggest that the range, cross-range, and altitude covered by the flight vehicle should not be judged by the body-fixed axes but by appropriate projections along the Earth-fixed axes explained below.

1.1.2 Earth-Fixed Axis System

The origin of the Earth-fixed axis system is located on the surface of the Earth with the Z^E axis pointing toward the center of the Earth. The local altitude of an aircraft is therefore measured along the negative Z^E axis. The plane $X^E Y^E$ is locally tangent to the surface of the Earth, effectively approximating the Earth to be a flat surface. For instance, when the X^E axis points to the north and Y^E points to the east, it gives the right-handed orthogonal *NED* (*north–east–down*) axis system. Other choices of the Earth-fixed axes are possible.

Homework Exercise: Look up the Earth-centered, Earth-fixed (ECEF) axis system and under what circumstances would it be useful.

The flat Earth approximation is usually adequate for studying flight dynamics phenomena of the order of seconds to minutes. For phenomena involving longer flight duration, very high speeds, or extreme altitudes, a curved Earth model may be required, which impacts the way in which gravity is modeled, for example, or whether a Coriolis effect due to the Earth's rotation is considered.

Homework Exercise: What kind of axis system is likely to be useful in describing the motion of a satellite orbiting the Moon or a planet? Would one use an Earth-fixed axis or a Moon-fixed axis for instance in this case?

1.1.3 Wind-Fixed Axis System

The wind-fixed axes are also anchored at the airplane CG, same as the body-fixed axis system, but the X^W axis is always aligned along the resultant velocity vector (relative wind). The alignment of the Y^W and Z^W axes forming a right-handed orthogonal axis system is described a little later in this chapter.

It is worth pointing out the difference between the airplane inertial velocity vector V and the relative wind V_∞. The inertial velocity V is the velocity of the airplane (more properly, the velocity of the airplane CG) with respect to the Earth-fixed axis (i.e., as observed from the ground). Since the body- and wind-fixed axes are both pinned to the airplane CG and travel with it, they are also being carried along with the inertial velocity V. Consider the case depicted in Figure 1.5 of an aircraft flying with inertial velocity V in a crosswind v. As indicated in Figure 1.5, the relative wind V_∞ is the vector sum of V and v. The body X^B axis is aligned along the inertial velocity V, whereas the wind X^W axis is along the relative wind V_∞. When writing the equations of motion, it is the inertial velocity V that matters, whereas when modeling the aerodynamic forces acting on the airplane, the relative wind V_∞ is of interest. Of course, in the absence of wind (still atmosphere), the inertial velocity and the relative wind are numerically identical.

Homework Exercise: For a hovering vehicle, how would the wind-fixed axes be defined, if at all? What use would they serve in this case?

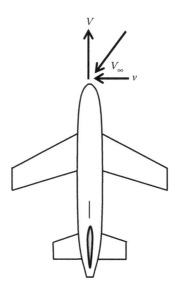

FIGURE 1.5   Airplane flying in crosswind, to illustrate the difference between inertial velocity and relative wind.

Another point is worth noting: since the X^W axis is always aligned along the relative wind, when the airplane maneuvers (e.g., climbs or descends, or turns left or right), the wind-fixed axes reorient with the relative wind. The orientation and *the angular rate* of the wind-fixed axes therefore are not necessarily the same as those of the body-fixed axes even though they share a common point of origin. To illustrate this, Figure 1.6 shows an airplane flying along a level, straight line flight path performing two different maneuvers—(a) the nose pitches up even as it continues to fly straight and level, and (b) the nose remains steady while the velocity vector pitches up putting the airplane into a climb. In case (a), the body axis reorients (with respect to the Earth-fixed axes) with an angular velocity while the wind axis holds fixed, whereas in case (b), it is the wind axis that is reorienting with an angular velocity while the body axis is unmoved.

There is sometimes a misconception that the wind axis is merely another form of the body axis with the X axis aligned differently—this is clearly not so. They are both collocated at the airplane CG; both are carried along with the airplane inertial velocity V with respect to the ground, but their orientation *and angular velocity* with respect to the Earth-fixed axes may be different. In simple words, the body-fixed axes reveal where the airplane is pointing, whereas the wind-fixed axes (in still atmosphere) show where the airplane is going—and these are not necessarily the same.

Having defined these axes, the translational and rotational motion of the airplane can be described in terms of the position and orientation of these axes and their rates. However, it is often necessary to transform quantities from one axis system to another. For example, the gravity

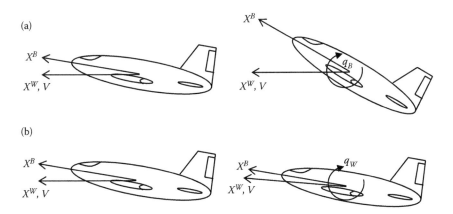

FIGURE 1.6   Different angular rates of the (a) body- and (b) wind-fixed axes.

vector is most conveniently defined in terms of the Earth-axis coordinates but, if the equations of translational motion are written in the body-fixed axes, the gravitational force will have to be transformed from Earth- to body-axis coordinates. So, we shall have to look at axis transformations, but before doing so we need to first define the variables that are required to describe the motion of an airplane in flight.

1.2 DEFINITION OF VARIABLES

In principle, the motion of the airplane can be fully described in terms of two vectors—the inertial velocity V of its CG and its inertial angular velocity ω about the CG; that is, the translational velocity V of the origin and angular velocity ω about the origin of the body-fixed axis with respect to the Earth-fixed axis. Integrating the inertial velocity V gives the instantaneous position and integration of the angular velocity ω yields the instantaneous orientation of the body-fixed axis (equivalently, the airplane) with regard to a ground observer.

Before proceeding further, it is instructive to clarify a point about a vector and its components about a given set of axes. For instance, the inertial velocity vector may be written in terms of its components along the body-fixed axis, or the Earth-fixed axis, or indeed the wind-fixed axis—the representation may change but its character remains unchanged. Thus, the body-axis angular velocity vector may be written in terms of its components along the wind axes, but it is still the body-axis angular velocity and not the wind-axis angular velocity. This may seem obvious, but it is not an uncommon mistake.

Figure 1.7 shows the Earth- and body-fixed axes and the vectors V and ω. The three components of V along the body-fixed X, Y, and Z axes are u, v, and w, respectively. The components u, v, and w are not physically very meaningful and may even be misleading; for instance, w by itself cannot be taken to suggest that the airplane has a downward/upward motion. Likewise, the components of ω about the body-fixed X, Y, and Z axes are p, q, and r, respectively. The components p, q, and r cover the three angular motions of the airplane—roll about the X^B axis, pitch about the Y^B axis, and yaw about the Z^B axis. Thus, p, the angular velocity about the X^B axis, is called the roll rate; q, the angular velocity about the Y^B axis, is called the pitch rate; and r, the angular velocity about the Z^B axis, is called the yaw rate. Positive sense of direction of the angular rotations follow the right-hand thumb rule and are as shown in Figure 1.7. The components p, q, r, likewise, do not provide a good representation of the airplane's

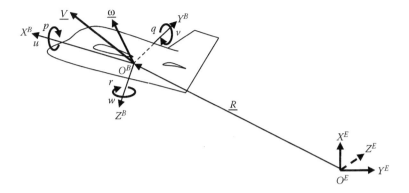

FIGURE 1.7 Sketch of an aircraft in flight with $X^B Y^B Z^B$ axes fixed to its center of gravity (CG) O^B, velocity vector V at CG, and angular velocity vector ω about the CG, axes $X^E Y^E Z^E$ fixed to the Earth, position vector $\underline{R}$ from origin (O^E) of $X^E Y^E Z^E$ to aircraft CG (O^B).

reorientation with respect to the Earth. For instance, it is a common misconception to imagine q in terms of a nose up/down motion; for example, as seen in Figure 1.8, an airplane performing a banked horizontal turn usually has a pretty large value of q though the nose is rock steady and not bobbing up/down at all.

Therefore, to describe the motion of the airplane as seen by an observer on the ground, it is necessary to represent the translational and rotational motion of the airplane with reference to the Earth-fixed axis. The Earth-fixed (inertial) axes are marked in Figure 1.7 as $X^E Y^E Z^E$ with origin at O^E. The position of the airplane is given by the position vector R in Figure 1.7 between the origin O^E of the Earth-fixed axis and the origin O^B of the

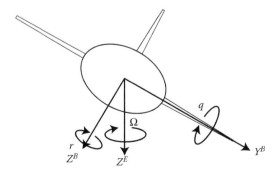

FIGURE 1.8 Example of an airplane in banked, horizontal turn having nonzero q but showing no obvious "pitching motion" of its nose.

body-fixed axis (airplane CG). The components of R along the Earth-fixed axes, x^E, y^E, and z^E, give the range, cross-range, and altitude of the airplane at any instant of time. The rate of change of these three components, that is, $\dot{x}^E$, $\dot{y}^E$, $\dot{z}^E$, are the inertial velocity components along the Earth axes. Assuming the X^B and X^E axes to be coplanar, these components are usually the forward, sideways, and up/down velocities of the airplane in flight.

Now, $\dot{x}^E$, $\dot{y}^E$, $\dot{z}^E$ and u, v, w are both components (along different axis systems) of the same vector, the inertial velocity V; hence they ought to be related. To relate them, we shall work out transformations between two sets of axes, in particular, between the Earth- and body-fixed axes shortly.

The orientation or attitude of the airplane as seen from the ground is described by a set of three angles (φ, θ, ψ). Thus, the angles (φ, θ, ψ), known as Euler angles, relate the inertial axis system $X^E Y^E Z^E$ and the body-fixed axis system $X^B Y^B Z^B$. Basically, one transports the Earth-fixed axes with no change in alignment to the CG of the airplane so that the origins O^E and O^B coincide, and then one performs a sequence of three rotations such that at the end of three rotations, the transported axes $X^E Y^E Z^E$ entirely coincide with the body-fixed axes $X^B Y^B Z^B$. In principle, any sequence of rotations such that two successive rotations are not about the same axis is acceptable, but conventionally a 3–2–1 sequence of rotations is used in airplane flight dynamics. Alternative sequences are common in other fields such as robotics and space flight dynamics. To perform a 3–2–1 sequence of rotations, first rotate the Earth axes about the Z^E axis by angle ψ, followed by a rotation about the intermediate Y axis by angle θ, and finally a third rotation about the X^B axis by an angle φ as shown in Figure 1.9.

Homework Exercise: Review the Euler angles and their rotation sequence as used in robotics and spacecraft dynamics. Does the sequence of

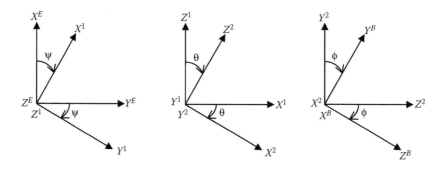

FIGURE 1.9 Rotations using 3–2–1 rule.

rotations really not matter at all, or does a particular sequence provide an advantage in any circumstance?

The angles ψ, θ, and φ are called the yaw, pitch, and roll angles, respectively. The positive sense of (φ, θ, ψ) is similar to that shown for p, q, r in Figure 1.7. Thus, for example, when θ is positive, it means the airplane nose is above the horizon, and when φ is positive, it means that the right (starboard) wing is banked down. The rates of change of the Euler angles, ($\dot{\varphi}, \dot{\theta}, \dot{\psi}$), are obviously related to the angular velocity vector ω and they provide a much more meaningful description of the airplane reorientation as seen from the ground. Thus, for instance, a positive $\dot{\theta}$ always means that the airplane nose is pitching up, and a positive $\dot{\varphi}$ suggests that the right wing is rolling down. Unfortunately, ($\dot{\varphi}, \dot{\theta}, \dot{\psi}$) are not called the roll, pitch, and yaw rates—that terminology is usurped by the body-axis components (p, q, r). As we shall see shortly, the Euler angle rates ($\dot{\varphi}, \dot{\theta}, \dot{\psi}$) can be related to the components (p, q, r)—after all, they represent the same vector ω.

At this stage, let us summarize in Table 1.1 the variables that have been defined so far. As can be seen, these cover the motion of the airplane relative to the ground (inertial reference). However, for calculation of aerodynamic forces, we need to determine the orientation and angular rate of the airplane with respect to the relative wind, that is, the relation between the body- and wind-fixed axes.

Figure 1.10 shows the body- and wind-axes, their origins collocated at the airplane CG, with the X^W axis aligned along the velocity vector as per its definition. The relative orientation of the body- and wind-fixed axes is defined in terms of two angles—the aerodynamic angles, β and α. Starting with the body-fixed axes, first rotate around the Y^B axis by an angle ($-\alpha$). As shown in Figure 1.10, this brings the $X^B Z^B$ axes along the $X^S Z^S$ axes, intermediate axes called the "stability" axes. Note that the $X^S Z^S$ axes still lie in the aircraft's plane of symmetry. Next, rotate by the angle β about

TABLE 1.1   Summary of Flight Dynamics Variables Defined so Far

Quantity	Vector	Components/Representation
Position of airplane CG w.r.t. Earth	R	(x^E, y^E, z^E)
Orientation of airplane body w.r.t. Earth	–	(φ, θ, ψ)
Inertial velocity of airplane CG	V	(u, v, w)
		$(\dot{x}^E, \dot{y}^E, \dot{z}^E)$
Inertial body angular velocity	ω	(p, q, r)
		$(\dot{\varphi}, \dot{\theta}, \dot{\psi})$

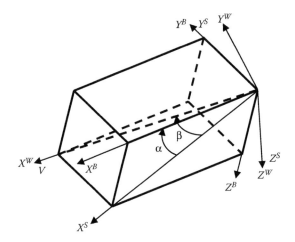

FIGURE 1.10 Body- and wind-fixed axis systems in aircraft dynamics.

the Z^S axis so that the X axis aligns with the relative wind. By definition, this then is the wind-fixed axis system. The Z^W axis is the same as the Z^S axis, which remains in the plane of symmetry ($X^B Z^B$ plane). The Y^W axis completes the right-handed orthogonal coordinate system. In the absence of wind (still atmosphere), the relative wind is identical in magnitude to the inertial velocity. The wind-fixed axis system, thus, gives yet another representation of the inertial velocity—in terms of its magnitude V and two angles, α and β.

The orientation of the wind-fixed axis with respect to the ground reveals the direction in which the airplane is flying, which could be different from the direction in which it is pointing (given by the orientation of the body-fixed axis). A set of three Euler angles (μ, γ, χ) can be defined through which the Earth-fixed axes may be rotated to coincide with the wind-fixed axes—exactly along similar lines as the Euler angles (φ, θ, ψ) used in case of the body axes. Following the same 3–2–1 convention, the Earth-fixed axis system is first rotated about the Z^E axis by an angle χ, then about the intermediate Y^1 axis by an angle γ, and finally about the X^W axis by an angle μ to coincide with the wind-fixed axis system. χ is called the heading angle—it shows the direction is which the airplane is heading. γ is called the flight path angle and indicates whether the airplane is climbing or descending. μ is the wind-axis roll angle, also called the velocity-vector roll angle because the wind X^W axis coincides with the velocity vector.

It is important to distinguish between the wind-axis Euler angles (μ, γ, χ) and the body-axis Euler angles (φ, θ, ψ) defined previously. For

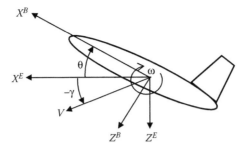

FIGURE 1.11 Airplane in landing approach highlighting the difference between flight path angle γ and pitch angle θ.

example, Figure 1.11 shows an airplane in landing approach—the velocity vector is below the horizon, that is, the flight path angle γ is negative and the airplane is descending. On the other hand, the nose is pointed above the horizon, that is, the pitch angle θ is positive.

The rates of change of the wind-axis Euler angles, $(\dot{\mu}, \dot{\gamma}, \dot{\chi})$, are obviously related to the angular velocity vector ω_W of the wind-fixed axes. For instance, an airplane rolling to the right would show a positive $\dot{\mu}$, one that is performing a vertical pull-up would have a positive $\dot{\gamma}$, and a horizontal turn to the right would mean a positive $\dot{\chi}$. As in the case of the body-axis angular velocity ω, the wind-axis angular velocity vector ω_W can be represented by its three components (p_W, q_W, r_W). The positive sense of (p_W, q_W, r_W) is similar to that shown in Figure 1.7 for (p, q, r). Again, the components (p_W, q_W, r_W) do not give a good impression of the airplane's maneuver as seen from the ground. A nonzero q_W simply means that there is a component of ω_W about the Y axis in question; it does not necessarily imply that the velocity vector is looping up/down with respect to the horizon. $\dot{\gamma}$ is a better variable to judge whether the airplane is doing a pull up/down.

As may be expected, since both $(\dot{\mu}, \dot{\gamma}, \dot{\chi})$ and (p_W, q_W, r_W) represent the angular velocity of the wind-fixed axis ω_W, they must be mutually related, as we shall indeed see shortly.

Finally, it may be worth emphasizing that the angular velocities of the wind- and body-fixed axes, ω_W and ω, respectively, are distinct quantities, and as illustrated in Figure 1.6 may change independently of each other. To sum up, it may be worthwhile to update Table 1.1 to include the wind-fixed axis variables and the aerodynamic angles. The final list of variables is in Table 1.2.

TABLE 1.2 Summary of All Flight Dynamics Variables

Quantity	Vector	Components/Representation
Position of airplane CG w.r.t. Earth	R	(x^E, y^E, z^E)
Orientation of airplane body w.r.t. Earth	–	(φ, θ, ψ)
Orientation of relative wind w.r.t. Earth	–	(μ, γ, χ)
Inertial velocity of airplane CG	V	(u, v, w)
		$(\dot{x}^E, \dot{y}^E, \dot{z}^E)$
		(V, α, β)
Inertial body angular velocity	ω	(p, q, r)
		$(\dot{\varphi}, \dot{\theta}, \dot{\psi})$
Inertial relative wind angular velocity	ω_w	(p_w, q_w, r_w)
		$(\dot{\mu}, \dot{\gamma}, \dot{\chi})$

1.3 3–2–1 TRANSFORMATION

Having defined the variables of interest for describing the flight of an airplane, we can get down to writing the equations that may be used to transform a variable from one axis system to another. It is important to remember that the variable itself is not altered in any manner during such a transformation—it is just that its components along the axes of one coordinate system are used to find a different set of components in another axis system. The magnitude of a vector, for instance, or the relative orientation of two vectors remain unchanged in this process.

To understand axis transformations, imagine an XYZ right-handed orthogonal axis system with the Z axis into the plane of the paper as sketched in Figure 1.12. Rotate this axis system by an angle δ about the Z axis in the clockwise sense. The rotated axis system labeled $X'Y'Z'$ (where Z' is the same as Z) is also sketched in Figure 1.12.

Clearly, the original and rotated axes are related to each other by the transformation in Equation 1.1. The matrix in Equation 1.1 is called the transformation matrix.

$$\begin{bmatrix} X \\ Y \\ Z \end{bmatrix} = \begin{bmatrix} \cos\delta & -\sin\delta & 0 \\ \sin\delta & \cos\delta & 0 \\ 0 & 0 & 1 \end{bmatrix} \begin{bmatrix} X' \\ Y' \\ Z' \end{bmatrix} \tag{1.1}$$

The equality relation in Equation 1.1 is the mathematical representation of Figure 1.12.

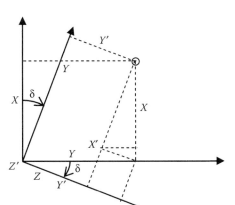

FIGURE 1.12   Rotation of XYZ axis system about the Z axis.

1.3.1 Earth- and Body-Fixed Axes

In case of the 3–2–1 transformation between Earth- and body-fixed axes as indicated in Figure 1.9, three such rotations are performed successively in the following sequence: (3) rotation about original Z (Z^E) axis by angle ψ, (2) rotation about intermediate Y (Y^1) axis by angle θ, and (1) rotation about final X (X^2 or X^B) axis by angle φ. Each rotation is represented by a transformation matrix as

$$\begin{bmatrix} X^E \\ Y^E \\ Z^E \end{bmatrix} = \underbrace{\begin{bmatrix} c\psi & -s\psi & 0 \\ s\psi & c\psi & 0 \\ 0 & 0 & 1 \end{bmatrix}}_{R_\psi} \begin{bmatrix} X^1 \\ Y^1 \\ Z^1 \end{bmatrix} \qquad \begin{bmatrix} X^1 \\ Y^1 \\ Z^1 \end{bmatrix} = \underbrace{\begin{bmatrix} c\theta & 0 & s\theta \\ 0 & 1 & 0 \\ -s\theta & 0 & c\theta \end{bmatrix}}_{R_\theta} \begin{bmatrix} X^2 \\ Y^2 \\ Z^2 \end{bmatrix}$$

$$\begin{bmatrix} X^2 \\ Y^2 \\ Z^2 \end{bmatrix} = \underbrace{\begin{bmatrix} 1 & 0 & 0 \\ 0 & c\varphi & -s\varphi \\ 0 & s\varphi & c\varphi \end{bmatrix}}_{R_\varphi} \begin{bmatrix} X^B \\ Y^B \\ Z^B \end{bmatrix} \tag{1.2}$$

where $c(\cdot) = \cos(\cdot)$ and $s(\cdot) = \sin(\cdot)$.

The transformation matrices are named R_ψ, R_θ, R_φ. As with all transformation matrices, R_ψ, R_θ, R_φ are orthogonal and satisfy the property of orthogonality:

$$R_\psi R_\psi^T = R_\psi^T R_\psi = I, \quad R_\theta R_\theta^T = R_\theta^T R_\theta = I, \quad R_\varphi R_\varphi^T = R_\varphi^T R_\varphi = I,$$

where I is the identity matrix. The net transformation between $X^E Y^E Z^E$ and $X^B Y^B Z^B$ is given by the matrix multiplication of the three transformation matrices in Equation 1.27, as follows:

$$
\begin{bmatrix} X^E \\ Y^E \\ Z^E \end{bmatrix} = \underbrace{\begin{bmatrix} c\psi & -s\psi & 0 \\ s\psi & c\psi & 0 \\ 0 & 0 & 1 \end{bmatrix}}_{R_\psi} \underbrace{\begin{bmatrix} c\theta & 0 & s\theta \\ 0 & 1 & 0 \\ -s\theta & 0 & c\theta \end{bmatrix}}_{R_\theta} \underbrace{\begin{bmatrix} 1 & 0 & 0 \\ 0 & c\varphi & -s\varphi \\ 0 & s\varphi & c\varphi \end{bmatrix}}_{R_\varphi} \begin{bmatrix} X^B \\ Y^B \\ Z^B \end{bmatrix} = R_\psi R_\theta R_\varphi \begin{bmatrix} X^B \\ Y^B \\ Z^B \end{bmatrix}
$$

(1.3)

Multiplying both sides of Equation 1.3 by $R_\varphi^T R_\theta^T R_\psi^T$ and using the orthogonality property of the transformation matrices, the reverse transformation between body- and Earth-fixed axes is

$$
\begin{bmatrix} X^B \\ Y^B \\ Z^B \end{bmatrix} = R_\varphi^T R_\theta^T R_\psi^T \begin{bmatrix} X^E \\ Y^E \\ Z^E \end{bmatrix}
$$

(1.4)

Let us illustrate the use of the transformations in Equations 1.3 and 1.4 by some examples. First, we find the components of aircraft weight along the body-fixed axes. The aircraft weight acts along the Z^E axis, so in the Earth-fixed axis, the weight vector may be written as $[0, 0, mg]^T$. Then using Equation 1.4, the components of the gravitational force along the three body-fixed axes are

$$
\begin{bmatrix} X^G \\ Y^G \\ Z^G \end{bmatrix} = R_\varphi^T R_\theta^T R_\psi^T \begin{bmatrix} 0 \\ 0 \\ mg \end{bmatrix}
$$

$$
= \begin{bmatrix} 1 & 0 & 0 \\ 0 & \cos\varphi & \sin\varphi \\ 0 & -\sin\varphi & \cos\varphi \end{bmatrix} \begin{bmatrix} \cos\theta & 0 & -\sin\theta \\ 0 & 1 & 0 \\ \sin\theta & 0 & \cos\theta \end{bmatrix} \begin{bmatrix} \cos\psi & \sin\psi & 0 \\ -\sin\psi & \cos\psi & 0 \\ 0 & 0 & 1 \end{bmatrix} \begin{bmatrix} 0 \\ 0 \\ mg \end{bmatrix}
$$

$$
= \begin{bmatrix} -mg\sin\theta \\ mg\sin\varphi\cos\theta \\ mg\cos\varphi\cos\theta \end{bmatrix}
$$

(1.5)

Homework Exercise: Try out the components of the weight vector along the body axis for different combinations of the Euler angles, θ and φ. For example, check out (a) $\theta = 90$ deg, $\varphi = 0$ and (b) $\theta = 0$, $\varphi = 90$ deg.

For an example of the reverse transformation, assume we know the components u, v, w of the airplane inertial velocity along the body-fixed axes and would like to find the airplane ground velocity components $\dot{x}^E$, $\dot{y}^E$, $\dot{z}^E$. Using Equation 1.3,

$$\begin{bmatrix} \dot{x}^E \\ \dot{y}^E \\ \dot{z}^E \end{bmatrix} = R_\psi R_\theta R_\varphi \begin{bmatrix} u \\ v \\ w \end{bmatrix} = \begin{bmatrix} c\psi & -s\psi & 0 \\ s\psi & c\psi & 0 \\ 0 & 0 & 1 \end{bmatrix} \begin{bmatrix} c\theta & 0 & s\theta \\ 0 & 1 & 0 \\ -s\theta & 0 & c\theta \end{bmatrix} \begin{bmatrix} 1 & 0 & 0 \\ 0 & c\varphi & -s\varphi \\ 0 & s\varphi & c\varphi \end{bmatrix} \begin{bmatrix} u \\ v \\ w \end{bmatrix}$$

$$= \begin{bmatrix} c\psi c\theta & c\psi s\theta s\varphi - s\psi c\varphi & c\psi s\theta c\varphi + s\psi s\varphi \\ s\psi c\theta & s\psi s\theta s\varphi + c\psi c\varphi & s\psi s\theta c\varphi - c\psi s\varphi \\ -s\theta & c\theta s\varphi & c\theta c\varphi \end{bmatrix} \begin{bmatrix} u \\ v \\ w \end{bmatrix}$$

$$(1.6)$$

The velocity components in Equation 1.6 can be integrated to determine the position of the airplane with respect to an Earth-fixed reference; hence Equation 1.6 is referred to as the *navigational equation*.

1.3.2 Earth- and Wind-Fixed Axes

Similarly, the conversion between Earth- and wind-fixed axes through the Euler angles μ, γ, χ may be represented using the transformation matrices as follows:

$$\begin{bmatrix} X^E \\ Y^E \\ Z^E \end{bmatrix} = \underbrace{\begin{bmatrix} c\chi & -s\chi & 0 \\ s\chi & c\chi & 0 \\ 0 & 0 & 1 \end{bmatrix}}_{R_\chi} \begin{bmatrix} X^1 \\ Y^1 \\ Z^1 \end{bmatrix} \qquad \begin{bmatrix} X^1 \\ Y^1 \\ Z^1 \end{bmatrix} = \underbrace{\begin{bmatrix} c\gamma & 0 & s\gamma \\ 0 & 1 & 0 \\ -s\gamma & 0 & c\gamma \end{bmatrix}}_{R_\gamma} \begin{bmatrix} X^2 \\ Y^2 \\ Z^2 \end{bmatrix}$$

$$\begin{bmatrix} X^2 \\ Y^2 \\ Z^2 \end{bmatrix} = \underbrace{\begin{bmatrix} 1 & 0 & 0 \\ 0 & c\mu & -s\mu \\ 0 & s\mu & c\mu \end{bmatrix}}_{R_\mu} \begin{bmatrix} X^W \\ Y^W \\ Z^W \end{bmatrix} \qquad (1.7)$$

The transformation matrices are named R_χ, R_γ, R_μ and satisfy the orthogonality property:

$$R_\chi R_\chi^T = R_\chi^T R_\chi = I; \quad R_\gamma R_\gamma^T = R_\gamma^T R_\gamma = I; \quad R_\mu R_\mu^T = R_\mu^T R_\mu = I$$

The net transformation between the $X^E Y^E Z^E$ and the $X^W Y^W Z^W$ axes is given by the matrix multiplication of the three transformation matrices in Equation 1.7, as follows:

$$
\begin{bmatrix} X^E \\ Y^E \\ Z^E \end{bmatrix} = \underbrace{\begin{bmatrix} c\chi & -s\chi & 0 \\ s\chi & c\chi & 0 \\ 0 & 0 & 1 \end{bmatrix}}_{R_\chi} \underbrace{\begin{bmatrix} c\gamma & 0 & s\gamma \\ 0 & 1 & 0 \\ -s\gamma & 0 & c\gamma \end{bmatrix}}_{R_\gamma} \underbrace{\begin{bmatrix} 1 & 0 & 0 \\ 0 & c\mu & -s\mu \\ 0 & s\mu & c\mu \end{bmatrix}}_{R_\mu} \begin{bmatrix} X^W \\ Y^W \\ Z^W \end{bmatrix} = R_\chi R_\gamma R_\mu \begin{bmatrix} X^W \\ Y^W \\ Z^W \end{bmatrix}
$$

(1.8)

And the reverse transformation is written as

$$
\begin{bmatrix} X^W \\ Y^W \\ Z^W \end{bmatrix} = R_\mu^T R_\gamma^T R_\chi^T \begin{bmatrix} X^E \\ Y^E \\ Z^E \end{bmatrix}
$$

(1.9)

In the absence of wind, the inertial velocity vector in wind-fixed axes is simply $[V\ 0\ 0]^T$. Let us transform this to the Earth-fixed axes using Equation 1.8 to find the airplane ground velocity components $\dot{x}^E$, $\dot{y}^E$, $\dot{z}^E$:

$$
\begin{bmatrix} \dot{x}^E \\ \dot{y}^E \\ \dot{z}^E \end{bmatrix} = \begin{bmatrix} c\chi & -s\chi & 0 \\ s\chi & c\chi & 0 \\ 0 & 0 & 1 \end{bmatrix} \begin{bmatrix} c\gamma & 0 & s\gamma \\ 0 & 1 & 0 \\ -s\gamma & 0 & c\gamma \end{bmatrix} \begin{bmatrix} 1 & 0 & 0 \\ 0 & c\mu & -s\mu \\ 0 & s\mu & c\mu \end{bmatrix} \begin{bmatrix} V \\ 0 \\ 0 \end{bmatrix} = \begin{bmatrix} V\cos\gamma\cos\chi \\ V\cos\gamma\sin\chi \\ -V\sin\gamma \end{bmatrix}
$$

(1.10)

Equation 1.10 is also called the *navigational equation*. In fact, this is the preferred form of the navigational equation (over that in Equation 1.6) as it is simpler and the variables V, γ, χ are more accessible as compared to u, v, w.

Homework Exercise: The ground track is the set of points on the Earth's surface that an airplane passes directly over. In other words, it is the projection of the flight trajectory on the Earth's surface. The heading angle χ is defined as 0 along north and progresses clockwise back to 360 deg along north—that is, 90 deg along east, 180 deg along south, and 270 deg along west. Find the ground track for a flight vehicle with (a) $\gamma = 0$, $\chi = 90$ deg and (b) $\gamma = 90$ deg, $\chi = 0$ (Notice something odd about case b)?)

1.3.3 Wind- and Body-Fixed Axes

The transformation between the wind- and body-fixed axes involves the angles α, β as indicated in Figure 1.10. The angles α and β are called aerodynamic angles, angle of attack and sideslip, respectively. Though they are not usually referred to as "Euler angles," the transformation between wind- and body-fixed axes follows a similar pattern to those between Earth-body and Earth-wind axes.

The wind-fixed $X^W Y^W Z^W$ axes can be rotated to coincide with the body-fixed axis system using two rotations as follows: First rotate around Z^W axis by angle $-\beta$ to match with the $X^S Y^S Z^S$ stability axis system; then rotate about Y^S by angle α to coincide with the $X^B Y^B Z^B$ axes. These rotations are represented by the following transformation matrices:

$$
\begin{bmatrix} X^W \\ Y^W \\ Z^W \end{bmatrix} = \underbrace{\begin{bmatrix} c\beta & s\beta & 0 \\ -s\beta & c\beta & 0 \\ 0 & 0 & 1 \end{bmatrix}}_{R_{-\beta}} \begin{bmatrix} X^S \\ Y^S \\ Z^S \end{bmatrix} \quad \begin{bmatrix} X^S \\ Y^S \\ Z^S \end{bmatrix} = \underbrace{\begin{bmatrix} c\alpha & 0 & s\alpha \\ 0 & 1 & 0 \\ -s\alpha & 0 & c\alpha \end{bmatrix}}_{R_\alpha} \begin{bmatrix} X^B \\ Y^B \\ Z^B \end{bmatrix} \quad (1.11)
$$

The net transformation in either sense (wind to body and body to wind) is given by the multiplication of the two matrices in Equation 1.11:

$$
\begin{bmatrix} X^W \\ Y^W \\ Z^W \end{bmatrix} = R_{-\beta} R_\alpha \begin{bmatrix} X^B \\ Y^B \\ Z^B \end{bmatrix} \Rightarrow \begin{bmatrix} X^B \\ Y^B \\ Z^B \end{bmatrix} = R_\alpha^T R_{-\beta}^T \begin{bmatrix} X^W \\ Y^W \\ Z^W \end{bmatrix} = R_{-\alpha} R_\beta \begin{bmatrix} X^W \\ Y^W \\ Z^W \end{bmatrix} \quad (1.12)
$$

With the usual orthogonality property of the transformation matrices,

$$
R_{-\beta} R_{-\beta}^T = R_{-\beta}^T R_{-\beta} = I \text{ and } R_\alpha R_\alpha^T = R_\alpha^T R_\alpha = I.
$$

To illustrate the use of Equation 1.12, let us transform the inertial velocity along the wind axes, given as $V = [V\ 0\ 0]^T$, into components along the body-fixed axes, (u, v, w).

$$\begin{bmatrix} u \\ v \\ w \end{bmatrix} = \begin{bmatrix} c\alpha & 0 & -s\alpha \\ 0 & 1 & 0 \\ s\alpha & 0 & c\alpha \end{bmatrix} \begin{bmatrix} c\beta & -s\beta & 0 \\ s\beta & c\beta & 0 \\ 0 & 0 & 1 \end{bmatrix} \begin{bmatrix} V \\ 0 \\ 0 \end{bmatrix} = \begin{bmatrix} V\cos\beta\cos\alpha \\ V\sin\beta \\ V\cos\beta\sin\alpha \end{bmatrix} \qquad (1.13)$$

From Equation 1.13, the definitions of the aerodynamic angles α, β are obtained as follows:

$$\alpha = \tan^{-1}\left(\frac{w}{u}\right) \quad \beta = \sin^{-1}\left(\frac{v}{V}\right) \quad \text{where } V = \sqrt{(u^2 + v^2 + w^2)} \qquad (1.14)$$

Homework Exercise: An alternative way of describing the aerodynamic angles is sometimes used in missile aerodynamics, especially in case of bodies that are dominantly axisymmetric. In these cases, due to rotation symmetry, the angles α and β are usually identical. Then, it may be convenient to define a total angle of attack α_t and an aerodynamic roll angle φ, which may be used to describe the orientation of the vehicle relative to the wind. See the Exercise Problems for more detail.

1.3.4 Relation between the Body-Axis and Wind-Axis Euler Angles

As seen earlier, the Earth-fixed axes can be rotated by a sequence of three Euler angles χ, γ, μ to reorient them along the wind-fixed axes. Further, as seen above, the wind-fixed axes when rotated by the sequence of angles $(-\beta, \alpha)$ will coincide with the body-fixed axes. This sequence of five consecutive rotations can be represented by the concatenation of the matrices $R_\chi R_\gamma R_\mu R_{-\beta} R_\alpha$. However, rotation through the three Euler angles ψ, θ, φ, represented by the matrix product $R_\psi R_\theta R_\varphi$, will also reorient the Earth-fixed axes along the body-fixed axes. Thus, there are two routes to transform from the Earth-fixed to the body-fixed axes: one is direct; the other is via the wind-fixed axes as an intermediate. Needless to say, both these transformations are equivalent—the starting and end points are one and the same. Hence the product of the rotation matrices either way must also be equivalent; that is,

$$R_\chi R_\gamma R_\mu R_{-\beta} R_\alpha = R_\psi R_\theta R_\varphi \quad \text{equivalently } R_\chi R_\gamma R_\mu = R_\psi R_\theta R_\varphi R_\alpha^T R_{-\beta}^T$$

$$(1.15)$$

Equation 1.15 relates the body-axis Euler angles ψ, θ, φ, and the wind-axis Euler angles χ, γ, μ, in terms of the aerodynamic angles β, α. On expansion, the matrix Equation 1.15 appears as

$$
\underbrace{\begin{bmatrix}
c_\chi c_\gamma & c_\chi s_\gamma s_\mu - s_\chi c_\mu & c_\chi s_\gamma c_\mu + s_\chi s_\mu \\
s_\chi c_\gamma & s_\chi s_\gamma s_\mu + c_\chi c_\gamma & s_\chi s_\gamma c_\mu - c_\chi s_\mu \\
-s_\gamma & c_\gamma s_\mu & c_\gamma c_\mu
\end{bmatrix}}_{R_\chi R_\gamma R_\mu}
=
\underbrace{\begin{bmatrix}
c_\psi c_\theta & c_\psi s_\theta s_\varphi - s_\psi c_\varphi & c_\psi s_\theta c_\varphi + s_\psi s_\varphi \\
s_\psi c_\theta & s_\psi s_\theta s_\varphi + c_\psi c_\theta & s_\psi s_\theta c_\varphi - c_\psi s_\varphi \\
-s_\theta & c_\theta s_\varphi & c_\theta c_\varphi
\end{bmatrix}}_{R_\psi R_\theta R_\varphi}
$$

$$
\underbrace{\begin{bmatrix}
c_\alpha c_\beta & -c_\alpha s_\beta & -s_\alpha \\
s_\beta & c_\beta & 0 \\
s_\alpha c_\beta & -s_\alpha s_\beta & c_\alpha
\end{bmatrix}}_{R_\alpha^T R_{-\beta}^T}
\tag{1.16}
$$

Further expanding the matrix product on the right-hand side of Equation 1.16 results in the following set of nine equalities relating ψ, θ, φ and χ, γ, μ with the angles β, α:

$$
\begin{bmatrix}
c_\chi c_\gamma & c_\chi s_\gamma s_\mu - s_\chi c_\mu & c_\chi s_\gamma c_\mu + s_\chi s_\mu \\
s_\chi c_\gamma & s_\chi s_\gamma s_\mu + c_\chi c_\gamma & s_\chi s_\gamma c_\mu - c_\chi s_\mu \\
-s_\gamma & c_\gamma s_\mu & c_\gamma c_\mu
\end{bmatrix}
$$

$$
=
\begin{bmatrix}
\begin{array}{l} c_\psi c_\theta c_\alpha c_\beta \\ +s_\beta(c_\psi s_\theta s_\varphi - s_\psi c_\varphi) \\ +s_\alpha c_\beta(c_\psi s_\theta c_\varphi + s_\psi s_\varphi) \end{array} &
\begin{array}{l} -c_\alpha s_\beta c_\psi c_\theta \\ +c_\beta(c_\psi s_\theta s_\varphi - s_\psi c_\varphi) \\ -s_\alpha s_\beta(c_\psi s_\theta c_\varphi + s_\psi s_\varphi) \end{array} &
\begin{array}{l} -s_\alpha c_\psi c_\theta \\ +c_\alpha(c_\psi s_\theta c_\varphi + s_\psi s_\varphi) \end{array} \\[2em]
\begin{array}{l} s_\psi c_\theta c_\alpha c_\beta \\ +s_\beta(s_\psi s_\theta s_\varphi + c_\psi c_\theta) \\ +s_\alpha c_\beta(s_\psi s_\theta c_\varphi - c_\psi s_\varphi) \end{array} &
\begin{array}{l} -c_\alpha s_\beta s_\psi c_\theta \\ +c_\beta(s_\psi s_\theta s_\varphi + c_\psi c_\theta) \\ -s_\alpha s_\beta(s_\psi s_\theta c_\varphi - c_\psi s_\varphi) \end{array} &
\begin{array}{l} -s_\alpha s_\psi c_\theta \\ +c_\alpha(s_\psi s_\theta c_\varphi - c_\psi s_\varphi) \end{array} \\[2em]
\begin{array}{l} -s_\theta c_\alpha c_\beta \\ +c_\theta s_\varphi s_\beta + c_\theta c_\varphi s_\alpha c_\beta \end{array} &
\begin{array}{l} s_\theta c_\alpha s_\beta \\ +c_\theta s_\varphi c_\beta - c_\theta c_\varphi s_\alpha s_\beta \end{array} &
\begin{array}{l} s_\alpha s_\theta \\ +c_\theta c_\varphi c_\alpha \end{array}
\end{bmatrix}
\tag{1.17}
$$

Clearly, not all the nine equations comprising Equation 1.17 are independent. Usually, the following set of three equations obtained by matching the last row of the respective matrices in Equation 1.17 are sufficient for most purposes; however, all the nine equations are valid and any of them may be used as per convenience.

$$\sin\gamma = \cos\alpha\cos\beta\sin\theta - \sin\beta\sin\varphi\cos\theta - \sin\alpha\cos\beta\cos\varphi\cos\theta$$
$$\sin\mu\cos\gamma = \sin\theta\cos\alpha\sin\beta + \sin\varphi\cos\theta\cos\beta - \sin\alpha\sin\beta\cos\varphi\cos\theta$$
$$\cos\mu\cos\gamma = \sin\theta\sin\alpha + \cos\alpha\cos\varphi\cos\theta$$

(1.18)

1.4 RELATION BETWEEN ANGULAR VELOCITY VECTOR AND EULER ANGLE RATES

We have seen that the angular velocity components (p, q, r) and the Euler angle rates $(\dot{\varphi}, \dot{\theta}, \dot{\psi})$ refer to the same vector ω, hence they ought to be related. However, the Euler rates $(\dot{\varphi}, \dot{\theta}, \dot{\psi})$ are not the "components" of ω along the Earth-fixed axes, so they cannot be related to (p, q, r) by the rotation matrices in the previous section. A relationship between the two is derived below using the notion that infinitesimal rotations may be added as vectors.

1.4.1 Relation between Body-Axis Angular Velocity Components (p, q, r) and Euler Angle Rates $(\dot{\varphi}, \dot{\theta}, \dot{\psi})$

p, q, r are the components of the body-axis angular velocity vector ω about the body axes $X^B Y^B Z^B$. To understand $\dot{\varphi}, \dot{\theta}, \dot{\psi}$, begin with the Earth-fixed axes and give a small rotation $\Delta\psi$ about the Z^E axis. This is indicated in Figure 1.13. Next, following the standard Euler angle sequence, give a rotation $\Delta\theta$—this will be about the intermediate Y^1 axis. The final rotation is $\Delta\varphi$ about the X^2 axis, which is the same as X^B. This sequence of infinitesimal rotations may be added vectorially with the appropriate unit vectors about each axis, as below:

$$\Delta\underline{\sigma} = \Delta\psi\,\hat{k}_E + \Delta\theta\,\hat{j}_1 + \Delta\varphi\,\hat{i}_2$$

(1.19)

where $\hat{k}_E$, $\hat{j}_1$, and $\hat{i}_2$ are unit vectors along Z^E, Y^1, and X^2 (same as X^B) axes, respectively. Taking the differential with time of the angles in Equation 1.19 yields the vector sum of the Euler angle rates:

$$\dot{\underline{\sigma}} = \dot{\psi}\,\hat{k}_E + \dot{\theta}\,\hat{j}_1 + \dot{\varphi}\,\hat{i}_2$$

(1.20)

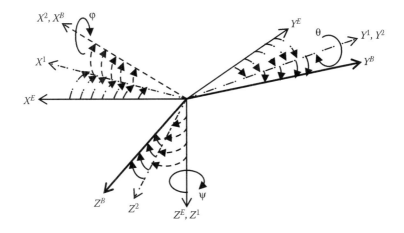

FIGURE 1.13   Infinitesimal rotations ψ (about Z^E), θ (about Y^1), and φ (about X^2).

Of course, $\underline{\dot\sigma}$ is the same as the body-axis angular velocity ω with components (p, q, r) about the body axes $X^B Y^B Z^B$. So, equating the two gives

$$
\begin{bmatrix} p \\ q \\ r \end{bmatrix}_B = \begin{bmatrix} 0 \\ 0 \\ \dot\psi \end{bmatrix}_E + \begin{bmatrix} 0 \\ \dot\theta \\ 0 \end{bmatrix}_1 + \begin{bmatrix} \dot\varphi \\ 0 \\ 0 \end{bmatrix}_B \tag{1.21}
$$

However, note that the vectors in Equation 1.21 are represented in different axis systems. To sum them up, they all need to be converted to a common axis system, which is chosen to be the body $X^B Y^B Z^B$ axes. This is possible using the rotation matrices in Equation 1.2. Thus,

$$
\begin{bmatrix} 0 \\ 0 \\ \dot\psi \end{bmatrix}_E = \begin{bmatrix} 1 & 0 & 0 \\ 0 & \cos\varphi & \sin\varphi \\ 0 & -\sin\varphi & \cos\varphi \end{bmatrix} \begin{bmatrix} \cos\theta & 0 & -\sin\theta \\ 0 & 1 & 0 \\ \sin\theta & 0 & \cos\theta \end{bmatrix} \begin{bmatrix} \cos\psi & \sin\psi & 0 \\ -\sin\psi & \cos\psi & 0 \\ 0 & 0 & 1 \end{bmatrix} \begin{bmatrix} 0 \\ 0 \\ \dot\psi \end{bmatrix}_B \tag{1.22}
$$

and

$$
\begin{bmatrix} 0 \\ \dot\theta \\ 0 \end{bmatrix}_1 = \begin{bmatrix} 1 & 0 & 0 \\ 0 & \cos\varphi & \sin\varphi \\ 0 & -\sin\varphi & \cos\varphi \end{bmatrix} \begin{bmatrix} \cos\theta & 0 & -\sin\theta \\ 0 & 1 & 0 \\ \sin\theta & 0 & \cos\theta \end{bmatrix} \begin{bmatrix} 0 \\ \dot\theta \\ 0 \end{bmatrix}_B \tag{1.23}
$$

After some algebra, Equation 1.21 can be written out in the form:

$$\begin{bmatrix} p \\ q \\ r \end{bmatrix} = \begin{bmatrix} 1 & 0 & -\sin\theta \\ 0 & \cos\varphi & \sin\varphi\cos\theta \\ 0 & -\sin\varphi & \cos\varphi\cos\theta \end{bmatrix} \begin{bmatrix} \dot\varphi \\ \dot\theta \\ \dot\psi \end{bmatrix} \tag{1.24}$$

Since the vectors on both sides of Equation 1.24 are in the body axes $X^B Y^B Z^B$, the subscript "B" has been dropped altogether.

The reverse transformation obtains by inverting the matrix in Equation 1.24 to yield

$$\begin{bmatrix} \dot\varphi \\ \dot\theta \\ \dot\psi \end{bmatrix} = \begin{bmatrix} 1 & \tan\theta\sin\varphi & \tan\theta\cos\varphi \\ 0 & \cos\varphi & -\sin\varphi \\ 0 & \sec\theta\sin\varphi & \sec\theta\cos\varphi \end{bmatrix} \begin{bmatrix} p \\ q \\ r \end{bmatrix} \tag{1.25}$$

Thus, for example, the second of Equation 1.25,

$$\dot\theta = q\cos\varphi - r\sin\varphi \tag{1.26}$$

reveals that in a horizontal turn where $q = \omega \sin\varphi$ and $r = \omega \cos\varphi$, the pitch angle rate $\dot\theta = 0$. In other words, the body X^B axis (the airplane nose) does not bob up/down relative to the horizon. This is an important point to note— even though there is a nonzero value of pitch rate q, there is no relative angular motion between the airplane and the free-stream velocity vector which happens to lie in the horizontal plane in this case.

There is a significant observation about the matrix in Equation 1.24 and its inverse in Equation 1.25 that we need to make here. Notice that in case $\theta = \pi/2$, the matrix in Equation 1.24 becomes

$$\begin{bmatrix} 1 & 0 & -1 \\ 0 & \cos\varphi & 0 \\ 0 & -\sin\varphi & 0 \end{bmatrix}$$

which is clearly singular (i.e., its determinant is 0). Therefore, its inverse, the matrix in Equation 1.25, is not defined. Hence, the relations in Equation 1.25 are not defined when $\theta = \pi/2$, which is a bit of a problem since Equation 1.25 is one of the set of equations used to solve for the motion of the airplane. However, this is a common problem with the use of Euler angles to define the orientation—every choice of Euler angle sequence will have a singular point. For most airplanes, $\theta = \pi/2$ is not expected to be part of a normal flight routine. Nevertheless, when using Equation 1.25 to simulate the motion of an airplane, it is usual practice to put in a check for the condition $\theta = \pi/2$ just in case.

Homework Exercise: Look up references to find alternative ways of describing the orientation of an airplane that can avoid the singularity problem (e.g., quaternions or direction cosines). Can you find the alternatives to Equation 1.25 in terms of quaternions or direction cosines (or any other choice of description)? How do those equations appear as against Equation 1.25 for the Euler angles?

1.4.2 Relation between Wind-Axis Angular Velocity Components (p_w, q_w, r_w) and Euler Angle Rates $(\dot{\mu}, \dot{\gamma}, \dot{\chi})$

Along exactly similar lines as in the case of the body-axis angular velocity, the relation between the wind-axis ω_w components in the wind axes, (p_w, q_w, r_w), and the wind-axis Euler angle rates $(\dot{\mu}, \dot{\gamma}, \dot{\chi})$ may be derived. Without further ado, these equations are written below:

$$\begin{bmatrix} p_w \\ q_w \\ r_w \end{bmatrix} = \begin{bmatrix} 1 & 0 & -\sin\gamma \\ 0 & \cos\mu & \sin\mu\cos\gamma \\ 0 & -\sin\mu & \cos\gamma\cos\mu \end{bmatrix} \begin{bmatrix} \dot{\mu} \\ \dot{\gamma} \\ \dot{\chi} \end{bmatrix} \qquad (1.27)$$

$$\begin{bmatrix} \dot{\mu} \\ \dot{\gamma} \\ \dot{\chi} \end{bmatrix} = \begin{bmatrix} 1 & \tan\gamma\sin\mu & \tan\gamma\cos\mu \\ 0 & \cos\mu & -\sin\mu \\ 0 & \sec\gamma\sin\mu & \sec\gamma\cos\mu \end{bmatrix} \begin{bmatrix} p_w \\ q_w \\ r_w \end{bmatrix} \qquad (1.28)$$

Homework Exercise: Discuss the singularity problem with reference to Equation 1.27. Does it mean that the motion of an airplane flying vertically upward cannot be studied? Or should (some of) the equations of motion in that case be formulated differently?

1.4.3 Difference between the Body-Axis Angular Velocity (p, q, r) and Wind-Axis Angular Velocity (p_w, q_w, r_w)

Since the angles separating the body- and wind-fixed axes are the aerodynamic angles α and β, whenever these two axes do not rotate in sync (i.e., with the same magnitude of angular velocity components), the angles

α and β will change. In other words, any difference between ω and ω_w will be seen as a rate of change of α and β. Just as the angles α and β decide the *static* aerodynamic forces on the airplane, their rates of change determine the *dynamic* aerodynamic forces acting on the airplane. Therefore, the difference $\omega - \omega_w$ is a significant quantity aerodynamically.

Note that ω and ω_w are different vectors, so one is not looking for a relationship between them. What we require is an expression for the *difference*, $\omega - \omega_w$, that may be of use in modeling the aerodynamic forces.

With reference to Figure 1.10, two infinitesimal rotations, the first by $-\Delta\beta$ and the second by $\Delta\alpha$, will transform the $X^W Y^W Z^W$ axes into coinciding with the $X^B Y^B Z^B$ axes. The first rotation takes place about the Z^W axis (which is same as the stability Z^S axis) with unit vector $\hat{k}_W$. The second rotation is about the stability Y^S axis (same as body Y^B axis) with unit vector $\hat{j}_B$. The vector sum of these two infinitesimal rotations is

$$\Delta\underline{\sigma} = (-\Delta\beta)\,\hat{k}_W + \Delta\alpha\,\hat{j}_B \tag{1.29}$$

Taking the differential with time gives

$$\dot{\underline{\sigma}} = (-\dot{\beta})\hat{k}_W + \dot{\alpha}\,\hat{j}_B \tag{1.30}$$

where $\dot{\underline{\sigma}}$ is the difference between the body- and wind-axis angular velocities, that is, $\omega - \omega_w$. Therefore, in terms of their components, we can write

$$\begin{bmatrix} p \\ q \\ r \end{bmatrix}_B - \begin{bmatrix} p_w \\ q_w \\ r_w \end{bmatrix}_W = (-\dot{\beta})\hat{k}_W + \dot{\alpha}\,\hat{j}_B \tag{1.31}$$

where the subscripts indicate that the body-axis angular velocity components are taken about the $X^B Y^B Z^B$ axes and the wind-axis angular velocity components are about the $X^W Y^W Z^W$ axes. Using the rotation matrices in Equation 1.11, we shall convert all the entities in Equation 1.31 to a common axis system, namely, the body $X^B Y^B Z^B$ axes. On carrying out this procedure, Equation 1.31 may be written out as

$$\begin{bmatrix} p_b^b - p_w^b \\ q_b^b - q_w^b \\ r_b^b - r_w^b \end{bmatrix} = \begin{bmatrix} \dot{\beta}\sin\alpha \\ \dot{\alpha} \\ -\dot{\beta}\cos\alpha \end{bmatrix} \tag{1.32}$$

TABLE 1.3 Summary of Airplane Kinematic Equations

Variable	Equation
Position (*body-axis*)	$$\begin{bmatrix} \dot{x}^E \\ \dot{y}^E \\ \dot{z}^E \end{bmatrix} = \begin{bmatrix} c\psi c\theta & c\psi s\theta s\varphi - s\psi c\varphi & c\psi s\theta c\varphi + s\psi s\varphi \\ s\psi c\theta & s\psi s\theta s\varphi + c\psi c\varphi & s\psi s\theta c\varphi - c\psi s\varphi \\ -s\theta & c\theta s\varphi & c\theta c\varphi \end{bmatrix} \begin{bmatrix} u \\ v \\ w \end{bmatrix} \quad (1.6)$$
Position (*wind-axis*)	$$\begin{bmatrix} \dot{x}^E \\ \dot{y}^E \\ \dot{z}^E \end{bmatrix} = \begin{bmatrix} V\cos\gamma\cos\chi \\ V\cos\gamma\sin\chi \\ -V\sin\gamma \end{bmatrix} \quad (1.10)$$
Orientation (*body-axis*)	$$\begin{bmatrix} \dot{\varphi} \\ \dot{\theta} \\ \dot{\psi} \end{bmatrix} = \begin{bmatrix} 1 & \tan\theta\sin\varphi & \tan\theta\cos\varphi \\ 0 & \cos\varphi & -\sin\varphi \\ 0 & \sec\theta\sin\varphi & \sec\theta\cos\varphi \end{bmatrix} \begin{bmatrix} p \\ q \\ r \end{bmatrix} \quad (1.25)$$
Orientation (*wind-axis*)	$$\begin{bmatrix} \dot{\mu} \\ \dot{\gamma} \\ \dot{\chi} \end{bmatrix} = \begin{bmatrix} 1 & \tan\gamma\sin\mu & \tan\gamma\cos\mu \\ 0 & \cos\mu & -\sin\mu \\ 0 & \sec\gamma\sin\mu & \sec\gamma\cos\mu \end{bmatrix} \begin{bmatrix} p_w \\ q_w \\ r_w \end{bmatrix} \quad (1.28)$$

The superscript "*b*" refers to components along body-fixed axes. Thus, the rates of change of the aerodynamic angles are related to the difference between the angular velocities of the body- and wind-fixed axes.

At this point we can pause to write out the airplane kinematic equations—these relate the airplane position to its inertial velocity and the orientation angles to the angular velocity. And, as we have seen, these may be written in terms of either body- or wind-axis variables. They are summarized in Table 1.3.

Now we have all the machinery required to derive the equations for the translational and rotational dynamics of an airplane in flight.

1.5 TRANSLATIONAL EQUATIONS OF MOTION

The translational equations of motion relate the rate of change of the inertial velocity V to the forces acting on the airplane. As discussed previously, we assume the airplane in flight to be a rigid body with six degrees of freedom with fixed mass m and fixed CG and MI. Additional effects, as in the examples in Section 1.1, may be taken into account as and when needed. A flat Earth is chosen as the inertial frame of reference.

Figure 1.14 shows the representation that we shall use of the aircraft for the purpose of deriving the equations of motion. The body-fixed axes $X^B Y^B Z^B$ are fixed to the CG labeled O^B. The Earth-fixed axes $X^E Y^E Z^E$

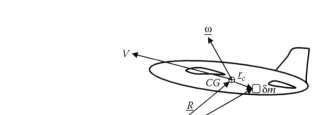

FIGURE 1.14 Aircraft representation for the purpose of deriving equations of motion.

provide an inertial reference frame. The airplane position is given by the position vector R from the origin O^E of the $X^E Y^E Z^E$ axes to the aircraft CG (O^B). The inertial velocity of the airplane is the velocity V of its CG relative to the Earth-fixed axes. Likewise its angular velocity relative to the inertial $X^E Y^E Z^E$ axes is denoted by ω. We shall assume the airplane geometry to be symmetric about the $X^B Z^B$ plane, which is called the longitudinal plane of the aircraft.

Consider an element of mass δm at a distance r_C from the CG O^B as marked in Figure 1.14. The velocity of this elemental mass as seen by an observer in the inertial frame of reference ($X^E Y^E Z^E$) is

$$\underline{V} = \underline{V}_C + \underline{\omega}_C \times \underline{r}_C \tag{1.33}$$

where $\underline{V}_C$ refers to the velocity of the CG, $\underline{\omega}_C$ refers to the angular velocity of the airplane about its CG, and $\underline{r}_C$ is the distance from the CG to the elemental mass. Linear momentum of the elemental mass δm is given by

$$\delta m \underline{V} = \delta m \underline{V}_C + \delta m (\underline{\omega}_C \times \underline{r}_C) \tag{1.34}$$

Summing this over the whole body of the airplane, the total linear momentum can be found as

$$\sum \delta m \underline{V} = \sum \delta m \underline{V}_C + \sum \delta m (\underline{\omega}_C \times \underline{r}_C) = \underline{V}_C \sum \delta m + \underline{\omega}_C \times \sum \delta m \underline{r}_C \tag{1.35}$$

since, being a rigid body, the angular velocity experienced by all the elements of the airplane is the same. In the limit, when an elemental mass is assumed to be very small, that is, $\delta m \to 0$, the summation can be written as an integral and therefore the total linear momentum appears as

$$\underline{p} = \underline{V}_C \int dm + \underline{\omega}_C \times \int \underline{r}_C dm \qquad (1.36)$$

Since all distances are being measured on the aircraft with respect to its CG (the origin of the body-fixed axes), which is also the center-of-mass (CM) of the airplane, one can further simplify Equation 1.36 using $\int \underline{r}_C dm = 0$ (definition of center-of-mass) to

$$\underline{p} = \underline{V}_C \int dm + \underline{\omega}_C \times \underbrace{\int \underline{r}_C dm}_{=0} = m\underline{V}_C \qquad (1.37)$$

In other words, the linear momentum of the airplane is simply the linear momentum of an equivalent point mass located at its CG.

The equation of translational motion can be written by differentiating Equation 1.37 with respect to time as

$$\underline{F} = \frac{d\underline{p}}{dt} = m\frac{d\underline{V}_C}{dt} \qquad (1.38)$$

where the rate of change d/dt is computed in the inertial frame. Note that Equation 1.38 is independent of which components of $\underline{V}_C$ are used to represent the vector. However, it is generally convenient to write the components as well as the rate of change d/dt in the same set of axes. In the present instance, we shall first choose the body-fixed axes to write the velocity components and at the same time convert the d/dt in the inertial frame in Equation 1.38 to the rate of change in the body-fixed frame.

The rate of change of a vector $\underline{A}$ in an inertial frame of reference (labeled "I") may be written in terms of its rate of change in another axis system (labeled "B") having angular velocity $\underline{\omega}_B$ with respect to the inertial frame as follows:

$$\left.\frac{d\underline{A}}{dt}\right|_I = \left.\frac{d\underline{A}}{dt}\right|_B + (\underline{\omega}_B \times \underline{A}) \qquad (1.39)$$

Note that the vector $\underline{A}$ has not yet been written in terms of its components; indeed, Equation 1.39 is valid no matter which choice of axes is made to write out these components. It is only convenient, not essential, to write the components of $\underline{A}$ in the same axis system (labeled "B").

If the axis system (labeled "B") is chosen to be the airplane body-fixed axis system, then $\underline{\omega}_B$, the angular velocity of the body-fixed axis relative to the inertial reference frame, is the same as the $\underline{\omega}_C$ used earlier where the subscript "C" was used to indicate that it was the angular velocity about the airplane CG.

Equation 1.38 in terms of the rate of change d/dt computed in the body-fixed axes is

$$F = m \frac{dV_C}{dt}\bigg|_B + m(\underline{\omega}_B \times \underline{V}_C) \tag{1.40}$$

Equation 1.38 may equally well be written in terms of the rate of change d/dt computed in the wind-fixed axes as follows:

$$F = m \frac{dV_C}{dt}\bigg|_W + m(\underline{\omega}_W \times \underline{V}_C) \tag{1.41}$$

where $\underline{\omega}_W$ is the angular velocity of the wind-fixed axis relative to the inertial reference frame. Either Equation 1.40 or 1.41 may be used to represent the translational dynamics of the airplane.

First, taking Equation 1.40 and writing $\underline{F}$, $\underline{V}_C$ in terms of their body-axis components, $\underline{V}_C = [u\ v\ w]^T$ and $\underline{\omega}_B = [p\ q\ r]^T$.

The cross product of vector $\underline{\omega}_B$ can be expressed in matrix form as

$$\underline{\omega}_B \times = \begin{bmatrix} 0 & -r & q \\ r & 0 & -p \\ -q & p & 0 \end{bmatrix} \tag{1.42}$$

Then, Equation 1.40 in the body-fixed axes system may be written out as

$$\underline{F}_B = m\begin{bmatrix} \dot{u} \\ \dot{v} \\ \dot{w} \end{bmatrix} + m\begin{bmatrix} 0 & -r & q \\ r & 0 & -p \\ -q & p & 0 \end{bmatrix}\begin{bmatrix} u \\ v \\ w \end{bmatrix} = m\begin{bmatrix} \dot{u} - rv + qw \\ \dot{v} + ru - pw \\ \dot{w} - qu + pv \end{bmatrix} \tag{1.43}$$

where the subscript "B" on the force vector indicates that its components are along the body-fixed axes and the d/dt have been replaced by overdots to indicate derivative with t.

Along very similar lines, writing $\underline{F}$, $\underline{V}_C$ in terms of their wind-axis components in Equation 1.41, $\underline{V}_C = [V \; 0 \; 0]^T$ and $\underline{\omega}_W = [p_w \; q_w \; r_w]^T$, the airplane translational dynamics equations now appear as

$$\underline{F}_W = m \begin{bmatrix} \dot{V} \\ 0 \\ 0 \end{bmatrix} + m \begin{bmatrix} 0 & -r_w & q_w \\ r_w & 0 & -p_w \\ -q_w & p_w & 0 \end{bmatrix} \begin{bmatrix} V \\ 0 \\ 0 \end{bmatrix} = m \begin{bmatrix} \dot{V} \\ r_w V \\ -q_w V \end{bmatrix} \qquad (1.44)$$

where $\underline{F}$ now carries a subscript "W" to suggest that its components are to be in the wind-fixed axes.

In case Equation 1.43 is used, the variables solved for are (u, v, w) and the usual variables of interest, namely, V, α, β, may be obtained by use of the relations in Equation 1.14. On the other hand, if Equation 1.44 is to be used, then further work is required to derive equations in terms of the variables α, β. First, from Equation 1.32, we have

$$\begin{bmatrix} p_w^b \\ q_w^b \\ r_w^b \end{bmatrix} = \begin{bmatrix} p_b^b - \dot{\beta}\sin\alpha \\ q_b^b - \dot{\alpha} \\ r_b^b + \dot{\beta}\cos\alpha \end{bmatrix} = \begin{bmatrix} p - \dot{\beta}\sin\alpha \\ q - \dot{\alpha} \\ r + \dot{\beta}\cos\alpha \end{bmatrix} \qquad (1.45)$$

where the subscript and superscript on the body-axis angular velocity components about the body-fixed axes (p, q, r) have been dropped. But note that the components of the wind-axis angular velocity in Equation 1.45 are in terms of its *body-axis* components. These need to be converted to the *wind-axis* components before use in Equation 1.44. Using Equations 1.11 and 1.12, this can be done as follows:

$$\begin{bmatrix} p_w^w \\ q_w^w \\ r_w^w \end{bmatrix} = R_{-\beta} R_\alpha \begin{bmatrix} p_w^b \\ q_w^b \\ r_w^b \end{bmatrix} = \underbrace{\begin{bmatrix} c\beta & s\beta & 0 \\ -s\beta & c\beta & 0 \\ 0 & 0 & 1 \end{bmatrix}}_{R_{-\beta}} \underbrace{\begin{bmatrix} c\alpha & 0 & s\alpha \\ 0 & 1 & 0 \\ -s\alpha & 0 & c\alpha \end{bmatrix}}_{R_\alpha} \begin{bmatrix} p - \dot{\beta}\sin\alpha \\ q - \dot{\alpha} \\ r + \dot{\beta}\cos\alpha \end{bmatrix} \qquad (1.46)$$

Equation 1.45 has been used in the second step. Expanding Equation 1.46, one gets the following relations for p_w, q_w, r_w:

$$
\begin{aligned}
p_w &= (p\cos\alpha + r\sin\alpha)\cos\beta + (q - \dot{\alpha})\sin\beta \\
q_w &= -(p\cos\alpha + r\sin\alpha)\sin\beta + (q - \dot{\alpha})\cos\beta \\
r_w &= (-p\sin\alpha + r\cos\alpha) + \dot{\beta}
\end{aligned}
\tag{1.47}
$$

These may now be inserted into Equation 1.44 to give the airplane translational dynamics equations in terms of the variables V, α, β:

$$
\underline{F}_W = m
\begin{bmatrix}
\dot{V} \\
\left((-p\sin\alpha + r\cos\alpha) + \dot{\beta}\right)V \\
\left((p\cos\alpha + r\sin\alpha)\sin\beta - (q - \dot{\alpha})\cos\beta\right)V
\end{bmatrix}
\tag{1.48}
$$

1.6 REPRESENTATION OF FORCES ACTING ON THE AIRPLANE

To complete the equations of translational dynamics, Equation 1.43 or 1.48, we need to write out the forces acting on the airplane. Typically, three sources of the forces are considered: gravity ("G"), aerodynamic ("A"), and propulsive ("P"). The net force is the sum of these three:

$$
\underline{F} = \underline{F}^G + \underline{F}^A + \underline{F}^P
\tag{1.49}
$$

Each component of force may usually be expressed most conveniently in a particular axis system, as explained below. Then it is necessary to transform them to the "B" or "W" axis system before being inserted into Equation 1.43 or 1.48, respectively.

1.6.1 Gravity

The gravitational force is best expressed in the Earth-fixed axes, as follows:

$$
\underline{F}^G =
\begin{bmatrix}
0 \\
0 \\
mg
\end{bmatrix}_E
\tag{1.50}
$$

As already worked out in Equation 1.5, the components of the gravitational force in body-fixed axes are

$$F_B^G = \begin{bmatrix} -mg\sin\theta \\ mg\sin\varphi\cos\theta \\ mg\cos\varphi\cos\theta \end{bmatrix} \tag{1.51}$$

From the Exercise Problem 4, the gravitational force components along the wind-fixed axes work out to

$$F_W^G = \begin{bmatrix} -mg\sin\gamma \\ mg\sin\mu\cos\gamma \\ mg\cos\mu\cos\gamma \end{bmatrix} \tag{1.52}$$

1.6.2 Aerodynamic Force

The aerodynamic force acting on the airplane is best expressed in the stability axis system. The stability axis is the intermediate axis system that arises in conversion between the wind- and body-fixed axes, as shown in Figure 1.10.

The different axis systems and the forces acting on the airplane are marked in Figure 1.15. The net aerodynamic force is first resolved into an

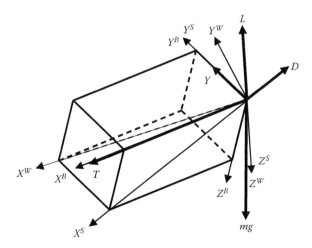

FIGURE 1.15 Sense of the forces acting on the aircraft—thrust, weight, and aerodynamics.

in-plane component in the $X^S Z^S$ plane (same as the body $X^B Z^B$ plane) and an out-of-plane component along the Y^S axis (same as Y^B axis). The out-of-plane component acting along positive Y^S axis is called the *side force*, Y. The in-plane component is then resolved into a *lift force*, L, acting along the negative Z^S direction and a *drag force*, D, acting along the negative X^S direction. Thus

$$\underline{F}^A = \begin{bmatrix} -D \\ Y \\ -L \end{bmatrix}_S \tag{1.53}$$

To transform Equation 1.53 to the body- and wind-fixed axes, we take recourse to the rotation matrices already introduced in Equation 1.11.

$$\underline{F}^A_B = \underbrace{\begin{bmatrix} c\alpha & 0 & -s\alpha \\ 0 & 1 & 0 \\ s\alpha & 0 & c\alpha \end{bmatrix}}_{R_{-\alpha}} \begin{bmatrix} -D \\ Y \\ -L \end{bmatrix}_S = \begin{bmatrix} -D\cos\alpha + L\sin\alpha \\ Y \\ -D\sin\alpha - L\cos\alpha \end{bmatrix} \tag{1.54}$$

$$\underline{F}^A_W = \underbrace{\begin{bmatrix} c\beta & s\beta & 0 \\ -s\beta & c\beta & 0 \\ 0 & 0 & 1 \end{bmatrix}}_{R_{-\beta}} \begin{bmatrix} -D \\ Y \\ -L \end{bmatrix}_S = \begin{bmatrix} -D\cos\beta + Y\sin\beta \\ D\sin\beta + Y\cos\beta \\ -L \end{bmatrix} \tag{1.55}$$

A comment here would perhaps be in order. Most elementary texts present a two-dimensional picture where they resolve the aerodynamic force into a component along the relative velocity vector called "drag" and a normal component called "lift." From this point on, "drag" tends to be defined as the component of aerodynamic force *in the direction* of the relative velocity vector. When this definition is blindly carried over to the more general three-dimensional case where the velocity vector does not lie in the $X^B Z^B$ plane, then the "drag" force so defined along the direction of V would not lie in the $X^B Z^B$ plane either. Equation 1.54 shows that this is not the case. Instead, it is appropriate to define the "drag" force as the component in the direction of the *component of the relative velocity vector in the $X^B Z^B$ plane.*

1.6.3 Propulsive Force

The usual propulsive force is the thrust that usually acts along a fixed direction relative to the body. However, in the interest of simplicity, it is assumed that the entire thrust acts along the body X^B axis. Therefore

$$\underline{F}_B^T = \begin{bmatrix} T_x \\ T_y \\ T_z \end{bmatrix}_B \approx \begin{bmatrix} T \\ 0 \\ 0 \end{bmatrix}_B \tag{1.56}$$

However, note that in case a thrust vectoring nozzle is in use, there may be significant components of the thrust force along the Y^B and Z^B directions.

Equation 1.56 can be transformed into the wind-fixed axis as below using Equations 1.11 and 1.12:

$$\underline{F}_W^T = \underbrace{\begin{bmatrix} c\beta & s\beta & 0 \\ -s\beta & c\beta & 0 \\ 0 & 0 & 1 \end{bmatrix}}_{R_{-\beta}} \underbrace{\begin{bmatrix} c\alpha & 0 & s\alpha \\ 0 & 1 & 0 \\ -s\alpha & 0 & c\alpha \end{bmatrix}}_{R_\alpha} \begin{bmatrix} T \\ 0 \\ 0 \end{bmatrix}_B = T \begin{bmatrix} \cos\beta\cos\alpha \\ -\sin\beta\cos\alpha \\ -\sin\alpha \end{bmatrix} \tag{1.57}$$

On collecting the appropriate terms, Equations 1.43 and 1.48 can each be written out with the force terms in full as follows:

$$\begin{bmatrix} -mg\sin\theta \\ mg\sin\varphi\cos\theta \\ mg\cos\varphi\cos\theta \end{bmatrix} + \begin{bmatrix} -D\cos\alpha + L\sin\alpha \\ Y \\ -D\sin\alpha - L\cos\alpha \end{bmatrix} + \begin{bmatrix} T \\ 0 \\ 0 \end{bmatrix} = m \begin{bmatrix} \dot{u} - rv + qw \\ \dot{v} + ru - pw \\ \dot{w} - qu + pv \end{bmatrix} \tag{1.58}$$

$$\begin{bmatrix} -mg\sin\gamma \\ mg\sin\mu\cos\gamma \\ mg\cos\mu\cos\gamma \end{bmatrix} + \begin{bmatrix} -D\cos\beta + Y\sin\beta \\ D\sin\beta + Y\cos\beta \\ -L \end{bmatrix} + \begin{bmatrix} T\cos\beta\cos\alpha \\ -T\sin\beta\cos\alpha \\ -T\sin\alpha \end{bmatrix}$$
$$= m \begin{bmatrix} \dot{V} \\ \left((-p\sin\alpha + r\cos\alpha) + \dot{\beta}\right)V \\ \left((p\cos\alpha + r\sin\alpha)\sin\beta - (q - \dot{\alpha})\cos\beta\right)V \end{bmatrix} \tag{1.59}$$

Equations 1.58 and 1.59 are the final, complete forms of the equations for the translational dynamics of airplanes. Their integration yields the components of the airplane inertial velocity vector V.

Homework Exercise: Compare between the body-axis form of the equations of translational motion in terms of the variables (u, v, w) in Equation 1.58 and the wind-axis form in Equation 1.59 written in terms of the variables (V, α, β). Which one appears easier to compute? Which one provides results that make more sense physically?

1.7 ROTATIONAL EQUATIONS OF MOTION

Referring back to Figure 1.14, the equations for rotational motion of aircraft can be arrived at by again considering the elemental mass δm. In this instance, we evaluate the angular momentum of this elemental mass about the CG as reference point as follows:

$$\delta \underline{h} = \underline{r}_C \times \delta m \underline{V} = \underline{r}_C \times \delta m(\underline{V}_C + \underline{\omega}_B \times \underline{r}_C) \tag{1.60}$$

Summing over the whole body of the aircraft, the total angular momentum comes out to be

$$\sum \delta \underline{h} = \sum \underline{r}_C \times \delta m(\underline{V}_C + \underline{\omega}_B \times \underline{r}_C)$$
$$= \sum \underline{r}_C \delta m \times \underline{V}_C + \sum \delta m\{\underline{r}_C \times (\underline{\omega}_B \times \underline{r}_C)\} \tag{1.61}$$

As before we take the limit $\delta m \rightarrow 0$, and the net airplane angular momentum takes the form

$$\underline{h} = \int \delta \underline{h} = \underbrace{\int \underline{r}_C dm \times \underline{V}_C}_{1} + \underbrace{\int dm\{\underline{r}_C \times (\underline{\omega}_B \times \underline{r}_C)\}}_{2} \tag{1.62}$$

Term "1" in Equation 1.62 is zero because we measure all distances from the center of mass of the aircraft, so that $\int \underline{r}_C dm = 0$ as before. Therefore, the net angular momentum of the whole body is

$$\underline{h} = \int dm\{\underline{r}_C \times (\underline{\omega}_B \times \underline{r}_C)\} \tag{1.63}$$

The equation of rotational dynamics of the airplane is obtained by equating the rate of change of its angular momentum to the net moment acting on the body about its CG. That is

$$\underline{M} = \frac{d\underline{h}}{dt} \tag{1.64}$$

where the derivative d/dt is calculated in the inertial reference frame. In case of the moment equations, invariably the body-fixed axes are used to write both the components of the angular momentum as well as the derivative d/dt. This is because the moments of inertia usually remain constant in the body-fixed axes, which helps to simplify the $d\underline{h}/dt$ term.

By the standard rule for converting the derivative d/dt from inertial (Earth-fixed) axis to body-fixed axis,

$$\underline{M} = \frac{d\underline{h}}{dt}\bigg|_I = \frac{d\underline{h}}{dt}\bigg|_B + \underline{\omega}_B \times \underline{h} \tag{1.65}$$

where the components of $\underline{M}$ and $\underline{h}$ are to be taken in the same axis system, usually the body-fixed axes. Expanding the expression for $\underline{h}$ in Equation 1.63 in terms of body-axis components of all the variables,

$$
\begin{aligned}
\underline{h} &= \int dm\{(x\hat{i} + y\hat{j} + z\hat{k}) \times [(p\hat{i} + q\hat{j} + r\hat{k}) \times (x\hat{i} + y\hat{j} + r\hat{k})]\\
&= \int dm\{(x\hat{i} + y\hat{j} + z\hat{k}) \times [(qz - yr)\hat{i} + (rx - pz)\hat{j} + (py - qx)\hat{k}]\\
&= [p\int(y^2 + z^2)dm - q\int xy\delta m - r\int xz\delta m]\hat{i}\\
&\quad + \left[-p\int yxdm + q\int(x^2 + z^2)dm - r\int yzdm\right]\hat{j}\\
&\quad + \left[-p\int zxdm - q\int zydm + r\int(x^2 + y^2)dm\right]\hat{j}\\
&= [pI_{xx} - qI_{xy} - rI_{xz}]\hat{i} + [-pI_{yx} + qI_{yy} - rI_{yz}]\hat{j}\\
&\quad + [-pI_{zx} - qI_{zy} + rI_{zz}]\hat{k} \tag{1.66}
\end{aligned}
$$

where the moments and products of inertia have their usual definitions as below:

$$I_{xx} = \int (y^2 + z^2)dm, \quad I_{yy} = \int (x^2 + z^2)dm, \quad I_{zz} = \int (x^2 + y^2)dm;$$

$$I_{xy} = I_{yx} = \int xy\,dm, \quad I_{xz} = I_{zx} = \int xz\,dm, \quad I_{yz} = I_{zy} = \int yz\,dm$$

$$(1.67)$$

Equation 1.66 can be written in matrix form as

$$\underline{h} = \begin{bmatrix} h_x \\ h_y \\ h_z \end{bmatrix} = \begin{bmatrix} I_{xx} & -I_{xy} & -I_{xz} \\ -I_{yx} & I_{yy} & -I_{yz} \\ -I_{zx} & -I_{zy} & I_{zz} \end{bmatrix} \begin{bmatrix} p \\ q \\ r \end{bmatrix} = \underline{\underline{I}} \cdot \omega_B \qquad (1.68)$$

where $\underline{\underline{I}}$ is the inertia tensor.

Inserting the form of the angular momentum vector in Equation 1.68 into the rotational dynamics Equation 1.65,

$$\underline{M} = \left.\frac{dh}{dt}\right|_B + \omega_B \times \underline{h} = \left.\frac{d}{dt}\right|_B \begin{bmatrix} h_x \\ h_y \\ h_z \end{bmatrix} + \omega_B \times \begin{bmatrix} h_x \\ h_y \\ h_z \end{bmatrix}$$

$$= \begin{bmatrix} I_{xx} & -I_{xy} & -I_{xz} \\ -I_{yx} & I_{yy} & -I_{yz} \\ -I_{zx} & -I_{zy} & I_{zz} \end{bmatrix} \left.\frac{d}{dt}\right|_B \begin{bmatrix} p \\ q \\ r \end{bmatrix} + \omega_B \times \begin{bmatrix} I_{xx} & -I_{xy} & -I_{xz} \\ -I_{yx} & I_{yy} & -I_{yz} \\ -I_{zx} & -I_{zy} & I_{zz} \end{bmatrix} \begin{bmatrix} p \\ q \\ r \end{bmatrix}$$

$$(1.69)$$

For most airplanes, the longitudinal plane $X^B Z^B$ is a plane of symmetry, which means that the corresponding cross products of inertia are zero; that is, $I_{yx} = I_{yz} = 0$. Equation 1.69 then simplifies a little to

$$\underline{M} = \begin{bmatrix} I_{xx} & 0 & -I_{xz} \\ 0 & I_{yy} & 0 \\ -I_{zx} & 0 & I_{zz} \end{bmatrix} \left.\frac{d}{dt}\right|_B \begin{bmatrix} p \\ q \\ r \end{bmatrix} + \omega_B \times \begin{bmatrix} I_{xx} & 0 & -I_{xz} \\ 0 & I_{yy} & 0 \\ -I_{zx} & 0 & I_{zz} \end{bmatrix} \begin{bmatrix} p \\ q \\ r \end{bmatrix} \qquad (1.70)$$

From Equation 1.42, we have

$$\underline{\omega}_B \times = \begin{bmatrix} 0 & -r & q \\ r & 0 & -p \\ -q & p & 0 \end{bmatrix} \tag{1.71}$$

Therefore, Equation 1.70 in full appears as

$$
\begin{aligned}
\underline{M} &= \begin{bmatrix} I_{xx} & 0 & -I_{xz} \\ 0 & I_{yy} & 0 \\ -I_{zx} & 0 & I_{zz} \end{bmatrix} \frac{d}{dt}\bigg|_B \begin{bmatrix} p \\ q \\ r \end{bmatrix} + \begin{bmatrix} 0 & -r & q \\ r & 0 & -p \\ -q & p & 0 \end{bmatrix} \begin{bmatrix} I_{xx} & 0 & -I_{xz} \\ 0 & I_{yy} & 0 \\ -I_{zx} & 0 & I_{zz} \end{bmatrix} \begin{bmatrix} p \\ q \\ r \end{bmatrix} \\
&= \begin{bmatrix} I_{xx}\dot{p} - I_{xz}\dot{r} \\ I_{yy}\dot{q} \\ -I_{zx}\dot{p} + I_{zz}\dot{r} \end{bmatrix} + \begin{bmatrix} -qI_{zx} & -rI_{yy} & qI_{zz} \\ rI_{xx} + pI_{zx} & 0 & -rI_{xz} - pI_{zz} \\ -qI_{xx} & pI_{yy} & qI_{xz} \end{bmatrix} \begin{bmatrix} p \\ q \\ r \end{bmatrix} \\
&= \begin{bmatrix} I_{xx}\dot{p} - I_{xz}\dot{r} \\ I_{yy}\dot{q} \\ -I_{zx}\dot{p} + I_{zz}\dot{r} \end{bmatrix} + \begin{bmatrix} -pqI_{zx} - qr(I_{yy} - I_{zz}) \\ (p^2 - r^2)I_{zx} - pr(I_{zz} - I_{xx}) \\ qrI_{xz} - pq(I_{xx} - I_{yy}) \end{bmatrix}
\end{aligned}
\tag{1.72}
$$

where the d/dt have been replaced by overdots to indicate derivative with t. Equation 1.72 is the standard form of the rotational dynamics equation for most airplanes (ones with a longitudinal plane of symmetry).

For the special case where the body-fixed axes are chosen to coincide with the airplane principal axes, the cross product of inertia $I_{zx} = I_{xz}$ is zero as well; then Equation 1.72 simplifies further to

$$\underline{M} = \begin{bmatrix} I_{xx}\dot{p} \\ I_{yy}\dot{q} \\ I_{zz}\dot{r} \end{bmatrix} + \begin{bmatrix} -qr(I_{yy} - I_{zz}) \\ -pr(I_{zz} - I_{xx}) \\ -pq(I_{xx} - I_{yy}) \end{bmatrix} \tag{1.73}$$

$\underline{M}$ in the preceding equations is, of course, the net moment acting about the airplane CG with components along the body-fixed axes.

1.8 REPRESENTATION OF MOMENTS ACTING ON THE AIRPLANE

Of the three sources of force acting on the airplane discussed in Section 1.6, gravity acting at the CG obviously does not produce a moment about the CG. That leaves the other two—aerodynamic and propulsive. Thus,

$$M = \underline{M}^A + \underline{M}^T \tag{1.74}$$

1.8.1 Aerodynamic Moment

The aerodynamic forces discussed in Section 1.7 and depicted in Figure 1.15 usually act at a point distinct from the airplane CG. Thus, when they are transferred to the CG, a moment is concurrently produced at the CG. In addition, there may also be a moment due to a pure couple.

The net moment about the CG is usually specified directly in terms of body-axis components, $\mathcal{L}$, M, and N, as marked in Figure 1.16. $\mathcal{L}$ acting about the X^B axis is called the rolling moment, M about the Y^B axis is the pitching moment, and N is the yawing moment about the Z^B axis. Their positive sense is as marked in Figure 1.16 and is similar to the positive sense of the corresponding angular velocity components, p, q, r, respectively. Thus,

$$\underline{M}^A_B = \begin{bmatrix} \mathcal{L} \\ M \\ N \end{bmatrix} \tag{1.75}$$

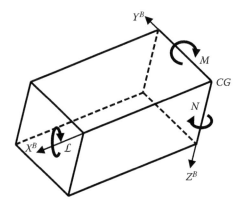

FIGURE 1.16 Sense of the moments acting on the aircraft due to aerodynamics.

1.8.2 Propulsive Moment

The propulsive force has been taken in Equation 1.56 to be a thrust T acting along the X^B axis. Depending on the offset h between the body-fixed axes and the thrust line, a pitching moment may be produced.

For example, as shown in Figure 1.17, the thrust line is offset above the X^B axis by a distance h marked as positive upward. In this case,

$$\underline{M}_B^T = \begin{bmatrix} 0 \\ -T.h \\ 0 \end{bmatrix} \tag{1.76}$$

For another example, consider a multiengine commercial airliner with wing-mounted engines as sketched in Figure 1.18. If either of the engines is turned off and only the other engine operates, then the sole operating engine produces a yawing moment about the airplane CG. In that case,

$$\underline{M}_B^T = \begin{bmatrix} 0 \\ 0 \\ \pm T.r_y \end{bmatrix} \tag{1.77}$$

where r_y is the moment arm from the engine thrust line to the airplane CG.

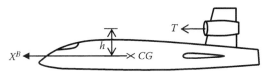

FIGURE 1.17   Moment acting on the aircraft due to offset thrust line.

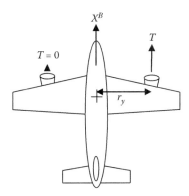

FIGURE 1.18   Moment acting on the aircraft due to asymmetric engine thrust.

In case of the thrust vectoring nozzle, the deflection of the nozzle will create a pitching/yawing moment. Thus, the form and value of the propulsive moment should be considered on a case-by-case basis. Usually, the engines are so placed on the airplane that the propulsive moment is as small as possible. However, in some cases, this may not be possible or desirable. For most of this text, we shall assume $\underline{M}^T = 0$ and consider only the aerodynamic moments in Equation 1.75 to act at the airplane CG. The propulsive moment will be additionally included as and when warranted.

Therefore, the final form of the airplane rotational dynamics Equation 1.72 with the moment components in Equation 1.75 is as below:

$$\begin{bmatrix} \mathcal{L} \\ M \\ N \end{bmatrix} = \begin{bmatrix} I_{xx}\dot{p} - I_{xz}\dot{r} \\ I_{yy}\dot{q} \\ -I_{zx}\dot{p} + I_{zz}\dot{r} \end{bmatrix} + \begin{bmatrix} -pqI_{zx} - qr(I_{yy} - I_{zz}) \\ (p^2 - r^2)I_{zx} - pr(I_{zz} - I_{xx}) \\ qrI_{xz} - pq(I_{xx} - I_{yy}) \end{bmatrix} \qquad (1.78)$$

At this point, we can gather the airplane dynamics (translation and rotation) equations into Table 1.4.

The kinematics equations in Table 1.3 and the dynamics equations in Table 1.4 constitute all the relations that are usually required for a 6 DOF analysis of the airplane dynamics as a rigid body. Obviously not all of them are simultaneously required for solution of a particular problem since there are multiple options of components/equations for some variables. Is there a preferred choice of equations? Usually, it depends on the specific problem at hand or even on the personal preference of the user. In the next section, we present certain selections of equations from Tables 1.3 and 1.4 used for the analysis of particular problems of airplane flight dynamics.

1.9 SELECTION OF EQUATIONS FOR SPECIFIC PROBLEMS

Sometimes a particular selection of variables or a reduced set of equations is convenient or necessary to address a specific problem of airplane flight dynamics. In case of constrained or restricted flight, only a few variables may be of interest dynamically, the others being set to fixed values. In such cases, a reduced set of equations should be studied, as suggested in the following.

TABLE 1.4 Summary of Airplane Translational and Rotational Dynamics Equations

Variable	Equation
Velocity (body-axis)	$$\begin{bmatrix} -mg\sin\theta \\ mg\sin\varphi\cos\theta \\ mg\cos\varphi\cos\theta \end{bmatrix} + \begin{bmatrix} -D\cos\alpha + L\sin\alpha \\ Y \\ -D\sin\alpha - L\cos\alpha \end{bmatrix} + \begin{bmatrix} T \\ 0 \\ 0 \end{bmatrix} = m \begin{bmatrix} \dot{u} - rv + qw \\ \dot{v} + ru - pw \\ \dot{w} - qu + pv \end{bmatrix} \quad (1.58)$$
Velocity (wind-axis)	$$\begin{bmatrix} -mg\sin\gamma \\ mg\sin\mu\cos\gamma \\ mg\cos\mu\cos\gamma \end{bmatrix} + \begin{bmatrix} -D\cos\beta + Y\sin\beta \\ D\sin\beta + Y\cos\beta \\ -L \end{bmatrix} + \begin{bmatrix} T\cos\beta\cos\alpha \\ -T\sin\beta\cos\alpha \\ -T\sin\alpha \end{bmatrix}$$ $$= m \begin{bmatrix} \dot{V} \\ ((-p\sin\alpha + r\cos\alpha) + \dot{\beta})V \\ ((p\cos\alpha + r\sin\alpha)\sin\beta - (q - \dot{\alpha})\cos\beta)V \end{bmatrix} \quad (1.59)$$
Angular velocity (body-axis)	$$\begin{bmatrix} \mathcal{L} \\ M \\ N \end{bmatrix} = \begin{bmatrix} I_{xx}\dot{p} - I_{xz}\dot{r} \\ I_{yy}\dot{q} \\ -I_{zx}\dot{p} + I_{zz}\dot{r} \end{bmatrix} + \begin{bmatrix} -pqI_{zx} - qr(I_{yy} - I_{zz}) \\ (p^2 - r^2)I_{zx} - pr(I_{zz} - I_{xx}) \\ qrI_{xz} - pq(I_{xx} - I_{yy}) \end{bmatrix} \quad (1.78)$$

1.9.1 Simulation of Arbitrary Maneuver

For time simulation of an arbitrary airplane maneuver in flight, the following set of equations in Table 1.5 is often used. Additionally, Equation 1.25 may also be solved if the orientation of the airplane body-fixed axes is of interest. The equations in Table 1.5 are mixed—the translational equations are in the wind axis and the rotational ones in the body axis with the body-axis (p, q, r) appearing in the translational dynamics. Since the wind-axis Euler angles are used in the translational equations, it makes sense to use the orientation equations in terms of (μ, γ, χ). It is not uncommon, and certainly not incorrect, to mix the wind- and body-axis variables and equations as long as they form a complete, consistent set. Most times, it is a matter of convenience or based on the requirements.

1.9.2 Steady States Such as Level Flight, Shallow Climb/Descent, Horizontal Turn, Spin, Etc.

Steady states, as we shall see later in greater detail, correspond to flights with fixed values of the variables. The three position variables and the yaw attitude (heading angle) are usually not of interest in these cases because values of these four variables usually do not cause any change in the other equations. Then a reduced set of eight equations listed in Table 1.6 is

TABLE 1.5 Airplane Equations for Simulation of Arbitrary Maneuver

Variable	Equation
Velocity (wind-axis)	$$\begin{bmatrix} -mg\sin\gamma \\ mg\sin\mu\cos\gamma \\ mg\cos\mu\cos\gamma \end{bmatrix} + \begin{bmatrix} -D\cos\beta+Y\sin\beta \\ D\sin\beta+Y\cos\beta \\ -L \end{bmatrix} + \begin{bmatrix} T\cos\beta\cos\alpha \\ -T\sin\beta\cos\alpha \\ -T\sin\alpha \end{bmatrix}$$ $$= m\begin{bmatrix} \dot{V} \\ ((-p\sin\alpha+r\cos\alpha)+\dot{\beta})V \\ ((p\cos\alpha+r\sin\alpha)\sin\beta-(q-\dot{\alpha})\cos\beta)V \end{bmatrix}$$ (1.59)
Angular velocity (body-axis)	$$\begin{bmatrix} \mathcal{L} \\ M \\ N \end{bmatrix} = \begin{bmatrix} I_{xx}\dot{p}-I_{xz}\dot{r} \\ I_{yy}\dot{q} \\ -I_{zx}\dot{p}+I_{zz}\dot{r} \end{bmatrix} + \begin{bmatrix} -pqI_{zx}-qr(I_{yy}-I_{zz}) \\ (p^2-r^2)I_{zx}-pr(I_{zz}-I_{xx}) \\ qrI_{xz}-pq(I_{xx}-I_{yy}) \end{bmatrix}$$ (1.78)
Position (wind-axis)	$$\begin{bmatrix} \dot{x}^E \\ \dot{y}^E \\ \dot{z}^E \end{bmatrix} = \begin{bmatrix} V\cos\gamma\cos\chi \\ V\cos\gamma\sin\chi \\ -V\sin\gamma \end{bmatrix}$$ (1.10)
Orientation (wind-axis)	$$\begin{bmatrix} \dot{\mu} \\ \dot{\gamma} \\ \dot{\chi} \end{bmatrix} = \begin{bmatrix} 1 & \tan\gamma\sin\mu & \tan\gamma\cos\mu \\ 0 & \cos\mu & -\sin\mu \\ 0 & \sec\gamma\sin\mu & \sec\gamma\cos\mu \end{bmatrix}\begin{bmatrix} p_w \\ q_w \\ r_w \end{bmatrix}$$ (1.28) $$p_w = (p\cos\alpha+r\sin\alpha)\cos\beta+(q-\dot{\alpha})\sin\beta$$ $$q_w = -(p\cos\alpha+r\sin\alpha)\sin\beta+(q-\dot{\alpha})\cos\beta$$ $$r_w = (-p\sin\alpha+r\cos\alpha)+\dot{\beta}$$ (1.47)

TABLE 1.6 Airplane Equations for Steady-State Flight Analysis

Variable	Equation
Velocity (body-axis)	$$\begin{bmatrix} -mg\sin\theta \\ mg\sin\varphi\cos\theta \\ mg\cos\varphi\cos\theta \end{bmatrix} + \begin{bmatrix} -D\cos\alpha+L\sin\alpha \\ Y \\ -D\sin\alpha-L\cos\alpha \end{bmatrix} + \begin{bmatrix} T \\ 0 \\ 0 \end{bmatrix} = m\begin{bmatrix} \dot{u}-rv+qw \\ \dot{v}+ru-pw \\ \dot{w}-qu+pv \end{bmatrix}$$ (1.58)
Angular velocity (body-axis)	$$\begin{bmatrix} \mathcal{L} \\ M \\ N \end{bmatrix} = \begin{bmatrix} I_{xx}\dot{p}-I_{xz}\dot{r} \\ I_{yy}\dot{q} \\ -I_{zx}\dot{p}+I_{zz}\dot{r} \end{bmatrix} + \begin{bmatrix} -pqI_{zx}-qr(I_{yy}-I_{zz}) \\ (p^2-r^2)I_{zx}-pr(I_{zz}-I_{xx}) \\ qrI_{xz}-pq(I_{xx}-I_{yy}) \end{bmatrix}$$ (1.78)
Orientation (body-axis)	$$\begin{bmatrix} \dot{\varphi} \\ \dot{\theta} \end{bmatrix} = \begin{bmatrix} 1 & \tan\theta\sin\varphi & \tan\theta\cos\varphi \\ 0 & \cos\varphi & -\sin\varphi \end{bmatrix}\begin{bmatrix} p \\ q \\ r \end{bmatrix}$$ (1.25)

TABLE 1.7   Airplane Equations for Steady-State Longitudinal Flight Analysis

Variable	Equation	
Velocity (*body-axis*)	$$\begin{bmatrix} -mg\sin\theta \\ mg\cos\theta \end{bmatrix} + \begin{bmatrix} -D\cos\alpha + L\sin\alpha \\ -D\sin\alpha - L\cos\alpha \end{bmatrix} + \begin{bmatrix} T \\ 0 \end{bmatrix} = m\begin{bmatrix} \dot{u} + qw \\ \dot{w} - qu \end{bmatrix}$$	(1.58)
Angular velocity (*body-axis*)	$$\begin{bmatrix} M \end{bmatrix} = \begin{bmatrix} I_{yy}\dot{q} \end{bmatrix} + \begin{bmatrix} \ \end{bmatrix}$$	(1.78)
Orientation (*body-axis*)	$$\begin{bmatrix} \dot{\theta} \end{bmatrix} = \begin{bmatrix} 1 \end{bmatrix}\begin{bmatrix} q \end{bmatrix}$$	(1.25)

used for analysis. However, note that in certain instances such as a steep climb/dive, ideally changes in atmospheric density will influence the thrust and the aerodynamic forces/moments and therefore the dynamic equations for velocity and angular velocity; hence, approximations must always be used judiciously.

Additionally, the relations in Equation 1.14 may be needed:

$$\alpha = \tan^{-1}\left(\frac{w}{u}\right) \quad \beta = \sin^{-1}\left(\frac{v}{V}\right) \quad \text{where } V = \sqrt{(u^2 + v^2 + w^2)} \quad (1.14)$$

1.9.3 Longitudinal Flight Steady States

Longitudinal flight refers to flight maneuvers restricted to the airplane longitudinal $X^B Z^B$ plane. In that case, only variables concerned with flight in the longitudinal plane are of interest; other variables are constrained, normally to zero. Table 1.7 lists the equations required.

Additionally, the relations in Equation 1.14 may also be used, and the flight path angle γ in case of flight restricted to the longitudinal plane is easily found as $\gamma = \theta - \alpha$.

1.9.4 Constant-Velocity Flight in the Longitudinal Plane

In case we are interested only in flights restricted to the longitudinal plane with the additional constraint that the inertial velocity has a constant magnitude, then it is better to use the wind-axis form of the velocity equations in Equation 1.59. The set of equations is presented in Table 1.8.

TABLE 1.8   Airplane Equations for Constant-Velocity Flight in the Longitudinal Plane

Variable	Equation	
Velocity (*wind-axis*)	$$\begin{bmatrix} -mg\sin\gamma \\ mg\cos\gamma \end{bmatrix} + \begin{bmatrix} -D \\ -L \end{bmatrix} + \begin{bmatrix} T\cos\alpha \\ -T\sin\alpha \end{bmatrix} = m\begin{bmatrix} 0 \\ \left(-(q-\dot\alpha)\right)V \end{bmatrix}$$	(1.59)
Angular velocity (*body-axis*)	$$\begin{bmatrix} M \end{bmatrix} = \begin{bmatrix} I_{yy}\dot q \end{bmatrix} + \begin{bmatrix} \\ \end{bmatrix}$$	(1.78)
Orientation (*body-axis*)	$$\begin{bmatrix} \dot\theta \end{bmatrix} = \begin{bmatrix} 1 \end{bmatrix}\begin{bmatrix} q \end{bmatrix}$$	(1.25)

Note that the equation in the first row of Equation 1.59 is now an algebraic relation, not a dynamic one, because of the constant-velocity constraint. As we shall see later, there is a more elegant way of imposing constraints on selected variables without having to write out a different set of equations for each choice of constrained variables.

1.9.5 Constant-Velocity Rolling Maneuvers

In this case, because the aircraft is in continuous rolling motion, the gravity vector continuously changes its orientation relative to the body-fixed axes; hence no true steady state is possible. It is, therefore, usual to consider the limit of $g/V \to 0$, where the gravity terms drop out of the equations of motion. Consequently, the Euler angles are rendered irrelevant since they appear in the velocity equations only through the gravity-related terms. The resulting set of equations is listed in Table 1.9.

Since the steady states of the equation set in Table 1.9 are not true steady states, they are usually called *pseudo-steady states*. Also note that the derivative of velocity on the right-hand side of Equation 1.59 in Table 1.9 has been set to zero because of the constant-velocity constraint.

1.10 EQUATIONS OF MOTION IN THE PRESENCE OF WIND

Figure 1.19 shows an aircraft in flight in the presence of an arbitrary wind represented by the inertial wind velocity vector W. V is the inertial aircraft velocity, that is, the velocity with respect to the Earth-fixed axes or the ground; hence it is also known as the ground velocity V_E. Since the

TABLE 1.9　Airplane Equations for Constant-Velocity Rolling Flight Analysis

Variable	Equation
Velocity (*wind-axis*)	$$\begin{bmatrix} 0 \end{bmatrix} + \begin{bmatrix} -D\cos\beta + Y\sin\beta \\ D\sin\beta + Y\cos\beta \\ -L \end{bmatrix} + \begin{bmatrix} T\cos\beta\cos\alpha \\ -T\sin\beta\cos\alpha \\ -T\sin\alpha \end{bmatrix}$$ $$= m \begin{bmatrix} 0 \\ \left((-p\sin\alpha + r\cos\alpha) + \dot{\beta}\right)V \\ \left((p\cos\alpha + r\sin\alpha)\sin\beta - (q-\dot{\alpha})\cos\beta\right)V \end{bmatrix} \qquad (1.59)$$
Angular velocity (*body-axis*)	$$\begin{bmatrix} \mathcal{L} \\ M \\ N \end{bmatrix} = \begin{bmatrix} I_{xx}\dot{p} - I_{xz}\dot{r} \\ I_{yy}\dot{q} \\ -I_{zx}\dot{p} + I_{zz}\dot{r} \end{bmatrix} + \begin{bmatrix} -pqI_{zx} - qr(I_{yy} - I_{zz}) \\ (p^2 - r^2)I_{zx} - pr(I_{zz} - I_{xx}) \\ qrI_{xz} - pq(I_{xx} - I_{yy}) \end{bmatrix} \qquad (1.78)$$

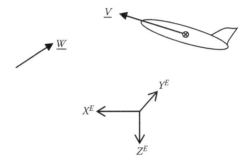

FIGURE 1.19　Aircraft in the presence of external wind.

equations of motion are written with respect to the inertial frame, the inertial velocity V (also called V_E) is the variable of interest for the airplane kinematics. Thus, for example, to gauge the distance traversed by the airplane with respect to the ground, that is, the navigation equation, the inertial velocity V is employed. However, in the presence of external wind W, the aircraft velocity with respect to atmosphere is given by $V_\infty = V - W$. Therefore, the airplane's relative wind velocity used in the aerodynamic modeling is V_∞.

Let the components of airplane inertial velocity V in the body-fixed axes be given by u, v, and w, as before, and let the components of the inertial wind velocity vector W along the body axes be represented by W_{xB}, W_{yB}, and W_{zB}, respectively. Then, the airplane's relative wind velocity V_∞, in the presence of external wind, has body axis components $(u - W_{xB})$, $(v - W_{yB})$, and $(w - W_{zB})$.

In practice, it is the Earth-axis components W_{xE}, W_{yE}, and W_{zE} of the inertial wind vector W that are likely to be known; in that case, the body-axis components can be found by the transformation law of Equation 1.4 as follows:

$$\begin{bmatrix} W_{xB} \\ W_{yB} \\ W_{zB} \end{bmatrix} = R_\varphi^T R_\theta^T R_\psi^T \begin{bmatrix} W_{xE} \\ W_{yE} \\ W_{zE} \end{bmatrix} \tag{1.79}$$

Of the airplane equations of motion in the body-fixed axes, the position and orientation equations remain unchanged as follows:

$$\begin{bmatrix} \dot{x}^E \\ \dot{y}^E \\ \dot{z}^E \end{bmatrix} = \begin{bmatrix} c\psi c\theta & c\psi s\theta s\varphi - s\psi c\varphi & c\psi s\theta c\varphi + s\psi s\varphi \\ s\psi c\theta & s\psi s\theta s\varphi + c\psi c\varphi & s\psi s\theta c\varphi - c\psi s\varphi \\ -s\theta & c\theta s\varphi & c\theta c\varphi \end{bmatrix} \begin{bmatrix} u \\ v \\ w \end{bmatrix} \tag{1.6}$$

$$\begin{bmatrix} \dot{\varphi} \\ \dot{\theta} \\ \dot{\psi} \end{bmatrix} = \begin{bmatrix} 1 & \tan\theta \sin\varphi & \tan\theta \cos\varphi \\ 0 & \cos\varphi & -\sin\varphi \\ 0 & \sec\theta \sin\varphi & \sec\theta \cos\varphi \end{bmatrix} \begin{bmatrix} p \\ q \\ r \end{bmatrix} \tag{1.25}$$

The wind relative velocity $V_\infty = V - W$ in terms of its body-axis components appears as below in vector format:

$$V_\infty = \begin{bmatrix} u \\ v \\ w \end{bmatrix} - \begin{bmatrix} W_{xB} \\ W_{yB} \\ W_{zB} \end{bmatrix} = \begin{bmatrix} u - W_{xB} \\ v - W_{yB} \\ w - W_{zB} \end{bmatrix} \tag{1.80}$$

By definition, the wind axis system is such that the X^W axis is aligned along the total relative wind velocity vector V_∞—note that it is no longer the same (or equal to) the inertial velocity V. The wind-axis Euler angles μ, γ, χ now reveal the orientation of V_∞ with respect to the ground. Thus, in Equation 1.10, for instance, V on the right-hand side must now be replaced by V_∞. With this change, Equation 1.10 cannot be used as it is as the *navigational equation* since it no longer tracks the airplane inertial velocity V (now different from V_∞).

In the wind axis, the components of V_∞ are $[V_\infty\ 0\ 0]^T$. Transforming between its components in wind and body axis, as in Equation 1.13, yields

$$\begin{bmatrix} u - W_{xB} \\ v - W_{yB} \\ w - W_{zB} \end{bmatrix} = \begin{bmatrix} c\alpha & 0 & -s\alpha \\ 0 & 1 & 0 \\ s\alpha & 0 & c\alpha \end{bmatrix} \begin{bmatrix} c\beta & -s\beta & 0 \\ s\beta & c\beta & 0 \\ 0 & 0 & 1 \end{bmatrix} \begin{bmatrix} V_\infty \\ 0 \\ 0 \end{bmatrix} = \begin{bmatrix} V_\infty \cos\beta\cos\alpha \\ V_\infty \sin\beta \\ V_\infty \cos\beta\sin\alpha \end{bmatrix} \quad (1.81)$$

From Equation 1.81, the updated definitions of the aerodynamic angles α, β in the presence of wind are obtained as follows:

$$\alpha = \tan^{-1}\left(\frac{w - W_{zB}}{u - W_{xB}}\right) \quad \beta = \sin^{-1}\left(\frac{v - W_{yB}}{V_\infty}\right)$$

$$\text{where } V_\infty = \sqrt{(u - W_{xB})^2 + (v - W_{yB})^2 + (w - W_{zB})^2} \quad (1.82)$$

In the absence of wind, the wind velocity components in Equation 1.82 may be set to zero, and the original definitions in Equation 1.14 are recovered.

Instruments onboard the airplane are likely to measure the quantities in Equation 1.82 or equivalently the body-axis components of the relative wind velocity on the left-hand side of Equation 1.81.

In case of the velocity (translational dynamics) and angular velocity (rotational dynamics) equations, reproduced below, while the equations themselves remain unchanged, the aerodynamic forces and moments therein need to be examined afresh.

$$\begin{bmatrix} -mg\sin\theta \\ mg\sin\varphi\cos\theta \\ mg\cos\varphi\cos\theta \end{bmatrix} + \begin{bmatrix} -D\cos\alpha + L\sin\alpha \\ Y \\ -D\sin\alpha - L\cos\alpha \end{bmatrix} + \begin{bmatrix} T \\ 0 \\ 0 \end{bmatrix} = m\begin{bmatrix} \dot{u} - rv + qw \\ \dot{v} + ru - pw \\ \dot{w} - qu + pv \end{bmatrix} \quad (1.58)$$

$$\begin{bmatrix} \mathcal{L} \\ M \\ N \end{bmatrix} = \begin{bmatrix} I_{xx}\dot{p} - I_{xz}\dot{r} \\ I_{yy}\dot{q} \\ -I_{zx}\dot{p} + I_{zz}\dot{r} \end{bmatrix} + \begin{bmatrix} -pqI_{zx} - qr(I_{yy} - I_{zz}) \\ (p^2 - r^2)I_{zx} - pr(I_{zz} - I_{xx}) \\ qrI_{xz} - pq(I_{xx} - I_{yy}) \end{bmatrix} \quad (1.78)$$

Thankfully, the definition of the aerodynamic forces in the stability axis, as in Equation 1.54, and their conversion to body axis, as in Equation

1.55, continue to be valid even in the presence of wind. Only the quantities in Equation 1.82 must be now used in those calculations. Therefore, the dynamic pressure should now be taken as

$$\bar{q} = \frac{1}{2}\rho\left[(u-W_{xB})^2 + (v-W_{yB})^2 + (w-W_{zB})^2\right] \qquad (1.83)$$

Thus, the set of Equations 1.6, 1.25, 1.58, and 1.78, with the aerodynamic relations in Equations 1.82 and 1.83, constitute the complete set of equations for the aircraft dynamics in the presence of wind in terms of the body-axis variables. In general, the wind velocity field may be a function of space and time; at every instant, the wind velocity vector as "seen by the airplane," transformed as per Equation 1.79, is used in the modeling.

No purpose is served by attempting to write the force and navigation equations in the wind axis since the velocity along X^W in the presence of wind, V_∞, is no longer the same as the inertial velocity.

EXERCISE PROBLEMS

1. A thrust vectoring nozzle in its undeflected position gives thrust T along the body X^B axis. The nozzle deflection is defined by two angles—an azimuthal angle τ between 0 and 360 deg, and a deflection angle σ from the X^B axis, as shown in the figure below. Find the components of thrust along the three body axes in case of a vectored nozzle. [Answer: $T_x = T\cos\sigma$, $T_y = T\sin\sigma\cos\tau$, $T_z = T\sin\sigma\sin\tau$.]

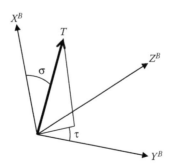

Thrust vectoring nozzle deflection angles

2. Imagine an airplane that maintains straight and level flight with an angle of attack α. That is, the X^B axis is at an angle α rotated upward

to the horizontal. Further, imagine that the airplane is flying in a uniform crosswind such that the relative wind makes an angle β with the direction of the inertial velocity. What will be the components of the inertial velocity along the body axis and along the wind axis? [Answer: ($V \cos \alpha, 0, V \sin \alpha$) and ($V \cos \beta, -V \sin \beta, 0$), respectively.]

3. Consider an airplane flying a vertical loop as shown with the velocity vector always tangential to the flight path. Define and distinguish between the angular velocity components q, q_W and the Euler angle rates $\dot{\theta}, \dot{\gamma}$.

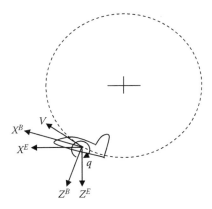

Airplane flying a vertical loop.

4. Transform the aircraft weight vector from Earth-fixed axis to wind-fixed axes and, hence, find the three components of the gravitational force along the wind-fixed axes.

5. Write out the remaining six equations in Equation 1.17 besides the three already listed in Equation 1.18.

6. Work out the steps between Equations 1.31 and 1.32 in full.

7. Verify that the matrix in Equation 1.42 is skew-symmetric. Study the properties of a skew-symmetric matrix.

8. Derive from first principles the relation between Earth- and body-axis rates:

$$\left.\frac{d\underline{A}}{dt}\right|_I = \left.\frac{d\underline{A}}{dt}\right|_B + (\underline{\omega}_B \times \underline{A})$$

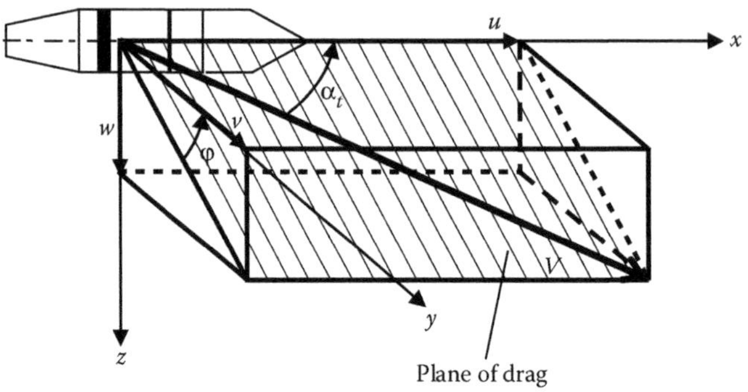

FIGURE 1.20 Sketch of body-axis and relative velocity vector marking the total angle of attack and aerodynamic roll angle. (Adapted from Baranowski, L., *Journal of Theoretical and Applied Mechanics*, 51(1), 235–246, 2013.)

9. Rewrite the equations of angular motion (attitude dynamics) for the case when the body-fixed axes are the airplane principal axes.

10. What is the significance of 3–2–1 rule of transformation between Earth- and body-fixed axes? Assuming aircraft axes to be initially coincident with Earth-fixed axes, draw the final orientation of aircraft after the three rotations.

11. For the equations of angular motion with the Euler angles as defined here to be usable, certain limits on the angles are to be assumed. Mathematically, $\theta \to \pi/2$ rad gives rise to a singularity in Equation 1.25, which is known as "Gimbal locking" in spacecraft dynamics. Study Equation 1.25 in this limit and discuss its implications.

12. The total angle of attack α_t and the aerodynamic roll angle φ used in case of an axisymmetric flight vehicle are defined as follows:

$$\alpha_t = \frac{\sqrt{w^2 + v^2}}{u}, \quad \varphi = \tan^{-1}\frac{w}{v}$$

Derive these relations with reference to Figure 1.20.

REFERENCE

1. Baranowski, L., Equations of motion of a spin-stabilized projectile for flight stability testing, *Journal of Theoretical and Applied Mechanics*, 51(1), 235–246, 2013.

CHAPTER 2

Modeling and Interpreting the Aerodynamics

A S REMARKED PREVIOUSLY, THE equations for the 6 DOF dynamics of a wide spectrum of flying vehicles are remarkably similar but for differences in mass and inertia properties, choice of propulsive device, structural flexibility, peculiarities in configuration, etc. Several such instances were highlighted at the beginning of Chapter 1. However, what really distinguishes the dynamics in flight of one airplane from another are its aerodynamic properties. Even between airplanes with a similar configuration flying under identical conditions, one airplane may recover its original flight state after being upset by a gust of wind, whereas another may enter a state of oscillatory motion. What distinguishes one from the other is the nature and magnitude of the aerodynamic forces and moments generated in flight. Therefore, accurate modeling and evaluation of the aerodynamic properties of airplanes is critical without which the results of a 6 DOF simulation may as well be meaningless.

2.1 DEFINITION OF AERODYNAMIC COEFFICIENTS

With reference to Figure 1.15, the aerodynamic forces acting on the airplane may be written in terms of their components along the *stability axis* as follows:

$$\underline{F}^A = \begin{bmatrix} -D \\ Y \\ -L \end{bmatrix}_S \tag{1.53}$$

Likewise, the aerodynamic moments, referenced to Figure 1.16, have the following components:

$$\underline{M}_B^A = \begin{bmatrix} \mathcal{L} \\ M \\ N \end{bmatrix} \tag{1.75}$$

Traditionally, these forces and moments have been expressed in terms of force/moment coefficients as below:

$$\underline{F}^A = \begin{bmatrix} -D \\ Y \\ -L \end{bmatrix}_S = \begin{bmatrix} -\bar{q}SC_D \\ \bar{q}SC_Y \\ -\bar{q}SC_L \end{bmatrix}_S \tag{2.1}$$

$$\underline{M}_B^A = \begin{bmatrix} \mathcal{L} \\ M \\ N \end{bmatrix} = \begin{bmatrix} \bar{q}SbC_l \\ \bar{q}ScC_m \\ \bar{q}SbC_n \end{bmatrix} \tag{2.2}$$

where $\bar{q} = (1/2)\rho V^2$ is the dynamic pressure, S is a reference area, usually taken to be the wing planform area; c is a reference length, usually the mean aerodynamic chord, used in case of variables defined in the longitudinal plane; and b is another reference length, usually the wing span, used in case of variables defined out of the longitudinal plane. Effectively, the force components in Equation 2.1 have been defined in terms of three force coefficients—C_D, C_Y, C_L—the drag, side force, and lift coefficients, respectively. Notice that dimensionally $\bar{q}S$ has dimensions of force and the force coefficients are all dimensionless. It is well known from dimensional analysis as employed in fluid mechanics that the force coefficients must only be functions of other dimensionless quantities. This is an important point that is often overlooked and it is not uncommon to find, for example, a force coefficient being modeled as a function of a dimensional quantity such as a velocity component.

Similarly, in Equation 2.2, the three moment components have been modeled in terms of a dimensional part—either $\bar{q}Sc$ or $\bar{q}Sb$, which have dimensions of moment—and a dimensionless coefficient each: rolling moment coefficient C_l, pitching moment coefficient C_m, and yawing moment coefficient C_n. Here too, the moment coefficients must only be functions of other dimensionless quantities.

There are a couple of obvious advantages to using the dimensionless force and moment coefficients. One, it becomes possible to compare aerodynamic data on a common scale for different airplanes and under different flight conditions. Two, it helps carry out experiments on scaled models under dynamically similar conditions and port the coefficient data directly to the full-scale airplane.

Homework Exercise: Look at the 3-view drawings of the airplanes in Figure 2.1. The B-707 entered service in the 1960s, the B-747 came in the 1970s, whereas the B-777 began flying commercially in the 1990s. Yet, configurationally, these airplanes appear very similar. So can one expect them to have similar aerodynamic characteristics? At first glance, it would appear that there has been little or no change in the state of the art where external airplane aerodynamics is concerned. Would that be a correct assessment?

2.2 MODELING OF AERODYNAMIC COEFFICIENTS

Traditionally, aerodynamic coefficients have been modeled in what is called a *quasi-steady* manner. That is, they are functions only of flight variables at *that* instant of time, not those at previous instants of time. The only exception is the modeling of the downwash lag effect, which we shall examine shortly. In effect, the history of the airplane maneuver—that is, how the airplane came to be in that particular state of flight—is of no consequence in evaluating the aerodynamic coefficients.

Without doubt, it is the relative flow incident on the airplane that is responsible for the generation of the aerodynamic forces and moments acting on the airplane, that is, the magnitude and orientation of the relative velocity and the angular velocity of the airplane with respect to the relative wind. Of the factors that certainly *do not* influence the aerodynamic coefficients—the position and orientation of the airplane relative to the ground (body-axis Euler angles) have no effect on the aerodynamic

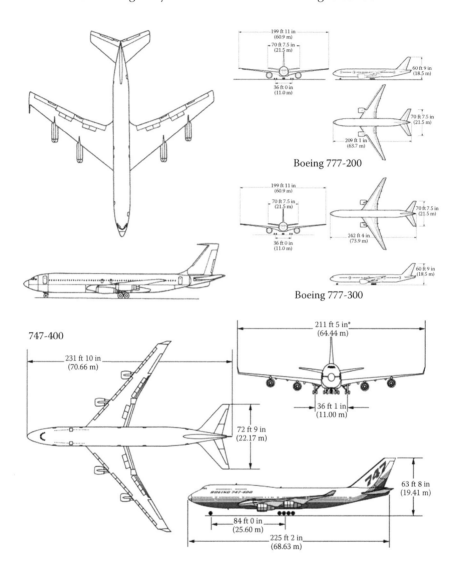

FIGURE 2.1 3-View drawings of different airplanes (clockwise from top left): B-707, B-777, B-747. (From www.aerospaceweb.org.)

coefficients. Atmospheric density does change with altitude from the ground, which affects the aerodynamic forces and moments, but that is accounted for in the factor $\bar{q} = (1/2)\rho V^2$ in Equations 2.1 and 2.2 as is the magnitude of the relative velocity V. Likewise, the angular velocity of the body-fixed axes relative to the ground has no bearing on the aerodynamic forces and moments either. It is the angular velocity of the body-fixed axes with respect to the relative wind that matters; that is, the relative angular velocity between the body- and wind-fixed axes. The effect of airplane size on the aerodynamic forces/moments is represented by the factors S, b, c in Equations 2.1 and 2.2 and they appear "outside" the aerodynamic coefficients as well.

Homework Exercise: How can the reference area S and reference lengths b, c be chosen for configurations that do not have a lifting surface such as a wing? For example, a missile or an airship. Note that with different choices of S, b, or c, the values of the aerodynamic coefficients can be grossly different, so it is important to make sure that the values reported and being compared use similar reference area and lengths.

With this background, the aerodynamic forces may be modeled as sum of four different effects, as follows:

$$
C_k(t) = \underbrace{C_{ksta}(Ma, \alpha, \beta, \delta)}_{\text{Static term}} + \underbrace{C_{kdyn}\left(\frac{(p_b - p_w)b}{2V}, \frac{(q_b - q_w)c}{2V}, \frac{(r_b - r_w)b}{2V}\right)}_{\text{Dynamic term}}
$$

$$
+ \underbrace{C_{kflo}\left(\frac{p_w b}{2V}, \frac{q_w c}{2V}, \frac{r_w b}{2V}\right)}_{\text{Flow curvature}} + \underbrace{C_{kdow}(\alpha(t - \tau), \beta(t - \tau))}_{\text{Downwash lag}} \qquad (2.3)
$$

where k represents any of the aerodynamic coefficients in Equations 2.1 and 2.2, and all variables are evaluated at the same time t (except where explicitly mentioned). The static term is a function of the relative flow Mach number Ma, the aerodynamic angles α, β, and the control surface deflections δ. The dynamic term is a function of the three components of the relative angular velocity between the body- and wind-fixed axes, suitably nondimensionalized. The third term brings in the *flow curvature* effect, which arises when the airplane is flying along a curved flight path—it is a function of the nondimensional wind-axis angular velocity components, which represent the flight path curvature. The downwash lag term is the only one that is strictly not a *quasi-steady* effect as it represents the aerodynamic forces/moments generated at the tail due to changes in α, β at the wing at a time $\tau \approx l_t/V$ earlier, where l_t is roughly the distance between the wing and tail aerodynamic centers.

The model in Equation 2.3 is fairly representative for most modern airplanes though certain additional effects may need to be added in special cases. For instance, in case of rapidly rolling vehicles, an additional Magnus moment effect should be included. For extremely low-speed flight, as in the case of micro air vehicles, the Reynolds number may become an important parameter for the aerodynamic coefficients, especially C_D. By and large, the *quasi-steady model* of Equation 2.3 is used to simulate airplane motion in both steady-state and unsteady flight conditions. However, certain unsteady aerodynamic modeling approaches, such as indicial functions and internal state variables, have been proposed though they are not used widely [1].

Homework Exercise: An added mass model becomes necessary for modeling the flight of aircraft such as airships and parachutes. Actually, a body accelerating in a fluid also displaces (accelerates) some of the fluid adjacent to it. This added mass of fluid displaced should be accounted for as an inertial effect in the dynamics equations. For heavier-than-air vehicles such as airplanes, the amount of air so displaced is of the order of 1% of the airplane mass, and so may be ignored. But for lighter-than-air (LTA) vehicles like airships or parachutes, the added mass is a significant factor. Find out how the added mass effect is modeled for such LTA systems.

In the following sections, we shall dwell on each of the terms in Equation 2.3 one by one, look at general trends or a specific example, and point out the main design features that impact each coefficient term.

2.3 STATIC AERODYNAMIC COEFFICIENT TERMS

2.3.1 Longitudinal Coefficients with Angle of Attack

Of the six aerodynamic coefficients, the longitudinal ones—namely, C_D, C_L, C_m—depend significantly on the angle of attack α. Figure 2.2 shows the typical variation of these three coefficients for a typical military airplane over the α range of $(-14, 90)$ deg for one fixed value of Mach number (in this case, fairly low subsonic). Figure 2.3 is a zoomed-in view of Figure 2.2, focusing on the low angle of attack range, roughly chosen here to be $(-14, 14)$ deg of α.

The variation of C_D at low angles of attack is close to parabolic and is usually well approximated by a quadratic function of α. As is well known, there are two sources of contribution to C_D—one due to the shear stress (skin friction) and the other due to the pressure. Around zero α, where $C_L \approx 0$, almost all the C_D is due to skin friction, which is nearly independent of α. A predominant source of pressure drag in case of a lifting wing, we know,

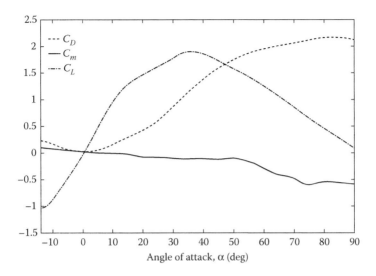

FIGURE 2.2   Longitudinal aerodynamic coefficients as function of α for a typical military airplane.

is the induced drag related to the mechanism of generation and shedding of wing-tip vortices. This effect is usually modeled as a quadratic function of C_L, hence the parabolic dependence of C_D on α. The typical parabolic bucket shape of C_D is quite clearly visible in Figure 2.3. Lower aspect ratio and a larger deviation from the ideal elliptical span-wise lift distribution

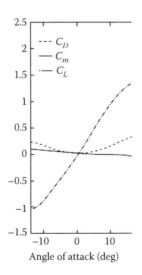

FIGURE 2.3   Zoom-in of the longitudinal aerodynamic coefficients as function of α for low angles of attack.

are two factors that tend to increase the induced drag. For the standard trapezoidal wing planform, a near-elliptical span-wise lift distribution can be obtained by judicious choice of the wing twist and the wing taper ratio (the ratio of tip chord to root chord), thereby minimizing the induced drag. For airplanes, such as gliders, for which maximizing endurance is a key objective, use of a large aspect ratio wing to cut down induced drag as far as possible is standard. However, for military airplanes, there are usually conflicting objectives and other requirements often that take precedence over C_D reduction in the selection of the aspect ratio. At higher angles of attack in Figure 2.2, C_D continues to increase, though not as sharply, before peaking at $\alpha \approx 90$ deg.

The lift coefficient C_L is very close to linear at low angles of attack, as exemplified in Figure 2.3 with $C_L \approx 0$ at zero α. As a rough estimate, one can assume a C_L of 0.08–0.1 per degree of α at low angles of attack. Thus, for example, at $\alpha \approx 10$ deg, one may expect a value of C_L around 0.8–1.0. The C_L per degree of α (called the wing lift curve slope) for a given airfoil section depends on the wing sweepback angle and aspect ratio. In general, the lift curve slope decreases for higher sweepback angle and lower aspect ratio, though in case of some low aspect ratio wings, there may be nonlinear effects that provide a sharp increase of the lift curve slope over some ranges of α. For larger α in Figure 2.2, note that the lift coefficient continues to increase, though with a smaller value of the slope. The peak value of C_L, approximately 1.8, occurring at $\alpha \approx 35$ deg corresponds to stall. However, in case of the airplane model in Figure 2.2, stall does not lead to a precipitous fall in C_L; rather, C_L decreases gradually and continues to be positive all the way to $\alpha \approx 90$ deg.

Coming to the pitching moment coefficient evaluated at the airplane CG, there is a small (usually positive) contribution to C_m from the fuselage and other nonlifting components of the airplane. Otherwise, a large proportion of C_m arises from the lifting components (wing and tail). A lifting force that acts at a point (center of pressure) separated from the airplane CG creates a pitching moment about the airplane CG. Thus, the variation of C_m with α is similar (nearly linear at low angles of attack) to that of the lift coefficient C_L. However, there may also be a smaller effect on C_m due to the shift in the location of the center of pressure relative to the airplane CG. In Figure 2.3, a positive C_L acting at a point aft of (behind) the airplane CG can be seen to produce a negative C_m at the airplane CG (and vice versa). As a result, if the slope of C_L with α is positive, then the slope of C_m with α is negative. At higher angles of attack in Figure 2.2, the value of

C_m remains largely negative though the slope fluctuates and even becomes positive over a small stretch of α around 75–80 deg.

The location of the airplane CG has a predominant impact on the value and trend of the coefficient C_m with angle of attack. For a vehicle like a hang glider, shifting of the CG by moving the flyer's weight is usually the only way to adjust the vehicle trim. For conventional airplanes, CG location and its movement in flight is normally part of the problem. Change in CG location due to configuration change (e.g., variable-sweep wings), fuel consumption or reallocation (e.g., Concorde), or payload arrangement/store drop can have a major impact on net C_m requiring a vehicle to be retrimmed in flight. Obviously, the tail usually being at a greater distance from the CG, changes to the tail lift have a greater impact on C_m than from the wing. Two design parameters that the designer has control over in this regard are the tail setting angle and the tail volume ratio. The latter, in case of the horizontal tail, is defined as

$$HTVR = \frac{S_t l_t}{Sc} \tag{2.4}$$

where *HTVR* stands for *horizontal tail volume ratio* and S_t and l_t are respectively the tail planform area and the distance between the tail and wing aerodynamic centers.* The tail setting angle i_t is used to set the tail chord at an angle with respect to the body X^B axis, to bias the tail lift as it were, in order to adjust the value of C_m at zero angle of attack. The *HTVR* is then used to decide the slope of the C_m curve with α—a larger value of *HTVR* should give a larger negative slope.

The trend of C_D, C_L, and C_m varying with α in the low-α range in Figure 2.3 is fairly typical of most conventional airplane configurations. The variations at higher angles of attack, however, are particular to each airplane's aerodynamic quirks and must be considered individually.

Homework Exercise: Many modern airplane wings feature devices such as flaps and slats, which are meant to increase the lift coefficient during particular phases of flight. Look up information on how the aero coefficient curves (such as those in Figure 2.2) will be altered when flaps/slats are deployed. Figure 2.4 gives one example of the lift coefficient for a DC-9 airplane. Notice the significant increase in C_L between the zero-flap and deflected flap cases—the lift curve shifts upward. On the other hand,

---

* Alternatively from the airplane CG, but that is a variable parameter.

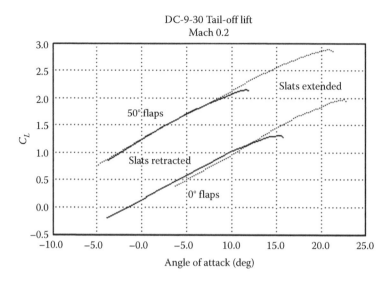

FIGURE 2.4 Effect of flap and slat deflection on lift coefficient, example of DC-9 airplane.

extending the slats shifts the C_L curve to the right and upward, yielding higher C_L but at a higher effective angle of attack.

2.3.2 Lateral Coefficients with Angle of Attack and Sideslip Angle

The lateral static aerodynamic coefficients—C_y, C_l, and C_n—are usually functions of both α and β. In many cases, the range of variation of β considered for the aerodynamic model is limited and these coefficients may justifiably be modeled as

$$C_k(\alpha,\beta) = \frac{\partial C_k}{\partial \beta}(\alpha) \cdot \beta = C_{k_\beta}(\alpha) \cdot \beta \qquad (2.5)$$

An example of the variation of $C_{l_\beta}(\alpha)$ and $C_{n_\beta}(\alpha)$ for a typical military airplane is shown in Figure 2.5.

Notice for instance that $C_{l_\beta}(\alpha)$ is negative and grows more negative with increasing α between 5 and 15 deg of angle of attack. The main design parameter that directly correlates with $C_{l_\beta}(\alpha)$ is the wing dihedral angle—the angle by which the wings are canted up (or down, called negative dihedral/anhedral) relative to a planar surface. Other configuration features that contribute to $C_{l_\beta}(\alpha)$ are the wing sweepback, which gives the variation with α referred to above, the wing position vis-à-vis the fuselage

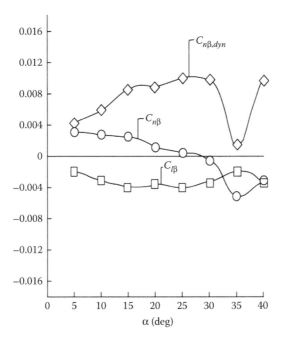

FIGURE 2.5   Typical variation of the lateral aerodynamic coefficients as function of α for a military airplane. (From NASA TP-1538, publ. 1979.)

(high, low, or mid), and the vertical tail volume ratio and the location of its aerodynamic center above the body X^B axis. However, the wing sweep-back and position, as well as the vertical tail size are usually determined by other factors, leaving the dihedral angle as the main weapon in the designer's arsenal to manipulate $C_{l_\beta}(\alpha)$. Too large a dihedral angle may give an acceptable $C_{l_\beta}(\alpha)$ at $\alpha = 5$ deg but too much $C_{l_\beta}(\alpha)$ at $\alpha = 15$ deg due to the cumulative effect of the sweepback angle. On the other hand, a small dihedral angle (or even a slight anhedral) may give an acceptable value of $C_{l_\beta}(\alpha)$ at $\alpha = 15$ deg at the cost of an undesirable value of $C_{l_\beta}(\alpha)$ at $\alpha = 5$ deg. The variation of $C_{l_\beta}(\alpha)$ at higher angles of attack is very con-figuration dependent.

For airplanes with a vertical tail, $C_{n_\beta}(\alpha)$ is expected to be positive at small values of α. In fact, this is one of the key requirements for sizing the vertical tail and the value of $C_{n_\beta}(\alpha)$ depends on the *vertical tail volume ratio (VTVR)* defined as

$$VTVR = \frac{S_V l_V}{Sc} \tag{2.6}$$

where S_V and l_V are the vertical tail planform area and distance of its aero-dynamic center from the wing aerodynamic center,[*] respectively.

With increasing angle of attack, as more of the vertical tail gets blanked by the flow separating from the fuselage, the vertical tail effectiveness decreases, and $C_{n_\beta}(\alpha)$ reduces correspondingly. This trend is clearly visible in Figure 2.5 where $C_{n_\beta}(\alpha)$ hits zero around $\alpha = 30$ deg and then reverses sign to go negative. This trend is fairly universal for a wide range of airplanes.

Homework Exercise: Figure 2.5 features a third curve—that labeled $Cn_{\beta,dyn}$, defined as follows:

$$Cn_{\beta,dyn} = Cn_\beta \cos\alpha - \frac{I_{zz}}{I_{xx}} Cl_\beta \sin\alpha$$

Look up references (e.g., Reference 2) for the significance and use of $Cn_{\beta,dyn}$.

2.3.3 Variation with Mach Number

The typical variation of drag coefficient C_D with Mach number is sketched in Figure 2.6. At low Mach numbers, there is little or no change in C_D with Mach number. Beyond the critical Mach number, Ma_{cr}, the drag coefficient begins to gradually increase and then increases steeply beyond M_{DD}, the drag divergence Mach number. This is due to the formation of shock waves over the airplane body, primarily the wing—the additional drag component is called *wave drag*, a form of pressure drag. Peak C_D usually occurs somewhere in the vicinity of Mach number 1 beyond which the drag coefficient falls off sharply. This trend of C_D with Mach number in the transonic regime is typical of many airplanes.

Figure 2.6 also shows the typical variation of the lift coefficient with Mach number in the subsonic and supersonic regimes. The increase in C_L with Mach number in the subsonic range, and the fall with Mach number in the supersonic range are both fairly standard. The variation in the transonic regime is somewhat more difficult to standardize and is overly dependent on the finer details of the airplane configuration, especially the key wing parameters such as aspect ratio, sweep angle, taper ratio, twist, airfoil shape, wing–fuselage interaction, possible leading-edge kinks, under-wing attachments, etc. However, for many airplane configurations,

---

[*] Alternatively from the airplane CG, but that is a variable parameter.

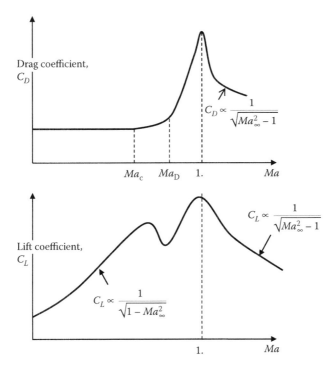

FIGURE 2.6 Typical variation of the drag and lift coefficients as function of Mach number.

there may be a dip in C_L around $Ma \approx 0.85$–0.9 before a rise in C_L peaking around Mach 1, as sketched in Figure 2.6.

The variation of the side-force coefficient C_Y with Mach number is in principle similar to that of C_L. And since the moment coefficients C_m and C_n arise primarily as a consequence of C_L and C_Y, respectively, the trend of their variation with Mach number may be expected to be similar as well. The variation of rolling moment coefficient C_l with Mach number is harder to pin down but, since much of the rolling moment in a conventional airplane configuration arises due to wing lift, the trend is likely to mirror the change of C_L with Ma to some extent.

In practice, it is not necessary to have functional relationships for the variation of the static aerodynamic coefficients with α, β, Ma—it is fairly common and acceptable to have tabular data that are looked up and interpolated during analysis. However, it is worthwhile and highly advisable to plot these variations and check the magnitude and trend for consistency. It may be possible to raise some red flags before even launching into a flight dynamics analysis.

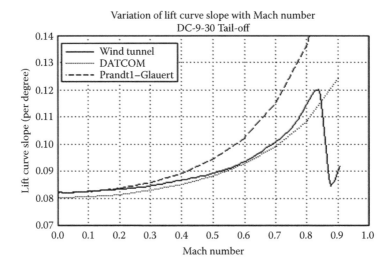

FIGURE 2.7   Lift curve slope variation with Mach number and theoretical predictions, example of DC-9 airplane.

Homework Exercise: Predicting the variation of the aero coefficients accurately in the transonic regime is a bit of a challenge. Most theoretical methods do not work well for Mach numbers around 1.0. For example, see the variation of lift curve slope for the DC-9 airplane plotted in Figure 2.7 and the comparison therein with the calculated values using the Prandtl–Glauert rule and an empirical prediction method using DATCOM.

2.3.4 Variation with Control Surface Deflection

The static aerodynamic coefficients may be modeled as a function of a control surface deflection in the following manner:

$$C_k(\delta) = \frac{\partial C_k}{\partial \delta} \cdot \delta = C_{k_\delta} \cdot \delta \qquad (2.7)$$

Each C_{k_δ} is called a *control effectiveness parameter* and may itself be a function of α, β, Ma. Typical values for the elevator control effectiveness parameter in case of the longitudinal coefficients have been presented in an Exercise Problem. Usually, $C_{L_{\delta e}}$ is positive and $C_{m_{\delta e}}$ is negative, suggesting that a positive (down) elevator deflection adds to the tail lift and hence the total airplane lift, and that the added tail lift acting aft of the airplane CG induces a negative (nose-down) pitching moment at the airplane CG.

Sample values of the aileron and rudder control effectiveness parameters in case of the lateral aerodynamic coefficients are as under (control deflection angle in deg):

$$C_{Y_{\delta a}} = \frac{1}{25}(-0.00227\alpha + 0.039)$$

$$C_{Y_{\delta r}} = \frac{1}{30}(-0.00265\alpha + 0.141)$$

$$C_{l_{\delta a}} = \frac{1}{25}(-0.00121\alpha - 0.0628)$$

$$C_{l_{\delta r}} = \frac{1}{30}(-0.000351\alpha + 0.0124)$$

$$C_{n_{\delta a}} = \frac{1}{25}(0.000213\alpha + 0.00128)$$

$$C_{n_{\delta r}} = \frac{1}{30}(0.000804\alpha - 0.0474)$$

Positive aileron deflection is defined as right aileron deflected down and left aileron deflected up. The net aileron deflection, δa, is then defined as

$$\delta a = \frac{\delta a_R + \delta a_L}{2} \tag{2.8}$$

Positive aileron deflection increases lift on the right wing and decreases lift on the left wing, producing a net negative rolling moment. Hence, $C_{l_{\delta a}}$ is usually negative.

Each deflected aileron creates additional drag and the drag differential between the left and right wing sections produces a yawing moment, which is the primary source of $C_{n_{\delta a}}$. In general, $C_{n_{\delta a}}$ may be either positive (called adverse yaw) or negative (proverse yaw). In the present instance, $C_{n_{\delta a}}$ is positive—that is, adverse yaw, which means that when the airplane rolls left due to aileron application, it also yaws right (and vice versa). This combination is considered preferable from the point of view of pilot handling. A common way to ensure adverse yaw is to deflect the up-going and down-going ailerons unequally such that the ensuing drag differential assures a positive $C_{n_{\delta a}}$. Usually, the up-going aileron is given a proportionately larger deflection for this purpose.

Rudder deflected to the left (as seen from the rear) is positive δr. Usually, this gives a positive side force, that is $C_{Y_{\delta r}} > 0$, which creates a positive rolling moment $C_{l_{\delta r}} > 0$ and a negative yawing moment $C_{n_{\delta r}} < 0$. With increasing angle of attack, all the rudder control effectiveness parameters are generally degraded, similar to the case of vertical tail effectiveness with α as seen earlier.

Homework Exercise: Several airplanes use alternative control surfaces, either exclusively or in addition to the regular control surfaces mentioned above. Some examples are: canards, all-moving tail-planes and elevons for pitch control, elevons, spoilers, tailerons, and flaperons for roll control. Investigate how these may be modeled and find some typical aero data for these control surface deflections. Note that most roll control devices also produce a yawing moment, so they may, in principle, also be used for yaw control.

2.4 DYNAMIC AERODYNAMIC COEFFICIENT TERMS

Just as the static aerodynamic terms depend on the relative orientation between the body- and wind-fixed axes, the dynamic coefficients represent the effect of the relative angular velocity (rate of change of orientation) between the body- and wind-fixed axes. The "dynamic term" in Equation 2.3 is modeled as

$$
\begin{aligned}
C_{kdyn} & \left(\frac{(p_b - p_w)b}{2V}, \frac{(q_b - q_w)c}{2V}, \frac{(r_b - r_w)b}{2V} \right) \\
&= \frac{\partial C_k}{\partial \frac{(p_b - p_w)b}{2V}} \frac{(p_b - p_w)b}{2V} + \frac{\partial C_k}{\partial \frac{(q_b - q_w)c}{2V}} \frac{(q_b - q_w)c}{2V} + \frac{\partial C_k}{\partial \frac{(r_b - r_w)b}{2V}} \frac{(r_b - r_w)b}{2V} \\
&= C_{k_{p1}} \frac{(p_b - p_w)b}{2V} + C_{k_{q1}} \frac{(q_b - q_w)c}{2V} + C_{k_{r1}} \frac{(r_b - r_w)b}{2V}
\end{aligned}
\tag{2.9}
$$

where the terms $C_{k_{p1}}$, etc. are called *rate derivatives* or *dynamic derivatives* and may themselves be functions of α, β, Ma. From Equation 1.32, we already have

$$
\begin{bmatrix} p_b^b - p_w^b \\ q_b^b - q_w^b \\ r_b^b - r_w^b \end{bmatrix} = \begin{bmatrix} \dot{\beta} \sin \alpha \\ \dot{\alpha} \\ -\dot{\beta} \cos \alpha \end{bmatrix}
\tag{1.32}
$$

So the expressions in Equation 2.9 may be further written using the relations in Equation 1.32 as follows:

$$C_{kdyn}\left(\frac{(p_b-p_w)b}{2V},\frac{(q_b-q_w)c}{2V},\frac{(r_b-r_w)b}{2V}\right)$$

$$=C_{k_{p1}}\frac{(\dot{\beta}\sin\alpha)b}{2V}+C_{k_{q1}}\frac{(\dot{\alpha})c}{2V}+C_{k_{r1}}\frac{(-\dot{\beta}\cos\alpha)b}{2V} \qquad (2.10)$$

Of these, the significant effects (at least at low values of angle of attack) are due to the relative pitch rate and the relative yaw rate (the second and third terms in Equation 2.10).

2.4.1 Pitching Moment Due to Relative Pitch Rate, $C_{m_{q1}}$

As sketched in Figure 2.8, $(q_b - q_w)$ is the relative angular velocity (relative pitch rate) of the body X^B axis with respect to the velocity vector V. If the horizontal tail chord is approximately aligned along the X^B axis, then the relative pitch rate induces a velocity $(q_b - q_w)l_t$ at the tail, more or less in the direction of the Z^B axis, as shown in Figure 2.8. From the tail point of view, the velocity components incident at the horizontal tail are marked in the inset to Figure 2.8. These are the free-stream velocity V_∞ and that induced by the relative pitch rate, $(q_b - q_w)l_t \cos \alpha$, normal to V_∞. For small α, the latter can be approximated as $(q_b - q_w)l_t$. The other component of the induced velocity, $(q_b - q_w)l_t \sin \alpha$, may be neglected, being the product of two small terms for small α. Thus, an additional angle of attack, $\Delta\alpha$, is induced at the horizontal tail, which may be approximated by

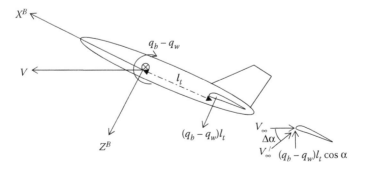

FIGURE 2.8 Sketch showing induced flow and hence induced angle of attack at the horizontal tail due to a relative pitch rate between body- and wind-fixed axes.

$$\Delta\alpha = \frac{(q_b - q_w)l_t}{V} \tag{2.11}$$

That creates an additional tail lift, which in turn produces a pitching moment at the airplane CG opposite in sense to that of the relative pitch rate, $(q_b - q_w)$. The corresponding *rate derivative* $C_{m_{q1}}$ is usually negative and proportional to the *HTVR*. Since this pitching moment opposes the relative pitch rate $(q_b - q_w)$, the $C_{m_{q1}}$ term is also called a pitch damping effect.

2.4.2 Yawing and Rolling Moment Due to Relative Yaw Rate, $C_{n_{r1}}$ and $C_{l_{r1}}$

As in the case of the relative pitch rate and the additional angle of attack induced at the horizontal tail, a relative yaw rate $(r_b - r_w)$ induces an additional sideslip angle at the vertical tail, as below:

$$\Delta\beta = -\frac{(r_b - r_w)l_V}{V} \tag{2.12}$$

which creates a side force acting at the vertical tail aerodynamic center, at a height h_V above the line of the body X^B axis. Consequently, this side force produces a rolling moment as well as a yawing moment at the airplane CG. These corresponding *rate* derivatives are labeled as $C_{n_{r1}}$ and $C_{l_{r1}}$, and they are both proportional to the *VTVR*. Usually, $C_{n_{r1}} < 0$ and $C_{l_{r1}} > 0$, and $C_{n_{r1}}$ is called the yaw damping effect since the yawing moment produced usually opposes the relative yaw rate $(r_b - r_w)$.

There is another source of yawing and rolling moment in response to a relative yaw rate $(r_b - r_w)$ and this is due to the wing. The relative yaw rate $(r_b - r_w)$ induces a velocity $\pm(r_b - r_w)y$ at every span-wise station y on the wing. This induced velocity either adds to or subtracts from the free-stream velocity incident on the wing at each span-wise station. Consequently, the effective free-stream velocity "seen" by every span-wise station is modified and hence the lift and drag at each such station. The additional lift due to this process creates a rolling moment, whereas the additional drag produces a yawing moment, both proportional to the relative yaw rate $(r_b - r_w)$. Normally, the resultant moments are such that the *rate derivatives* have the following signs: $C_{n_{r1}} < 0$ and $C_{l_{r1}} > 0$, which are the same as that due to the vertical tail contribution. Thus, the wing and vertical tail act cooperatively in this regard. To a first approximation, $C_{n_{r1}}$ due to the wing is proportional to the airplane drag coefficient C_D and $C_{l_{r1}}$ due to the wing is proportional to the lift coefficient C_L.

Typical values of these rate derivatives for an example airplane are given below:

$$C_{m_{q1}} = -4.7658/\text{rad}$$

$$C_{l_{r1}} = 0.1502/\text{rad}$$

$$C_{n_{r1}} = -0.1693/\text{rad}$$

2.5 FLOW CURVATURE COEFFICIENT TERMS

The flow curvature terms in Equation 2.3 arising due to the angular velocity of the wind-fixed axes are modeled as below:

$$C_{k\,flo}\left(\frac{p_w b}{2V}, \frac{q_w c}{2V}, \frac{r_w b}{2V}\right) = \frac{\partial C_k}{\partial(p_w b/2V)}\frac{p_w b}{2V} + \frac{\partial C_k}{\partial(q_w c/2V)}\frac{q_w c}{2V} + \frac{\partial C_k}{\partial(r_w b/2V)}\frac{r_w b}{2V}$$

$$= C_{k_{p2}}\frac{p_w b}{2V} + C_{k_{q2}}\frac{q_w c}{2V} + C_{k_{r2}}\frac{r_w b}{2V}$$

$$(2.13)$$

2.5.1 Yawing and Rolling Moment Due to Wind-Axis Yaw Rate, $C_{n_{r2}}$ and $C_{l_{r2}}$

Consider an airplane flying along a curved flight path in the horizontal plane as sketched in Figure 2.9. Assume that the wind- and body-axis yaw rates are identical; that is, $(r_b - r_w) = 0$, which implies that the airplane nose is always aligned with the velocity vector, which is tangent to the flight path at every instant. Clearly, the airplane velocity and the wind-axis

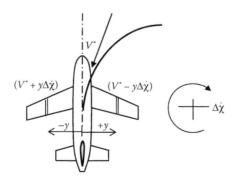

FIGURE 2.9 Sketch showing the differential velocity induced on the left and right wing halves when flying a curved flight path.

yaw rate (also called the turn rate in the horizontal plane) are related by $V = r_w R$, where R is the radius of the turn. Also, sections of the wing on the inside of the turn face a reduced relative flow velocity, lesser by $r_w y$, where y is the distance along the wing span as marked in Figure 2.9. Likewise, wing sections on the outside of the turn encounter an increased relative flow velocity. Consequently, there is a lift and drag differential between the two wings, which creates a rolling moment that rolls the airplane "into the turn" and yaws it "away from the turn." Therefore, the flow curvature derivatives due to wind-axis yaw rate usually have the following signs: $C_{n_{r2}} < 0$ and $C_{l_{r2}} > 0$. A similar effect but to a much lesser extent is visible due to the left and right halves of the horizontal tail as well.

2.5.2 Pitching Moment Due to Wind-Axis Pitch Rate, $C_{m_{q2}}$

Similarly, in case of an airplane flying a curved flight path in the vertical plane, there will be a radial variation of the relative flow velocity along the body Z^B axis. However, unless one of the lifting surfaces is located at a significant vertical distance from the airplane centerline (or from the X^B–Y^B plane), the effect on the airplane aerodynamically is negligible. So usually the flow curvature derivative $C_{m_{q2}}$ may be ignored.

2.5.3 Rolling Moment Due to Wind-Axis Roll Rate, $C_{l_{p2}}$

Figure 2.10 shows an airplane with velocity vector V pointing into the plane of the paper having a roll rate p_w about the wind X^W axis (aligned to the velocity vector). As sketched in the inset to Figure 2.10, the roll rate

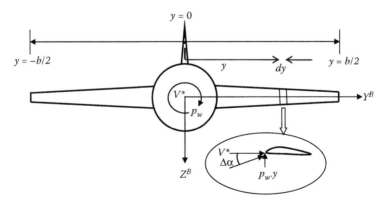

FIGURE 2.10 Sketch showing the additional angle of attack induced at a spanwise station of the wing due to a roll rate p_w about the velocity vector (pointing into the plane of the paper).

induces a relative velocity component $\pm p_w y$ normal to the free-stream velocity at every span-wise station of the wing. Hence, an additional angle of attack is induced as below, positive on the down-going wing and negative in case of the up-going one:

$$\Delta\alpha = \pm \frac{p_w y}{V} \tag{2.14}$$

There is therefore an increased lift on the down-going wing and decreased lift on the up-going one; this lift differential produces a rolling moment opposite in sense to the roll rate p_w. The derivative $C_{l_{p2}}$ is therefore usually negative and is called the roll damping effect. The value of the derivative $C_{l_{p2}}$ is largely dependent on the wing section (airfoil) lift curve slope and the wing geometry.

For example, a typical value of the derivative $C_{l_{p2}}$ for a military airplane with trapezoidal wing having aspect ratio about 3.5 is $C_{l_{p2}} = -0.3755/\mathrm{rad}$.

A second source of roll damping is the tail, both horizontal and vertical—the principle is quite similar to that of the wing above, but the usually small tail aspect ratio means that the tail contribution to $C_{l_{p2}}$ is relatively small. However, in case of finned missiles, for instance, which do not have a main lifting surface (wing), the tail is the primary contributor to $C_{l_{p2}}$.

Homework Exercise: Though not commonly modeled for airplanes, the Magnus effect is an important source of force/moment for spinning vehicles such as rockets, missiles, and even bullets. In brief, a spinning vehicle at an angle of incidence experiences a cross force and moment; that is, a pitching moment that depends on the sideslip angle and a yawing moment depending on the angle of attack. The Magnus effect is usually modeled as follows:

$$C_{L_{Mag}} = \frac{\partial C_L(\beta)}{\partial p_w}\frac{p_w b}{2V}, \quad C_{m_{Mag}} = \frac{\partial C_m(\beta)}{\partial p_w}\frac{p_w b}{2V}$$
$$C_{Y_{Mag}} = \frac{\partial C_Y(\alpha)}{\partial p_w}\frac{p_w b}{2V}, \quad C_{n_{Mag}} = \frac{\partial C_n(\alpha)}{\partial p_w}\frac{p_w b}{2V}$$

Note that for a Magnus force/moment to be created, two variables—roll rate and an incidence angle—are simultaneously required. For an understanding of the Magnus effect in flight dynamics, see, for instance, Reference 3.

2.6 DOWNWASH LAG TERMS

Downwash created by the flow over a lifting surface such as a wing modifies the flow-field downstream. As a result, any other surface, for example, a horizontal tail, located downstream of the wing experiences a slightly altered "free-stream" flow than seen by the wing. Since the aerodynamic forces and moments generated by a surface are dependent on the relative flow incident on that surface, the altered "free-stream" flow must be accounted for in the aerodynamic modeling of surfaces downstream of a lifting surface.

Figure 2.11 shows the flow with velocity V incident on a wing at angle of attack α_w and the resultant upwash ahead and downwash behind the wing created as a result of the lift created by the wing. The downwash at the tail is measured in terms of the downwash angle, $\varepsilon(t)$. Consequently, the angle of attack α_t at the tail is reduced by the downwash angle, $\varepsilon(t)$, and therefore α_t is given by

$$\alpha_t(t) = \alpha_w - \varepsilon(t) \tag{2.15}$$

The downwash angle itself depends on the wing angle of attack α_w (equivalently, the wing lift coefficient) but, since it takes a finite time for any change at the wing to travel downstream and affect the tail flow-field, not on its present value but its value at a time Δt previously. This effect is therefore known as the "downwash lag effect" and the lag time Δt is estimated as the time taken for a disturbance traveling with the free-stream velocity to cover a distance from the wing aerodynamic center to the tail aerodynamic center. That is, the time lag is estimated as $\Delta t = l_t/V$, where l_t is the distance between wing and tail aerodynamic centers, and V is the airplane speed.

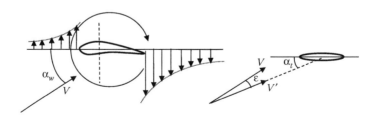

FIGURE 2.11   Sketch of downwash created by flow over a lifting surface (wing) and resultant change in the flow-field at a downstream surface (tail).

In practice, the downwash angle at the tail is modeled as

$$\varepsilon(t) = \frac{d\varepsilon}{d\alpha}\alpha(t - \Delta t) \tag{2.16}$$

where $\alpha(t - \Delta t)$ is the time-lagged angle of attack at the wing.

A change in angle of attack at the tail *due to downwash lag effect* is approximated as

$$\Delta\alpha_t = \frac{d\varepsilon}{d\alpha}\Delta\alpha = \frac{d\varepsilon}{d\alpha}\cdot\frac{\Delta\alpha}{\Delta t}\cdot\Delta t = \frac{d\varepsilon}{d\alpha}\cdot\dot{\alpha}\left(\frac{l_t}{V}\right) \tag{2.17}$$

The increment in tail lift due to change in tail angle of attack due to downwash lag effect is modeled as

$$\Delta C_{Lt} = C_{L\alpha_t}\Delta\alpha_t = C_{L\alpha_t}\frac{d\varepsilon}{d\alpha}\dot{\alpha}\left(\frac{l_t}{V}\right) \tag{2.18}$$

And the resultant change in pitching moment is given by

$$\Delta C_{mCG} = -\frac{S_t l_t}{Sc}C_{L\alpha_t}\frac{d\varepsilon}{d\alpha}\dot{\alpha}\frac{l_t}{V} = -V_H C_{L\alpha_t}\frac{d\varepsilon}{d\alpha}\dot{\alpha}\frac{l_t}{V} \tag{2.19}$$

Through these expressions, the downwash lag effect manifests itself as a damping in the direction of normal acceleration given by the derivative

$$C_{L\dot{\alpha}} = \frac{dC_L}{d(\dot{\alpha}c/2V)} = C_{L\alpha_t}\frac{S_t}{S}\frac{d\varepsilon}{d\alpha}\frac{l_t}{V}\cdot\frac{2V}{c} = 2V_H C_{L\alpha_t}\frac{d\varepsilon}{d\alpha} \tag{2.20}$$

And as a damping in pitch dynamics, in terms of the derivative

$$C_{m\dot{\alpha}} = -2\frac{l_t}{c}V_H C_{L\alpha_t}\frac{d\varepsilon}{d\alpha} \tag{2.21}$$

In the above expressions, $V_H = (S_t l_t/Sc)$ is horizontal tail volume ratio (HTVR) encountered earlier and $C_{L\alpha_t}$ is the lift curve slope of the tail.

Now all the components of the aero model in Equation 2.3—static, dynamic, flow curvature, downwash—are in place, the aerodynamic forces and moments acting on an airplane under all possible flight conditions of Mach number, altitude, flow incidence, control surface deflections, and body angular rates in any given maneuver can be estimated.

2.7 SAMPLE SIMULATION CASES

Combining three things—the equations of motion in Chapter 1, the aerodynamic model presented here, and a model for the engine thrust—one may, in principle, simulate the motion of the airplane to reveal its time history, that is, the changes in the variables describing its motion as a function of elapsed time. For our purposes, a simple model for the thrust as a function of altitude, Mach number, and throttle setting will usually suffice (see Box 2.1).

Analytical solutions to the airplane equations of motion are rare; it is common to carry out a numerical simulation to obtain the time history. A brief summary of standard numerical schemes used for time integration of the equations of aircraft dynamics is presented in Box 2.2.

Simulations may be launched with different choices of control inputs, either changing with time or fixed, and different atmospheric/wind conditions—these are called open-loop simulations. When the aircraft dynamics is simulated in tandem with a flight control law, then various control commands are given as input—those are called closed-loop simulations. Closed-loop simulations will be dealt with in a later chapter.

BOX 2.1 ENGINE THRUST MODEL

Thrust from aircraft engines (turbo-prop or turbo-jet) is usually a function of altitude (density of air changes with altitude), speed of flight, and throttle setting. The thrust at a given altitude h, Mach number Ma, and throttle setting η can be represented as

$$T(h, Ma, \eta) = \sigma C_T(Ma) T_{SL}(h_{SL}, Ma = 0, \eta) \qquad (2.22)$$

Subscript "SL" here represents a sea-level reference condition, $\sigma = \rho/\rho_{SL}$ is the ratio of air density at actual operating altitude to the air density at sea-level altitude, and $C_T(Ma)$ is the thrust coefficient that solely accounts for variation of thrust with Ma. Variation in $C_T(Ma)$ is quite significant for turbo-prop engines, whereas in case of turbojet engines, the thrust coefficient is fairly unchanged with Mach number.

BOX 2.2   STANDARD NUMERICAL SCHEMES FOR TIME INTEGRATION OF THE SET OF ORDINARY DIFFERENTIAL EQUATIONS USED IN AIRCRAFT FLIGHT DYNAMICS

Integration of aircraft equations of motion described by a set of ordinary differential equations, written in compact (vector) form as

$$\frac{d\underline{x}}{dt} = \underline{f}(\underline{x}, \underline{U}), \qquad \underline{f} : \Re^n \to \Re^n$$

requires an initial value problem solver. Numerical schemes such as Euler method for integration and Runge–Kutta methods are popular. These numerical schemes are based on time-marching from a given initial condition $\underline{x}(t = 0) = \underline{x}(0)$ for fixed values of the parameters $\underline{U}$. In MATLAB®, one can use an ordinary differential equation (ODE) solver function for computing aircraft trajectories from a starting point. For a quick introduction to ODE solvers in MATLAB and how to choose an appropriate one for your problem, see http://mathworks.com/help/matlab/math/choose-an-ode-solver.html.

Example: Let us say we want to simulate trajectories of the nonlinear second order ODE given by

$$\frac{d^2y}{dt^2} + 5\mu(1 - y^2)\frac{dy}{dt} + 9\sin y = 0$$

From the given initial condition

$$y(t = 0) = 0, \quad \dot{y}(t = 0) = 0.1$$

The above equation can be written as two first-order ODEs as below:

$$\text{Assuming } y = y_1; \frac{dy}{dt} = \dot{y} = y_2$$

$$\dot{y}_1 = y_2$$

$$\dot{y}_2 = -9\sin y_1 - 5\mu\left(1 - y_1^2\right)y_2$$

Create two files, namely, `afd.m` and `afdrun.m`, in MATLAB in the same directory. The contents of the two files, respectively, are the following:

```
%afd.m
function dydt = afd(t,y,Mu)
dydt = [y(2); -9*sin(y(1))-5*Mu*(1-y(1)^2)*y(2)];
```

```
%afdrun.m
tspan = [0, 20];
y0 = [0; 0.1];
Mu = 0.1;
ode = @(t,y) afd1(t,y,Mu);
[t,y] = ode45(ode, tspan, y0);

% Plot of the solution
plot(t,y(:,1))
xlabel('Time (s)')
ylabel('y(t)')
```

The output of the MATLAB code is plotted in Figure 2.12.

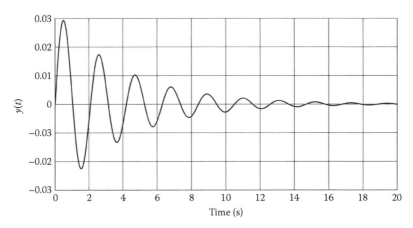

FIGURE 2.12 Result of numerical integration of example ODE showing time history.

Though the airplane's steady states may be found, in principle, without carrying out a time simulation by merely solving the algebraic trim equations, a time history may be used to confirm a particular steady state or check its stability. In the latter case, it is common to start a simulation from an initial state slightly disturbed from a steady state with controls held fixed and observe the variation of the state variables to conclude about the stability or otherwise of that steady state. Such simulations are usually carried out for short durations of the order of a few seconds to a minute to determine the open-loop natural response of an airplane to disturbances, either natural or pilot-induced or due to flight events such as a gust of wind or the dropping of a store.

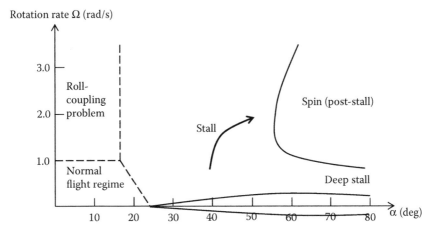

FIGURE 2.13 Short-term maneuvers distinguished based on angular rate and angle of attack (From Sinha, N. K., and Ananthkrishnan, N., *Elementary Flight Dynamics with an Introduction to Bifurcation Analysis and Continuation Methods*, CRC Press, Taylor & Francis publications, Boca Raton, Florida, USA, 2014.).

Time simulations may also be carried out to simulate maneuvers as flown by a pilot by suitable variation of the controls. The maneuvers may be placed in two categories depending on the time-scale—short-term and long-term. Short-term maneuvers deal with airplane attitudes and rates and are usually described in terms of two metrics—the rotation rate and angle of attack. As shown in Figure 2.13, one way in which different flight regimes may be distinguished is based on angle of attack—either low (well below stall) or controlled flight at moderate-to-high angles of attack (stall and beyond). On the other axis, the flight regime is qualified based on rotation rate—either low or high (rapid roll or large yaw-rate turns). A mix of high angle of attack and large rotation rates is obtained in case of spin, either inadvertent or deliberate.

Long-term maneuvers, on the other hand, are concerned with time of flight and distance (range and altitude) covered and are usually talked about in terms of Mach number and altitude with energy height and specific excess power (SEP) used as metrics (see Figure 2.14). The focus here is on optimal acceleration and climb profiles, minimizing time taken or fuel consumed, or maximizing distance covered. Changes in control input and the state variables are normally more gradual but the simulation time periods are long.

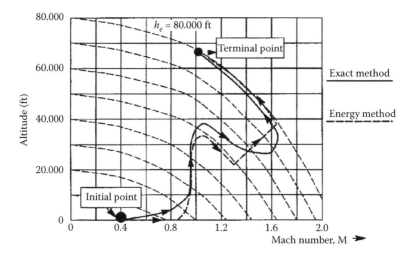

FIGURE 2.14 Long-term maneuvers shown on a plot of altitude and Mach number. (From http://forum.keypublishing.com/showthread.php?72673-Boyd-s-E-M-Theory.)

2.7.1 Example Airplane and Aerodynamic Models

Three aircraft models with their corresponding aerodynamic data are used in this text to illustrate the flight dynamic phenomena in the various flight regimes in Figure 2.13. These are

- The low-angle-of-attack F-18 model for the normal flight regime

- The F-18/HARV (High Angle of Attack Research Vehicle) for flight at high angles of attack

- The low-angle-of-attack, high-roll-rate pseudo-steady-state (PSS) model for roll coupling studies

These models are described below.

2.7.1.1 F-18 Low-Angle-of-Attack Model

Common geometry, mass, inertia, and thrust data for the F-18 aircraft are as listed in Table 2.1. The aerodynamic coefficients for the low-angle-of-attack model are reported in Table 2.2. The coefficients are a function of angle of attack α and sideslip angle β, body-axis pitch and yaw rates, q_b and r_b, respectively, and the wind-axis roll rate p_w, following the modeling scheme in Equation 2.3. Additionally, they are also functions of the control

TABLE 2.1   Geometry, Mass, Inertia, and Thrust Data
Common to Both F-18 Models

Quantity	Value	Units
Mass, m	15,118.35	kg
Wing span, b	11.405	m
Mean aerodynamic chord, c	3.511	m
Wing planform area, S	37.16	m^2
Roll moment of inertia, I_{xx}	31,181.88	kg-m^2
Pitch moment of inertia, I_{yy}	205,113.07	kg-m^2
Yaw moment of inertia, I_{zz}	230,400.22	kg-m^2
Maximum thrust, T_m	49,817.6	N

surface deflection angles. No Mach number dependence is modeled in Table 2.2. The longitudinal coefficients are limited to an angle of attack of 40 deg, whereas the lateral coefficients are modeled up to 35 deg angle of attack.

2.7.1.2 F-18/HARV High-Angle-of-Attack Model

The F-18/HARV model* consists of a tabulated database of the aerodynamic derivatives and coefficients as listed in Table 2.3 over the range of angle of attack −4 deg ≤ α ≤ 90 deg given at an interval of 4 deg. Each coefficient is assembled from a static term, a dynamic or flow curvature term, and appropriate control deflection terms. Each lateral-directional static term is a linear function of sideslip angle. Elevator derivative terms are specified separately for left and right elevator deflections. Note that the lateral-directional terms also depend on elevator deflection—this is because of the asymmetry in the aerodynamics due to left and right elevator deflection. Mach number dependence is not explicitly modeled.

For the convenience of readers, the data listed in Table 2.3 have been plotted in Figures 2.15 through 2.17. In practice, the tabular data are smoothened and curve-fitted before being used in the simulations.

The limits on the actuators for the different control deflections and rates are as listed in Table 2.4.

2.7.1.3 Pseudo-Steady-State Model for Rapid Rolling Maneuvers

The airplane model used for studying rapid rolling maneuvers using the pseudo-steady-state equations of motion as described in Chapter 1 is as follows. The mass and geometric data are provided in Table 2.5.

---

* Available at http://www.nasa.gov/centers/dryden/history/pastprojects/HARV/Work/NASA2/nasa2.html.

TABLE 2.2 Aerodynamic Coefficients for the Low-Angle-of-Attack F-18 Model

Coefficient	Functional Relation	Range
C_D	$0.0013\alpha^2 - 0.00438\alpha + 0.1423$	$-5 \le \alpha \le 20$
	$-0.0000348\alpha^2 + 0.0473\alpha - 0.358$	$20 \le \alpha \le 40$
C_L	$0.0751\alpha + 0.0144\delta e + 0.732$	$-5 \le \alpha \le 10$
	$-0.00148\alpha^2 + 0.106\alpha + 0.0144\delta e + 0.569$	$10 \le \alpha \le 40$
C_Y	$-0.0186\beta + \dfrac{\delta a}{25}(-0.00227\alpha + 0.039) + \dfrac{\delta r}{30}(-0.00265\alpha + 0.141)$	$-5 \le \alpha \le 35$
C_m	$-0.00437\alpha - 0.0196\delta e - 0.123q_b - 0.1885$	$-5 \le \alpha \le 40$
C_l	$C_l(\alpha,\beta) - 0.0315p_w + 0.0126r_b + \dfrac{\delta a}{25}(0.00121\alpha - 0.0628) - \dfrac{\delta r}{30}(0.000351\alpha - 0.0124)$	
	$C_l(\alpha,\beta) = \begin{cases} (-0.00012\alpha - 0.00092)\beta \\ (0.00022\alpha - 0.006)\beta \end{cases}$	$-5 \le \alpha \le 15$ $15 \le \alpha \le 25$
C_n	$C_n(\alpha,\beta) - 0.0142r_b + \dfrac{\delta a}{25}(0.000213\alpha + 0.00128) + \dfrac{\delta r}{30}(0.000804\alpha - 0.0474)$	
	$C_n(\alpha,\beta) = \begin{cases} 0.00125\beta \\ (-0.00022\alpha + 0.00342)\beta \\ -0.00201\beta \end{cases}$	$-5 \le \alpha \le 10$ $10 \le \alpha \le 25$ $25 \le \alpha \le 35$

Note: All angles and control deflections in deg and angular rates in rad/s.

TABLE 2.3   Tabulated Aerodynamic Coefficients for the High-Angle-of-Attack F-18 Model

| | **Tabulated Coefficients/Derivatives** | | | | |
| | | **Dynamic or Flow** | | | |
Force/Moment	**Static**	**Curvature**		**Elevator**	**Aileron**	**Rudder**
Drag	C_{D0}	C_{Dq1}		$C_{D\delta e,r}, C_{D\delta e,l}$		
Side force	$C_{Y\beta}$	C_{Yp2}	C_{Yr1}	$C_{Y\delta e,r}, C_{Y\delta e,l}$	$C_{Y\delta a}$	$C_{Y\delta r}$
Lift	C_{L0}	C_{Lq1}		$C_{L\delta e,r}, C_{L\delta e,l}$		
Rolling moment	$C_{l\beta}$	C_{lp2}	C_{lr1}	$C_{l\delta e,r}, C_{l\delta e,l}$	$C_{l\delta a}$	$C_{l\delta r}$
Pitching moment	C_{m0}	C_{mq1}		$C_{m\delta e,r}, C_{m\delta e,l}$		
Yawing moment	$C_{n\beta}$	C_{np2}	C_{nr1}	$C_{n\delta e,r}, C_{n\delta e,l}$	$C_{n\delta a}$	$C_{n\delta r}$

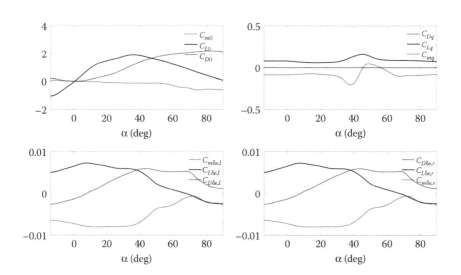

FIGURE 2.15   F-18/HARV longitudinal aerodynamic and control derivatives from Table 2.3.

The aerodynamic derivatives are as listed in Table 2.6. Drag and thrust are not used in the PSS model.

2.7.2 Example Simulation Results

A simulation is run by setting the initial conditions—initial values of the state variables—and prescribing the control inputs for the duration of the simulation. The numerical method is then supposed to calculate the time history of the state variables subject to the accuracy of the scheme (an issue that we will not deal with in this text). In principle, any set of initial conditions is acceptable; however, in practice, the initial condition is usually chosen to be (or close to) a steady state (also referred as trim

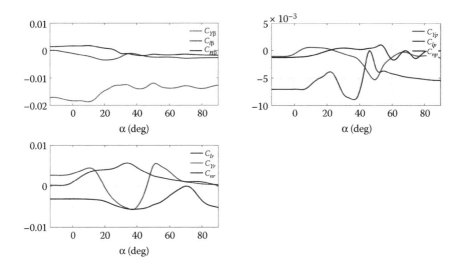

FIGURE 2.16   F-18/HARV lateral aerodynamic derivatives from Table 2.3.

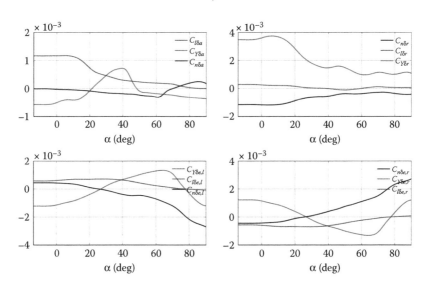

FIGURE 2.17   F-18/HARV lateral control derivatives from Table 2.3.

TABLE 2.4   Actuator Position and Rate Limits for F-18/HARV Model

Control Surface	Position Limit (deg)	Rate Limit (deg/s)
Elevator	(−25, 10)	±40
Aileron	(−35, 35)	±100
Rudder	(−30, 30)	±82

TABLE 2.5   Geometry, Mass, Inertia, and Thrust Data for the
PSS Airplane Model

Quantity	Value	Units
Mass, m	2,718	kg
Wing span, b	11.0	m
Mean aerodynamic chord, c	1.829	m
Wing planform area, S	20.07	m^2
Roll moment of inertia, I_{xx}	2,304.9	kg-m^2
Pitch moment of inertia, I_{yy}	16,809	kg-m^2
Yaw moment of inertia, I_{zz}	18,436	kg-m^2
Maximum thrust, T_m	Not used	N

condition). Simple trim states can often be obtained by manual solution of
the steady-state equations of motion; otherwise, a fairly elementary trim
routine can numerically do the job. In Chapter 3, we shall present a more
elegant method of obtaining trim states.

The following two straight and level flight trim states are selected for the
simulation examples in this section for the F-18/HARV data given in Tables
2.1 and 2.3. The equations being numerically integrated are those in Table 1.6.

Flight States	Trim State 1	Trim State 2
Mach number	$M = 0.206$	$M = 0.6$
Angle of attack and sideslip (rad)	$\alpha = 0.3, \beta = 0$	$\alpha = 0.035, \beta = 0$
Roll, pitch, and yaw rates (rad/s)	$p = 0, q = 0, r = 0$	$p = 0, q = 0, r = 0$
Body-axis roll and pitch Euler angle (rad)	$\varphi = 0, \theta = 0.3$	$\varphi = 0, \theta = 0.035$
Throttle setting (between 0 and 1)	$\eta = 0.54$	$\eta = 0.38$
Elevator, aileron, and rudder angle (rad)	$\delta e = -0.0522, \delta a = 0, \delta r = 0$	$\delta e = 0.006, \delta a = 0, \delta r = 0$

Incidentally both the trim states selected here are longitudinal flight
states, that is, the initial condition corresponds to motion in the longitudinal plane. The first trim state is at a low Mach number, nearly at landing angle of attack (but still in level flight), whereas the Trim State 2 is
at a moderately high subsonic Mach number corresponding to a cruise
condition.

Sample Simulation 1: Starting with the initial condition at Trim State 1,
an elevator pulse is applied for a period of 2 s. All other control parameters
are held unchanged. The prescribed control input and the time response
are plotted in Figure 2.18. As expected, the down elevator deflection at the

TABLE 2.6 Aerodynamic Coefficients for the PSS Airplane Model

		Tabulated Coefficients/Derivatives			
Force/Moment	**Static**	**Dynamic or Flow Curvature**	**Elevator**	**Aileron**	**Rudder**
Side force	$C_{y\beta} = -0.081/\text{rad}$	$C_{yp2} = 0$ $\quad$ $C_{yr1} = 0$	$C_{y\delta e} = 0$	$C_{y\delta a} = 0$	$C_{y\delta r} = 0$
Lift	$C_{L\alpha} = 4.35/\text{rad}$	$C_{Lq1} = 0$	$C_{L\delta e} = 0$		
Rolling moment	$C_{l\beta} = -0.081/\text{rad}$	$C_{lp2} = -0.442/\text{rad}$ $\quad$ $C_{lr1} = 0.0309/\text{rad}$	$C_{l\delta e} = 0$	$C_{l\delta a} = -1.24/\text{rad}$	$C_{l\delta r} = 0$
Pitching moment	$C_{m\alpha} = -0.435/\text{rad}$	$C_{mq1} = -9.73/\text{rad}$	$C_{m\delta e} = -1.07/\text{rad}$		
Yawing moment	$C_{n\beta} = 0.0218/\text{rad}$	$C_{np2} = 0$ $\quad$ $C_{nr1} = -0.0424/\text{rad}$	$C_{n\delta e} = 0$	$C_{n\delta a} = 0$	$C_{n\delta r} = -0.01/\text{rad}$

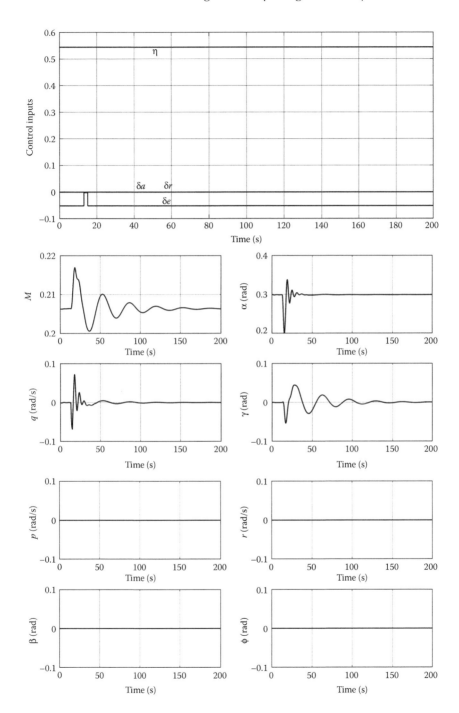

FIGURE 2.18 Response to elevator pulse input from level flight condition (δa, δe, δr in rad).

start of the pulse induces a nose-down pitching motion, decreasing the angle of attack and slightly raising the speed (Mach number). With the withdrawal of the elevator deflection at the end of the pulse, the perturbations in pitch rate, angle of attack, Mach number, and flight path angle die down and the original trim state is recovered—this suggests that Trim State 1 is a stable equilibrium. Of the variables, q and α predominantly show oscillations at a higher frequency and also damp out more rapidly than Mach number and γ, which vary at a lower frequency. This suggests that the dynamics in the longitudinal variables is made up of two (natural) modes—a high-frequency, well-damped mode (called the *Short* period) and a relatively low-frequency, lightly damped one (called the *Phugoid*). The lateral-directional variables are not notably affected by the elevator pulse input as observed from Figure 2.18.

This simulation has shown that Trim State 1 is indeed a steady-state condition of the airplane (for the data set considered). The trim state has been confirmed to be a stable one since all the state variables returned to their initial (trim) value after having been perturbed by the elevator pulse. The response of the longitudinal variables was as expected from our understanding of the physics of airplane flight. Lastly, the simulation revealed the existence of two distinct modes for the longitudinal flight dynamics.

Sample Simulation 2: Starting with the same initial conditions as in the previous simulation, in this case, a step input in elevator deflection is applied. All other control inputs are held fixed. The relevant plots are shown in Figure 2.19. As a result of the up-elevator deflection, the airplane settles to a new steady state at a higher angle of attack and correspondingly lower speed (Mach number). Obviously, this new trim state is also stable as seen from the simulation; however, its flight path angle is slightly negative, indicating a descending flight, different from the level condition of Trim State 1 at the beginning of the simulation. This is a consequence of having an unchanged thrust by keeping the throttle setting η fixed. No change in the lateral-directional variables is visible during this maneuver either and the two distinct longitudinal flight modes are apparent in Figure 2.19 as well. Thus, the primary consequence of a step elevator input is to change the airplane trim speed by virtue of a change in the angle of attack.

Sample Simulation 3: In this case, airplane response to a step decrease in thrust is simulated while keeping other control inputs at their equilibrium values. The resulting time history plots are shown in Figure 2.20.

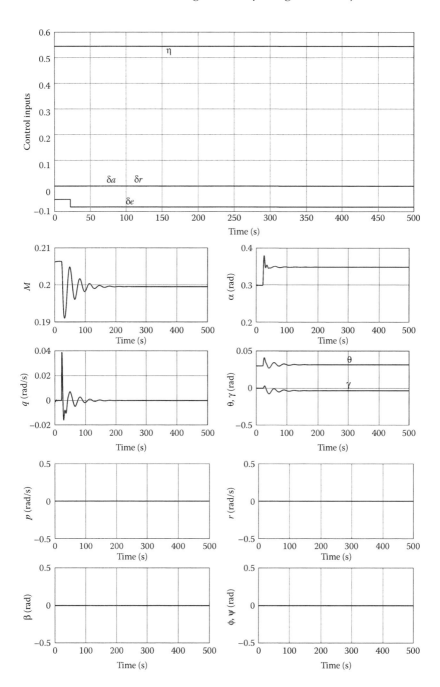

FIGURE 2.19   Response to elevator step input from level flight condition (δa, δe, δr in rad).

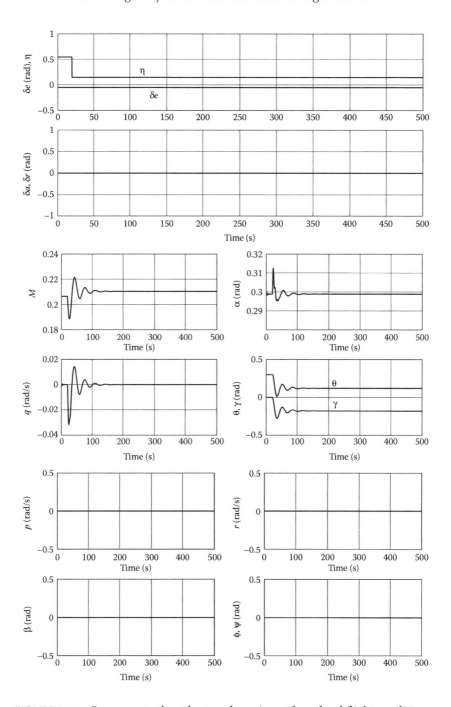

FIGURE 2.20 Response to throttle step-down input from level flight condition.

Interestingly, the final steady state is a descending flight at a slightly *higher* speed (Mach number). This may be contrary to the popular perception of an increase/decrease in thrust corresponding to an increase/decrease in steady-state speed. The steady-state angle of attack is nearly unchanged due to the elevator deflection being held fixed.

Actually, the drop in thrust as seen in Figure 2.20 does result initially in an immediate decrease in speed, which in turn reduces both lift and drag (each being proportional to V^2). The airplane then settles into a steady state satisfying the standard force equilibrium conditions ($(T − D)/W = \sin \gamma$, $L/W = \cos \gamma$), where γ is the flight path angle (climb angle, negative in case of descending flight). Figure 2.21 shows the loss in altitude due to flying in a straight, descending trajectory.

Sample Simulation 4: In this case, a step-up increase in thrust is applied while keeping the other control inputs at their equilibrium values. This is opposite to Case 3; here, a steady climb is observed (see time history plots in Figures 2.22 and 2.23).

The important conclusion from these simulation cases is that thrust (throttle input) is primarily meant to change the flight path angle (i.e., climb or descend) having only a marginal effect on the speed and trim angle of attack. On the other hand, elevator input is the primary means of adjusting the trim speed and angle of attack, having only a negligible effect on the flight path angle.

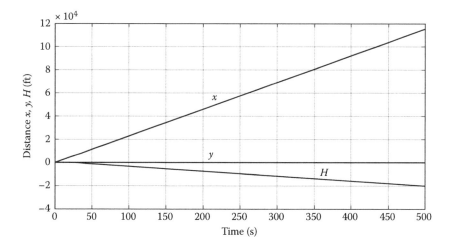

FIGURE 2.21 Descending airplane in response to decrease in thrust from level flight condition.

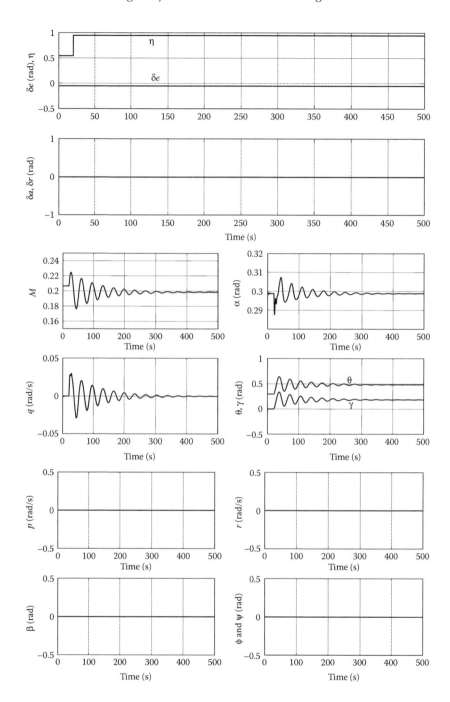

FIGURE 2.22   Response to throttle step-up input from level flight condition.

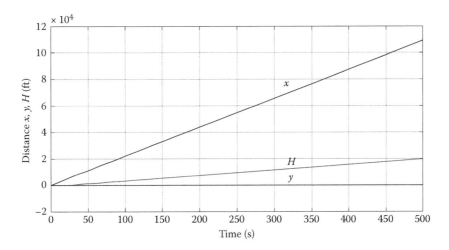

FIGURE 2.23 Airplane ascending in response to increase in thrust from level flight condition.

Homework Exercise: Imagine an airplane flying along a straight line, level flight path in trim. The pilot desires to quickly accelerate the airplane to a much higher flight speed, but not necessarily settle into another trim state. Which controls can he use and to what degree? What happens if he slams the throttle to increase the thrust appreciably? On the other hand, what happens if he pushes the stick forward (move the elevator down) to put the airplane into a dive?

Sample Simulation 5: Now we initiate a simulation from Trim State 1 keeping the throttle and elevator fixed but deploying the lateral-directional controls (aileron and rudder). In this instance, a small-magnitude aileron pulse of 3-s duration is applied with the rudder kept at neutral. The resulting time history is shown in Figure 2.24. As expected, the longitudinal variables hardly show any change whereas the lateral-directional variables are all perturbed from their trim values. The aileron input is able to excite both the rolling and yawing motions, albeit to different degrees. Over the time scale plotted in Figure 2.24, two modes of lateral-directional motion are visible—one is an oscillatory motion with moderately high frequency and damping (called the *Dutch Roll* mode) and the other is a slow first-order response (called the *Spiral* mode). Both the modes appear to be stable as the variables regain their trim values. (Note that the heading angle ψ and the position variables x, y, z are discounted when regarding stability).

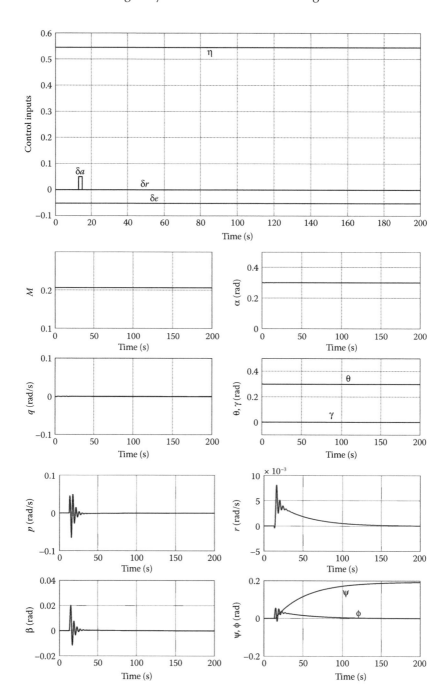

FIGURE 2.24 Response to aileron pulse input from level flight condition (δa, δe, δr in rad).

Sample Simulation 6: In this case, we initiate a simulation from Trim State 2 using only the aileron while keeping the throttle, elevator, and rudder fixed. The setup is similar to the previous instance in Sample Simulation 5 except that the aileron input is of a larger magnitude and is in the form of a doublet. The other controls are undisturbed from their trim values. The results of the simulation are plotted in Figure 2.25. It may be noticed that this input (aileron doublet of reasonably large magnitude) excites both the longitudinal and lateral-directional variables. Whereas the *Short Period* mode seems to damp out rapidly, the *Phugoid* mode persists and appears to be very lightly damped. Among the directional variables, the *Dutch Roll* motion is clearly seen in the oscillatory behavior of the lateral-directional variables.

The initial positive aileron input caused the airplane to bank left and hence also turn left. The second half of the doublet with negative aileron deflection nearly resets the bank angle to zero (wings level). However, the spiral mode in this case (Trim State 2) is found to be marginally unstable. The bank angle therefore very slowly veers off from the level flight condition and the airplane is seen to yaw to the right very slowly. In principle, Trim State 2 is unstable; however, a marginally unstable spiral mode is usually not of concern and for all practical purposes, the airplane behaves (in the other modes) as if it were flying at a stable trim.

From this simulation and the previous one, it is clear that aileron deflection primarily produces a roll rate p. To manipulate the bank angle requires additional integral action.

Homework Exercise: Work out a control schedule that will roll the airplane to a prescribed bank angle and hold the bank. Carry the exercise a step further by getting the airplane to fly a banked turn with zero sideslip. Finally, attempt to get a zero-sideslip banked turn while maintaining a constant velocity and unchanged altitude (steady, level flight).

Sample Simulation 7: For the last simulation in this set, we provide a step input to the rudder after trimming the airplane at Trim State 1. The input corresponds to a hard left rudder. The time history plots appear in Figure 2.26.

The airplane begins yawing to the left, banks left, and pitches down steeply. At steady state, the pitch angle is nearly 60 deg nose down. The angle of attack being about 17 deg, the airplane velocity vector is inclined at a descent angle of approximately 75 deg, making it a steep spin. The spiraling trajectory is clearly seen in the trajectory plot of Figure 2.26. The roll, pitch, and yaw rates are all significant—they are the components of the angular velocity vector in spin about the body-fixed axes; in

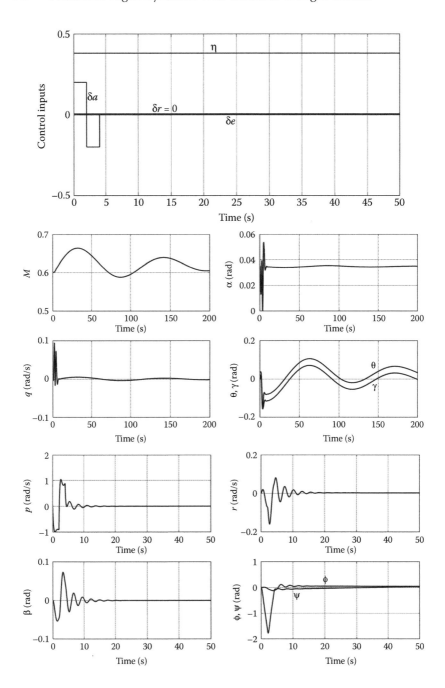

FIGURE 2.25 Response to aileron doublet input from level flight condition at low angle of attack showing both longitudinal and lateral-directional modes (δa, δe, δr in rad).

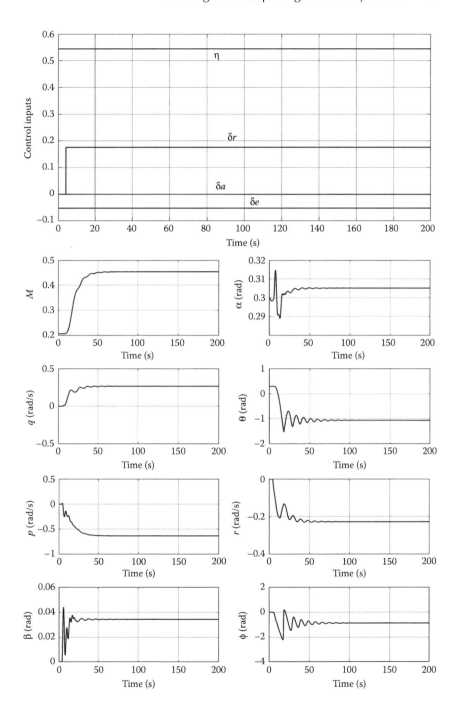

FIGURE 2.26 Response to rudder step input from level Trim State 1 showing the airplane enter into a spin motion (δa, δe, δr in rad). *(Continued)*

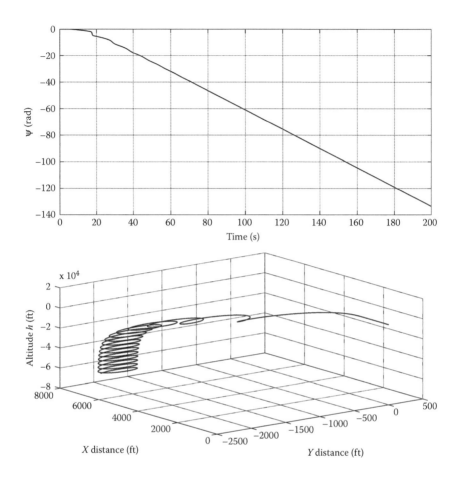

FIGURE 2.26 (Continued) Response to rudder step input from level Trim State 1 showing the airplane enter into a spin motion (δa, δe, δr in rad).

a steep spin, the X-component (roll rate p) is usually large. Finally, note that this spin is actually a steady state—all the variables have settled down to constant values.

Homework Exercise: Stall leading to spin is a leading cause of airplane accidents, especially in case of general aviation aircraft. Look up cases on the Internet or in your library of stall/spin-related air crashes.

EXERCISE PROBLEMS

1. Write out the force and moment components in terms of coefficients for a wingless missile of body diameter D. (Use D instead of b, c, and body cross-sectional area $S_c = \pi D^2/4$ in place of S.)

2. Model the aerodynamic moment coefficient due to the Magnus effect. Explicitly write out the contribution to the aerodynamic moments using the Magnus moment coefficient.

3. Plot the following airplane data and analyze the trend of the variation of the static longitudinal coefficients with angle of attack. (δe is the elevator deflection angle in deg, α in deg).

$$C_D = 0.0013\alpha^2 - 0.00438\alpha + 0.1423, \quad -5 \leq \alpha \leq 20°$$
$$-0.0000348\alpha^2 + 0.0473\alpha - 0.3580, \quad 20 \leq \alpha \leq 40°$$
$$C_L = 0.0751\alpha + 0.0144\delta e + 0.732, \quad -5 \leq \alpha \leq 10°$$
$$-0.00148\alpha^2 + 0.106\alpha + 0.0144\delta e + 0.569, \quad 10 \leq \alpha \leq 40°$$
$$C_m = -0.00437\alpha - 0.0196\delta e - 0.1885$$

4. Derive an approximate expression for the derivative C_{l_β} due to wing dihedral angle Γ assuming a rectangular wing of constant chord c and identical airfoil sections. ($C_{l_\beta} = -c_{l_\alpha}\Gamma/4$, where c_{l_α} is the wing section (airfoil) lift curve slope.)

5. Derive an approximate expression for the derivative C_{l_β} due to wing sweep angle Λ assuming a rectangular wing of constant chord c and identical airfoil sections and verify the trend of increasing (negative) C_{l_β} with increasing α (or C_L). ($C_{l_\beta} = -C_L \sin 2\Lambda/4$, where C_L is the wing (airplane) lift coefficient.)

6. Derive an estimate for the C_{n_β} contribution from the vertical tail and verify the dependence on $VTVR$. ($C_{n_\beta} = VTVR * C_{L\alpha_V}$, where $C_{L\alpha_V}$ is the vertical tail "lift" curve slope.)

7. Plot the following airplane data and analyze the trend of the variation of the static lateral coefficients with angle of attack (in deg).

$$C_{Y_\beta}(\alpha) = -0.0186$$
$$C_{l_\beta}(\alpha) = -0.00012\alpha - 0.00092, \quad -5 \leq \alpha \leq 15°$$
$$0.00022\alpha - 0.006, \quad 15 \leq \alpha \leq 25°$$
$$C_{n_\beta}(\alpha) = 0.00125, \quad -5 \leq \alpha \leq 10°$$
$$-0.00022\alpha + 0.00342, \quad 10 \leq \alpha \leq 25°$$
$$-0.00201, \quad 25 \leq \alpha \leq 35°$$

8. Evaluate the derivative $C_{m_{q2}}$ in case of a T-tailed airplane with the horizontal tail located at a height h_H from the airplane centerline.

9. For a trapezoidal wing with taper ratio λ, derive an expression for the derivative $C_{l_{p2}}$. ($C_{l_{p2}} = (-c_{l_\alpha}/12)((1+3\lambda)/(1+\lambda))$, where c_{l_α} is the wing section (airfoil) lift curve slope.)

10. Carry out a simulation by applying a step rudder input (as in Sample Simulation 7) for the low-angle-of-attack case of Trim State 2. Does the airplane enter a spin? Explain your result.

11. The example simulations in this chapter use the high-angle-of-attack F-18 aero data in Table 2.3. Write your own code with the low-angle-of-attack F-18 aero data in Table 2.2 and carry out sample simulation runs.

12. The sample simulations in this chapter output the aircraft motion by plotting the time history of the state variables. Another way of postprocessing the output data from such a simulation is to feed it to a *flight simulator*, which will then show the motion of the airplane on a computer screen. *FlightGear* (www.flightgear.org) is one such free, open-source flight simulator. Try to hook up *FlightGear* (or any other simulator) to your code and animate the airplane motion from your simulations.

REFERENCES

1. Klein, V., and Noderer, K. D., *Modeling of Aircraft Unsteady Aerodynamics Characteristics*, Part I—Postulated Models, NASA TM 109120, 1994.
2. Sinha, N. K., and Ananthkrishnan, N., *Elementary Flight Dynamics with an Introduction to Bifurcation Analysis and Continuation Methods*, CRC Press, Taylor & Francis publications, Boca Raton, Florida, USA, 2014.
3. Cayzac, R., Carette, E., Denis, P., and Guillen, P., Magnus effect: Physical origin and numerical prediction, *Journal of Applied Mechanics*, 78(5), 051005, 2011.

CHAPTER 3

Introduction to Dynamical Systems Theory

T HE EQUATIONS OF MOTION of rigid aircraft developed in Chapter 1, including aerodynamic and propulsive force/moment terms described in Chapter 2, can be compactly put in vector form as a set of first-order ordinary differential equations:

$$\frac{d\underline{x}}{dt} = \underline{f}(\underline{x},\underline{U}) \tag{3.1}$$

where $\underline{x} \in \Re^n$ is the vector of n state variables describing a system's state, $\underline{U} \in \Re^m$ is the vector of m control parameters, and $\underline{f} : \Re^{n+m} \to \Re^n$ represents the vector field comprising of nonlinear functions of state and control parameters. The right-hand side of Equation 3.1 has no explicit term indicating dependence on time, hence dynamical systems (systems evolving in time) represented in the form of Equation 3.1 are called autonomous; a nonautonomous system will have explicit dependence of parameters (other than state variables) with respect to time, for example, unsteady or time-varying wind. Equivalent autonomous form can be notionally obtained for a nonautonomous system by treating the time-dependent term as additional state variable, for example,

$$\frac{dx}{dt} = f(\underline{x}, \underline{U}) + A \sin \omega t \tag{3.2}$$

can be rewritten as

$$\frac{dx}{dt} = f(\underline{x}, \underline{U}) + A \sin \theta; \quad \frac{d\theta}{dt} = \omega \tag{3.3}$$

so that a new variable θ becomes a state variable and the new set of equations

$$\frac{dy}{dt} = g(\underline{y}, \underline{U}) \tag{3.4}$$

has apparently no explicit dependence on time, $y = [\underline{x}, \theta]'$. Solutions of a set of ordinary differential equations represented by Equation 3.1 or 3.4 are trajectories ($\underline{x}$ vs. t in case of Equation 3.1 or y vs. t in case of Equation 3.4), which can be obtained by numerically integrating the equations starting from a given initial condition ($\underline{x}(0)$ at $t = 0$, for instance) for a fixed set of parameter values ($\underline{U}$).

Homework Exercise: Look for simple examples of dynamical systems that can be put in the form of Equation 3.1 or 3.4. One example is the simple pendulum whose dynamical system representation is

$$\ddot{\theta} + \omega_n^2 \sin \theta = 0, \quad \text{where } \omega_n^2 = \frac{g}{l} \tag{3.5}$$

where g is the acceleration due to gravity and l is the length of the pendulum (see Figure 3.1), and $\ddot{\theta}$ indicates second derivative of θ with respect to time.

By defining the variables, $x_1 = \theta$, $x_2 = \dot{\theta}$, Equation 3.5 can be written as follows:

$$\begin{aligned} \dot{x}_1 &= x_2 \\ \dot{x}_2 &= -\omega_n^2 \sin x_1 \end{aligned} \tag{3.6}$$

So, $\underline{x} = [x_1, x_2]$ and $f(\underline{x}) = [x_2, -\omega_n^2 \sin x_1]$.

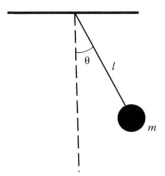

FIGURE 3.1   Sketch of the simple pendulum with the variables marked.

Homework Exercise: Consider the dynamical system consisting of a bead that can smoothly slide on a rotating hoop as pictured in Figure 3.2. The dynamics of this system can be modeled as

$$\ddot{\theta} + \sin\theta\left(\frac{g}{R} - \omega^2 \cos\theta\right) = 0 \tag{3.7}$$

where ω is the angular velocity of hoop rotation, R is the radius of the hoop, and g is the acceleration due to gravity. Write the model in Equation 3.7 in the dynamical system form of Equation 3.1.

In principle, a dynamical system of the form Equation 3.1 can be repeatedly solved (usually numerically) for several choices of initial

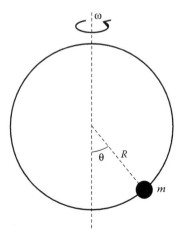

FIGURE 3.2   Sketch of the bead with rotating hoop with the variables marked.

conditions chosen either arbitrarily or in some systematic manner. For physical systems, one can expect that given an initial condition, there will be a unique trajectory starting from that initial condition. Using this approach, one may end up exploring a large number of initial conditions and examining individual trajectories to draw any useful conclusion about the behavior of the dynamical system under investigation. However, if one were to indeed obtain such a solution, how may it be represented? There are two usual ways in which the trajectories of a dynamical system of the form Equation 3.1 may be put out. One is to plot the variation of every state variable x_i as a function of time—this is called a "time series" or "time history." The other representation is a phase portrait which is useful but only for low-order dynamical systems—usually systems with just a couple of state variables. In a phase portrait, one plots two variables, say x_i and x_j, against each other with time as a parameter. For instance, in case of the simple pendulum that we just studied, the time series would be a plot of angular position θ and angular velocity $\dot{\theta}$ with time, as displayed in Figure 3.3. Also shown in Figure 3.3 is the phase portrait obtained by plotting angular position θ versus angular velocity $\dot{\theta}$. Several trajectories, each starting with a different initial condition, can be inked into a single phase portrait, giving a composite picture of the system dynamics.

Homework Exercise: Draw the phase portrait of the simple pendulum dynamics with several trajectories. What conclusions can be drawn from this picture about the dynamics of this system?

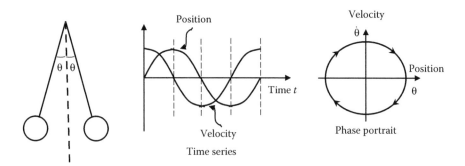

FIGURE 3.3 Illustration of time series and phase portrait representation of simple pendulum dynamics. (From https://commons.wikimedia.org/wiki/File: Pendulum_phase_portrait_illustration.svg.)

Homework Exercise: Simulate the rotating hoop with bead system to produce its phase portrait for different values of the parameter g/ω^2R. In particular, choose at least one value of this parameter greater than 1 and another less than 1.

Both the simple pendulum and the system with a bead on a rotating hoop are *nonlinear* and while their dynamics may be easy to understand, the behavior of a more general nonlinear dynamical system, such as an airplane in flight for instance, may be quite complex. By merely carrying out simulations from different initial conditions and looking at time series or phase portraits, there is a risk of not seeing the "big picture" or missing out some critical behavior. Therefore, for analysis of a general nonlinear dynamical system, it is common to use the *bifurcation and continuation method* and supplement it with a small number of carefully selected time simulations. In the bifurcation method, one primarily examines steady states of a dynamical system of the form of Equation 3.1 and their stability with a varying parameter. Typically, a continuation algorithm is used to "continue" from one initial steady state and compute all possible steady states as a parameter is varied. For example, in case of the rotating hoop with bead example, all possible steady (equilibrium type in this case) states with the parameter $\gamma = \omega^2R/g$ varied from 0 to 3 are plotted in Figure 3.4. Notice how the number of steady states changes with varying parameter. For $\gamma < 1$, there are only two distinct steady states—$\theta = 0$ at the bottom and $\theta = \pi$ at the top of the hoop. However, beyond $\gamma = 1$, two new steady states emerge at what is called a *supercritical pitchfork bifurcation*—we shall

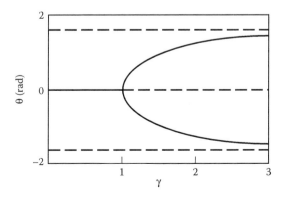

FIGURE 3.4 Bifurcation diagram of the rotating hoop with bead problem, $\gamma = \omega^2R/g$ (full line—stable; dashed line—unstable).

study such bifurcations in this chapter. For now, this is just a preview. Also note the change in stability at the bifurcation point—the steady state at $\theta = 0$ is stable up to $\gamma = 1$; thereafter, it becomes unstable and it is the newly created steady states that are stable. The other steady state at the top of the hoop ($\theta = \pi$) is unstable for all values of γ.

Homework Exercise: Interpret the phase portraits of the rotating hoop with bead system in a previous homework exercise in light of the bifurcation diagram in Figure 3.4.

In the following, we shall study steady and periodic states of *dynamical systems* and define their stability. We shall examine different types of bifurcations and the associated changes in the system dynamics. We shall present the rudiments of how a bifurcation and continuation algorithm works and how all of this may be put together to investigate a nonlinear dynamical system.

3.1 TYPES OF STEADY STATES

One can think of steady states of a dynamical system as some kind of fixed or periodically repetitive state that the system settles into and where it stays forever unless it is displaced from that state. The question of steady states of a given system can be approached in two possible ways: First, given the dynamical equations of the form Equation 3.1, one can determine all possible steady states by simply solving the set of algebraic equations, $f(x) = 0$. For example, in case of the simple pendulum dynamics given by Equation 3.6, this yields two possible steady states—$(\theta = 0, \dot{\theta} = 0)$ and $(\theta = \pi, \dot{\theta} = 0)$. A second way of establishing the steady states is to start a time simulation of the dynamics from some initial state, wait for the transients to die out and observe the state to which the system eventually settles down. For the example of the simple pendulum, if the initial state is either of the above two steady states, the system will remain at the initial state forever. However, for any other choice of the initial state, the simple pendulum behaves as in Figure 3.3, cyclically passing through the initial state at regular intervals. This periodic behavior is also a *steady state* of the simple pendulum dynamics. Thus, neither the first nor the second approach completely captures all steady states of a dynamical system. There may be steady states other than what are obtained by solving $f(x) = 0$. At the same time, there may be solutions of $f(x) = 0$ that are difficult to attain by the second method of starting a time simulation from one initial condition or another.

Homework Exercise: Consider the dynamics of the damped pendulum given by

$$\ddot{\theta} + c\dot{\theta} + \omega_n^2 \sin \theta = 0, \quad \text{where } \omega_n^2 = \frac{g}{l} \text{ and } c > 0 \qquad (3.8)$$

Solving $f(\underline{x}) = 0$ for the steady states of Equation 3.8 yields the same two steady states as for the simple (undamped) pendulum, viz. $(\theta = 0, \dot{\theta} = 0)$ and $(\theta = \pi, \dot{\theta} = 0)$. Due to the damping, the periodic steady states no longer exist.

Try to simulate the dynamics of Equation 3.8 with an initial condition that leads the system to the steady state $(\theta = \pi, \dot{\theta} = 0)$.

In the following, and later in this text, we shall describe and encounter different kinds of steady states of typical dynamical systems. The usual categories are listed below:

3.1.1 Equilibrium States

Steady states that do not change with time, that is states for which the system dynamics equations satisfy the condition $\dot{x} = 0$, can be identified by the first approach above—by solving $f(\underline{x}) = 0$. These are called equilibrium states. For example, the linear first-order system, $\dot{x} = ax$, $a \neq 0$, has a unique equilibrium state at $x = 0$ for all values of a. The nonlinear system $\dot{x} = ax + x^2$ has two unique equilibrium states, $x = 0$ and $x = -a$, for all values of a. Another nonlinear dynamical system, this one with a cubic nonlinearity, $\dot{x} = ax + x^3$, has an equilibrium state at $x = 0$ for all values of a and another pair at $x = \pm\sqrt{a}$ only for positive values of a.

Next, consider a linear second-order system of the form $\ddot{x} + c\dot{x} + kx = 0$. Taking $x = x_1$ as the first variable and $\dot{x} = x_2$ as the second variable, the second-order system can be converted to the dynamical system form of Equation 3.1, as below:

$$\begin{aligned} \dot{x}_1 &= x_2 \\ \dot{x}_2 &= -kx_1 - cx_2 \end{aligned} \qquad (3.9)$$

In case of mechanical systems, the state variables (x_1, x_2) typically represent displacement (x) and velocity $(\dot{x})$. Solutions of $f(\underline{x}) = 0$ in case of Equation 3.9 give the single equilibrium state $(x_1, x_2) = (0,0)$. However, a nonlinear second-order system may have multiple equilibrium states. For example, $\ddot{x} + c\dot{x} + x(1 - x^2) = 0$ has three equilibria: $(x_1 \ x_2) = (x \ \dot{x}) = \{(0,0) \ (-1, 0) \ (1, 0)\}$.

3.1.2 Periodic States

Periodic states are characterized by cyclic behavior, $x(t) = x(t + mT)$, m is a nonzero integer, and $T > 0$ is the minimum time period of the orbit; $m = 1$ corresponds to a period-1 or single period orbit, $m = 2$ corresponds to period-2 or period doubled orbit, and so on. A system should be at least of second-order to admit the existence of a periodic orbit; a first-order continuous-time system (having dynamics confined to a line) cannot exhibit a periodic state. An example of a periodic state of a dynamical system is the cyclic solution in Figure 3.3. However, that solution is not isolated in the sense that a different initial state will generally yield a different periodic cycle. In fact, there is an infinite number of distinct cyclic solutions in case of the simple pendulum. On the other hand, an isolated periodic orbit is called a limit cycle—there are no other closed orbits in the immediate vicinity of a limit cycle.

3.1.3 Quasi-Periodic States

Three-dimensional (and higher-order systems) admit periodic orbits with two frequencies. Closed orbits thus become possible in the form of "knots" when the ratio of two frequencies is rational. When the ratio of two frequencies present in a time signal is irrational, a quasi-periodic state becomes possible. These kinds of motion are usually visualized as trajectories on the surface of a torus where the two frequencies correspond to the two possible loops on the torus.

3.1.4 Chaotic States

Chaotic states are pictured in terms of strange attractors whose trajectories never exactly close in on themselves, hence they are not periodic, but remain confined to a finite volume nevertheless.

Figure 3.5 offers a depiction of the different kinds of steady states possible in dynamical systems. In the space of the state variables (state space), they either appear to be a point (equilibrium) or a closed loop (periodic state) or a toroidal surface (quasi-periodic state) or a strange attractor (chaotic state).

3.2 STABILITY OF STEADY STATES

Clearly, once a system is in one of the steady states discussed above, in the absence of any perturbation, it will continue to be in the same steady state, theoretically forever. The question is, what happens when it is slightly

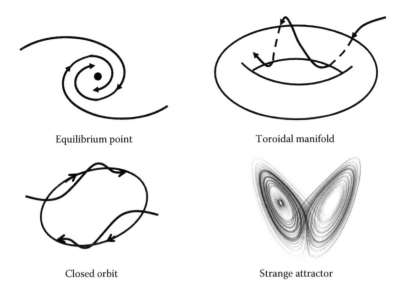

Equilibrium point   Toroidal manifold

Closed orbit   Strange attractor

FIGURE 3.5 Depiction of the different possible types of steady states of any dynamical system. (From http://www.cse.dmu.ac.uk.)

perturbed from a steady state? Imagine a system being moved to a state slightly away from an equilibrium state, or off a periodic orbit. Starting from that perturbed state, if the system evolves such that it remains near the original steady state in some sense, or even tends to move back to that steady state, then that steady state can be regarded as stable in some sense. We shall make this more precise in the following.

Note that the notion of stability introduced here refers to a steady state. In systems with multiple steady states, some of them may be *stable* as defined here, whereas some others may be *unstable*. It makes no sense, in general, to talk about *stability* of the dynamical system itself. *Stability* as a concept is usually applied to each individual steady state. However, in cases where the dynamical system has only a single steady state, and this always happens for linear systems, the notion of *stability* is sometimes loosely carried over from that of the lone steady state to that of the system itself. Thus one may come across statements about a linear dynamical system being *stable* or *unstable*. However, strictly, and to be safe, it is better to ensure that the notion of *stability* here is used only to describe a particular steady state.

3.2.1 Stability of Equilibrium States

The standard notion of *stability* widely in use is that introduced by Lyapunov [1]. As applied to an equilibrium state, it can be understood with reference to the sketch in Figure 3.6 as follows. Suppose $\underline{x}^*$ is an equilibrium state. Let us first fix a bound marked in Figure 3.6 by ε so that if ever a system trajectory $\underline{x}(t)$ breaches this bound, then the state may be declared to not be stable. Next, depending on our choice of ε, we choose a smaller region marked by δ in Figure 3.6. Now for every possible initial state $\underline{x}(0)$ within the region marked by δ, we need to confirm that during the evolution of the system dynamics, the trajectory does not go beyond the bound set by ε. If that is indeed the case, then the equilibrium state $\underline{x}^*$ can be said to be *stable in the sense of Lyapunov* [1].

Note that the choice of ε is arbitrary, we only need one choice of this domain. However, if this domain is too large, then one may presumably have to run the simulation from each initial state for a long time to rule out the possibility that it may breach the limits of this domain. So it would be prudent to select a reasonably small ε-region. Having done that, it is up to us to come up with the second, smaller domain marked by δ. If we happen to choose a δ-domain for which unfortunately one of the trajectories breaches the ε-domain, it only means that our choice of δ was unsuitable. Perhaps, another choice of δ could provide the conditions required to satisfy the requirement of *stability in the sense of Lyapunov*. Also, for a given choice of δ, we must test all possible initial conditions within that domain (actually, an infinite number) and check every trajectory for a very long time (ideally, infinite time) to ensure that the ε-bound is not breached. Finally, if we are successful in establishing *stability in the sense of Lyapunov* for an equilibrium state, it only means that perturbed

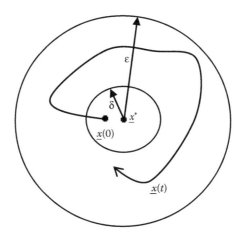

FIGURE 3.6 Sketch to describe Lypunov stability.

trajectories from that equilibrium state remain in the vicinity of that state and do not wander too far away from it. Note that this definition does not require the trajectories to actually tend back to the equilibrium state.

With so many caveats and conditions, one may imagine this definition of *stability* to be practically pretty much useless. But let us examine its application to the simple pendulum whose dynamics was shown in Figure 3.3. Consider the question of stability of the equilibrium state at $(\theta = 0, \dot{\theta} = 0)$. We know from Figure 3.3 that trajectories (other than the equilibrium states) are closed orbits. Select one such closed orbit as the ε-domain. Select the same domain as the δ-domain as well, though any smaller domain would work equally well. (This is an example to show that there is no unique choice of the δ-domain, it depends on the ingenuity of the reader!) It is easy to show that for every initial state within this δ-domain, the trajectories are closed orbits that lie within the common ε and δ domain, hence they will never breach the bound set by ε. One successful choice of δ is all that is needed to establish stability. Hence, we can state that the equilibrium state $(\theta = 0, \dot{\theta} = 0)$ for the simple pendulum dynamics is in fact *stable in the sense of Lyapunov.*

Homework Exercise: Similarly, try to examine the stability of the other equilibrium point at $(\theta = \pi, \dot{\theta} = 0)$ for the simple pendulum dynamics.

As you may have guessed, for an arbitrary dynamical system, the investigation of stability of an equilibrium point in the manner above may not be very convenient. Instead, there is a more practical way of testing for stability, which we shall describe shortly. However, the concept of stability as put forward by Lyapunov forms the basis of this test as well. It is quite remarkable that Lyapunov's notion of stability has survived virtually unscathed for over a century—a century of fantastic developments in science.

Before proceeding to the more practical matter of testing for stability of equilibrium states, there is another related concept put forward by Lyapunov—that is the notion of *asymptotic stability.* The explanation for the concept of *asymptotic stability* is literally identical to that of *stability in the sense of Lyapunov* except that in this case it is not enough for the trajectories to remain bounded by ε, they should also approach the equilibrium state with the passage of time. In other words, an additional condition:

$$\left\| x(t) - \underline{x}^* \right\| \to 0 \quad \text{as } t \to \infty \tag{3.10}$$

is necessary, where $\underline{x}^*$ is the equilibrium state in question.

Homework Exercise: For the dynamics of the simple pendulum, we have seen that the equilibrium state $(\theta = 0, \dot{\theta} = 0)$ is *stable in the sense of Lyapunov*. Now check whether it is also *asymptotically stable*. How about the same equilibrium state for the damped pendulum in Equation 3.9?

In practice, the way Lyapunov's definition of stability of an equilibrium state is actually applied is by obtaining a locally linearized approximation of the dynamical system about the equilibrium state in question. Since the ε in Lyapunov's definition of stability can be chosen to be arbitrarily small, it means that the perturbed initial states $\underline{x}(0)$ within the δ-region that are to be tested will also be correspondingly close to the equilibrium state $\underline{x}^*$ under investigation. Therefore, a small perturbation approximation is admissible.

Let $(\underline{x}^*, \underline{U}^*)$ be an equilibrium state of the system governed by Equation 3.1. Representing the disturbed state of the system from $(\underline{x}^*, \underline{U}^*)$ as $\underline{x} = \underline{x}^* + \Delta\underline{x}$ at fixed $\underline{U}^*$, Equation 3.1 can be written as

$$\underline{\dot{x}}^* + \Delta\underline{\dot{x}} = \underline{f}(\underline{x}^* + \Delta\underline{x}, \underline{U}^*) \tag{3.11}$$

Expanding Equation 3.11 in a Taylor series gives

$$\cancel{\underline{\dot{x}}^*} + \Delta\underline{\dot{x}} = \underline{f}(\underline{x}^* + \Delta\underline{x}, \underline{U}^*)$$

$$= \cancel{\underline{f}(\underline{x}^*, \underline{U}^*)} + \left.\frac{\partial \underline{f}}{\partial \underline{x}}\right|_{(\underline{x}^*, \underline{U}^*)} \Delta\underline{x} + \left.\frac{\partial^2 \underline{f}}{\partial \underline{x}^2}\right|_{(\underline{x}^*, \underline{U}^*)} \frac{\Delta\underline{x}^2}{2!} + \cdots \tag{3.12}$$

where the partial derivatives are to be evaluated at the equilibrium state in question. The terms marked by the arrows can be struck off from either side since they always satisfy the dynamical model given by Equation 3.1.

$$\Delta\underline{\dot{x}} = \left.\frac{\partial \underline{f}}{\partial \underline{x}}\right|_{(\underline{x}^*, \underline{U}^*)} \Delta\underline{x} + \left.\frac{\partial^2 \underline{f}}{\partial \underline{x}^2}\right|_{(\underline{x}^*, \underline{U}^*)} \frac{\Delta\underline{x}^2}{2!} + \cdots \tag{3.13}$$

Thus, the general perturbation dynamics equation in terms of $\Delta\underline{x}$ appears as in Equation 3.13. Up to this point, no approximation has been made. Equation 3.13 is entirely equivalent to the original Equation 3.1 except for a change of origin.

Now, if we assume the perturbation $\Delta \underline{x}$ to be small (small perturbation theory), terms with higher powers of $\Delta \underline{x}$ (negligibly small) in Equation 3.13 can be dropped, resulting in

$$\Delta \dot{\underline{x}} = \left. \frac{\partial \underline{f}}{\partial \underline{x}} \right|_{(\underline{x}^*, \underline{U}^*)} \Delta \underline{x} \tag{3.14}$$

which is a linear set of differential equations in the perturbed states represented by the vector $\Delta \underline{x}$. The solutions of Equation 3.14 do not depend on the perturbation $\Delta \underline{x}$ itself but only on the Jacobian matrix:

$$A = \left. \frac{\partial \underline{f}}{\partial \underline{x}} \right|_{(\underline{x}^*, \underline{U}^*)} \tag{3.15}$$

In fact, the solutions (i.e., the trajectories starting from a given $\Delta \underline{x}(t = 0)$ can be written out as

$$\Delta \underline{x}(t) = e^{At} \Delta \underline{x}(0) \tag{3.16}$$

In short, all the requirements for Lyapunov's test for stability of an equilibrium state can be checked by just examining the Jacobian matrix A.

Note that the Jacobian matrix is obtained by *evaluating* the partial derivatives in Equation 3.15 at the equilibrium state $(\underline{x}^*, \underline{U}^*)$ in question. For a given dynamical system, Equation 3.1, while the partial derivative $(\partial \underline{f}/\partial \underline{x})$ is functionally the same at all equilibrium states, when *evaluated* with the values $(\underline{x}^*, \underline{U}^*)$ of an equilibrium state, the Jacobian matrix will be evaluated differently at each equilibrium state. To establish the stability (or otherwise) of an equilibrium state according to Lyapunov's definition, we need to figure out whether all perturbations remain bounded (remember the ε-bound?) for all time. Further, for it to be asymptotically stable, we also need the perturbations to die down to zero with time. Both these possibilities can be checked by examining the eigenvalues of the Jacobian matrix A evaluated at an equilibrium state. A short primer on eigenvalues is provided in Box 3.1.

The conclusion in Box 3.1 for second-order (perturbed, linear) dynamical systems can be extended to higher-order dynamical systems as well. As long as all the eigenvalues of the Jacobian matrix lie in the left half-plane

BOX 3.1 EIGENVALUES OF THE JACOBIAN MATRIX

For a linear dynamical system, $\dot{\underline{x}} = A\underline{x}$, the eigenvalues λ of the matrix A are found by solving the characteristic equation: $\det(\lambda I - A) = 0$, where I is the identity matrix.

Possible arrangements of eigenvalues for a second-order system with distinct eigenvalues are shown in Figure 3.7. In the two cases where all (both)

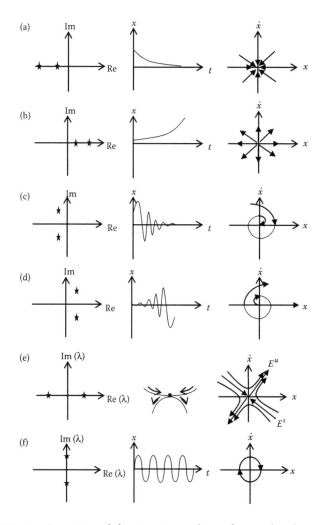

FIGURE 3.7 Location of distinct eigenvalues of second-order system, form of transient response, and corresponding phase portraits. (a) Stable node, (b) unstable node, (c) stable focus, (d) unstable focus, (e) saddle node, and (f) center.

eigenvalues lie in the left half-plane, the trajectory tends to the steady state at the origin indicating that the equilibrium point is stable. These cases are called *stable node/focus*. The reverse is the case where all (both) eigenvalues lie in the right half-plane—the trajectory diverges from the steady state at the origin suggesting that the equilibrium point is unstable. These are labeled *unstable node/focus*. However, even a single right half-plane eigenvalue is enough to make the equilibrium state unstable. Consider the case in Figure 3.7 where one real eigenvalue is located in the right half-plane. In this case, a pair of trajectories approach the equilibrium state (representing the left half-plane eigenvalue); however, all the other trajectories eventually diverge from the origin. The phase portrait appears like a saddle; hence, this case is called a *saddle*. Remember that Lyapunov's definition of stability requires *every* trajectory starting from the δ-region to remain bounded. Therefore, even though trajectories in the direction marked E^s do remain bounded (in fact, approach the origin), every other trajectory breaches the ε-bound, and so the *saddle* equilibrium state is unstable. The case with a pair of eigenvalues on the imaginary axis, called a *center*, gives a response similar to the simple pendulum dynamics seen in Figure 3.3—that is, it is stable but not asymptotically stable. For an equilibrium state to be asymptotically stable, all eigenvalues should lie in the left half-plane, or in other words, all eigenvalues should have a negative real part. For all cases where one or more eigenvalues lie on the right half-plane, the trajectory is not bounded; hence the equilibrium state is not stable in the sense of Lyapunov.

(have negative real part), the equilibrium state is asymptotically stable. Even one eigenvalue in the right half-plane is enough to adjudge the equilibrium state as unstable. Cases with no eigenvalue in the right half-plane but with one or more eigenvalues on the imaginary axis (zero real part) are the critical ones and usually signify a state of transition from stable to unstable or vice versa. However, note that certain even-dimensional systems may legitimately have pairs of eigenvalues located on the imaginary axis without signifying any "critical" behavior.

Homework Exercise: Try to picture a saddle in three-dimensional space with trajectories coming in along one dimension and spiraling out along the other two-dimensional surface.

This method of determining stability of an equilibrium state is often called *Lyapunov's first method*. Another method, usually referred to as *Lyapunov's second method*, involves the use of something called a *Lyapunov function* and is not discussed here. However, both methods have their basis in the same concept of stability as defined in the sense of Lyapunov.

3.2.2 Stability of Periodic Orbits

We have seen that periodic orbits form a closed loop in state space. That is, they traverse the same set of points over and over again with a periodicity T—the time period. Formally, $\underline{x}(t + T) = \underline{x}(t)$. Lyapunov's notion of stability discussed earlier deals with *points* and does not directly apply to periodic cycles. Imagine, for instance, a point on a periodic orbit, a ε-bounded region around it and a trajectory that starts within a δ-bounded region, as required by Lyapunov's statement. Almost certainly, the trajectory will leave the ε-bounded region, complete the loop, and then reenter the ε-bounded region—this happens repeatedly. Clearly, Lyapunov's definition needs some modification before being used for periodic states.

First of all, to adjudge the stability of a periodic orbit, it is not necessary to examine every point on that orbit—a single point is sufficient. Consider one such point and let the periodic orbit be initiated at that point at time $t = 0$. Then $\underline{x}(T) = \underline{x}(0)$. Next, consider a plane Σ transverse to the periodic orbit, as pictured in Figure 3.8. The plane Σ is usually called the Poincaré plane. The point on the periodic orbit with $\underline{x}(T) = \underline{x}(0)$ then turns out to be a *fixed point* on the Poincaré plane. That is, every time the periodic orbit intersects the Poincaré plane, it does so at the same point.

Now, Lyapunov's notion of stability can be applied to examine the *stability* of this *fixed point* on the Poincaré plane, and if this *fixed point* is found to be *stable* (in whatever sense), then the corresponding *periodic orbit* may also be adjudged to be stable (in the same sense). For this purpose, we need to select a ε-bounded region on the Poincaré plane, then find a δ-bounded region such that every "trajectory" starting from within the δ-bounded region remains within the ε bound for all time. However,

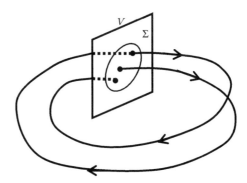

FIGURE 3.8   Depiction of Poincaré plane for use in defining stability of periodic orbit. (From http://www.mate.tue.nl.)

we first need to clarify about this "trajectory" and how it may remain bounded. Figure 3.8 shows a trajectory starting from an arbitrary point on the Poincaré plane and repeatedly intersecting with it. Note that successive intersections need not happen with exactly a time interval T. The time period T only applies to the *fixed point*. In general, trajectories from points slightly displaced from the *fixed point* will have slightly different time period for their next intersection with the Poincaré plane. That is, intersections with the Poincaré plane are not at regular intervals (they are not *stroboscopic*). Figure 3.8 shows a ε-bounded region labeled V. As long as successive intersections of the trajectory in Figure 3.8 lie within the V region, then it can be said to be ε-bounded for our purposes. If all such trajectories remain bounded, then the fixed point (and by extension the periodic orbit) can be called *stable*. Further, if successive intercepts tend to the *fixed point*, then it may be said to be *asymptotically stable* (and hence the periodic state is asymptotically state as well).

Homework Exercise: Look back at the dynamics of the simple pendulum—its phase portrait is reproduced in Figure 3.9. Identify the *fixed point* on the Poincaré plane that represents the periodic orbit. Take a "perturbed" point on the Poincaré plane and draw the trajectory passing through that point. Where does it next intersect the Poincaré plane and after what time interval? From an analysis of many such "perturbed" points, what can be inferred about the *stability* of the *fixed point* and hence of the periodic state of the simple pendulum dynamics?

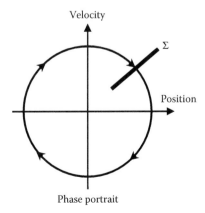

FIGURE 3.9   Phase portrait of periodic orbit of simple pendulum with Poincaré plane.

More formally, we need to first convert the original continuous-time dynamical system of the form Equation 3.1 into a discrete-time dynamical system (also called a map). For this purpose, pick a point on the periodic orbit, and erect a Poincaré plane through that point transverse to the periodic orbit. Then the dynamics of repeated intersections on the Poincaré plane may be written as

$$\underline{x}_{k+1} = P(\underline{x}_k) \tag{3.17}$$

where P is called the Poincaré map. Clearly, if the original dynamical system had n states, the Poincaré map represents dynamics with $(n-1)$ states. In writing the Poincaré map, the state in the direction normal to the Poincaré plane, that is, along the original periodic orbit, is discarded. This is sensible because a trajectory from a perturbed point that itself lies on the periodic orbit will simply return to precisely that point after a period T. That is, it neither tends toward nor away from the *fixed point*. By defining the $(n-1)$-dimensional Poincaré plane and limiting investigations to trajectories that originate from points on that plane, one dimension (that of perturbations along the periodic orbit itself) has been eliminated.

Let $\underline{x}^*$ be a *fixed point* on the Poincaré plane that belongs to a periodic orbit transverse to that plane in n-dimensional space. Then,

$$\underline{x}^* = P(\underline{x}^*) \tag{3.18}$$

Similar to Lyapunov's first method of determining stability, take an initial state on the Poincaré plane slightly perturbed from the *fixed point* as $\underline{x}^* + \Delta\underline{x}$. The trajectory from the perturbed state also satisfies the dynamics given by the Poincaré map in Equation 3.17. Hence,

$$(\underline{x}^* + \Delta\underline{x})_{k+1} = P((\underline{x}^* + \Delta\underline{x})_k) \tag{3.19}$$

Equation 3.19 can be expanded in a Taylor series as

$$\underline{x}^*_{k+1} + \Delta\underline{x}_{k+1} = P(\underline{x}^*_k) + \frac{\partial P}{\partial \underline{x}}\bigg|_{\underline{x}^*} \Delta\underline{x}_k + \text{higher-order terms} \tag{3.20}$$

The terms struck off in Equation 3.20 are identical by virtue of $\underline{x}^*$ being a fixed point. Then, assuming the perturbation $\Delta\underline{x}$ to be small, the higher-order terms may be ignored, leaving the small-perturbation dynamics equation as

$$\Delta \underline{x}_{k+1} = \left. \frac{\partial P}{\partial \underline{x}} \right|_{\underline{x}^*} \Delta \underline{x}_k \qquad (3.21)$$

For an originally n-dimensional dynamical system, the matrix $(\partial P/\partial \underline{x})|_{\underline{x}^*}$ is of size $(n-1) \times (n-1)$, so it has $(n-1)$ eigenvalues, which are usually referred to as the *Floquet multipliers*. Similar to the case of the equilibrium point that we have already seen, for the *fixed point* $\underline{x}^*$ to be merely *stable* (in the sense of Lyapunov), all the *Floquet multipliers* must lie on or within the unit circle in the complex plane. In that case, the corresponding periodic orbit is also merely *stable*. Further, if all the *Floquet multipliers* lie strictly within the unit circle, then the *fixed point* is asymptotically stable, and so is the corresponding periodic state. Thus, when a periodic state is asymptotically stable, trajectories from a point perturbed off the periodic orbit will approach the periodic orbit with time in the sense that successive intercepts on a Poincaré plane will tend toward the fixed point intercept of the periodic state.

Even if one *Floquet multiplier* lies outside the unit circle in the complex plane, the *fixed point* and hence the periodic state is unstable.

For a dynamical system in three-dimensional state space, the Poincaré plane is actually a two-dimensional surface and the linearization of the Poincaré map yields a matrix with two eigenvalues, λ_j. As long as $|\lambda_j| < 1$ for each of them, the periodic state is stable. Figure 3.10 shows a sketch of stable and unstable periodic orbits—the distinction is made depending on the behavior of the perturbed trajectory shown therein.

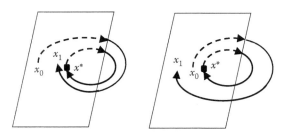

FIGURE 3.10   Sketch of periodic orbit and Poincaré plane showing stable case with perturbed trajectory (left) and unstable case (right).

3.3 BIFURCATIONS OF STEADY STATES

So far we have looked at two kinds of steady states—*equilibrium state* and *periodic state*—and their stability. In case of equilibrium states, we have seen that they can be identified either by solving the steady-state condition, $f(x) = 0$, or by simulating the dynamical system, $\dot{x} = f(x)$. However, as you may have realized after studying about stability, the method of simulating forward in time will only lead to *stable* steady states. Likewise, in case of periodic states, they may be located either by solving for the periodic condition, $x(t + T) = x(t)$, or by time simulation of the dynamical system; however, in the latter case, again only *stable* periodic states can be obtained.

We have seen that a dynamical system, $f(x) = 0$, may have multiple steady states—*equilibria* and *periodic* ones—and that some of them may be stable and others unstable. Therefore, the notion of *stability* should be applied only to a steady state, and not to the dynamical system as a whole.

Homework Exercise: Consider the dynamical system below, called the *van der Pol oscillator*:

$$\ddot{x} - \mu(1 - x^2)\dot{x} + x = 0 \tag{3.22}$$

with $\mu > 0$. Cast Equation 3.22 in the dynamical system form, $\dot{x} = f(x, u)$, $x = [x, \dot{x}]$, $u = \mu$, and determine its steady states and their stability by any of the methods discussed previously. Depending on your choice of the parameter μ, the phase portrait may appear as in Figure 3.11. Notice that the equilibrium state at (0, 0) is unstable and there is a single, isolated periodic state that appears to be stable. This is an example of a limit cycle.

We have seen one example of a dynamical system with a varying parameter (Figure 3.4) where the location as well as stability of the steady states can change with changing values of the parameter. Thus, in general, for a dynamical system of the form, $\dot{x} = f(x, U)$, where U is a vector of parameters, what is of interest is an entire family of steady states, such as x^*, and changes in stability, if any, as one or more of the parameters U are varied. At critical values of the parameters, there may be a qualitative change in the number and/or nature of stability of the steady states. These critical points are called *bifurcation points* and the corresponding phenomenon is identified as a particular kind of bifurcation. For example, in Figure 3.4, at the critical value of $\gamma = 1$, a single equilibrium state bifurcates into three, with the equilibrium state $x^* = 0$ changing from stable to unstable. Thus,

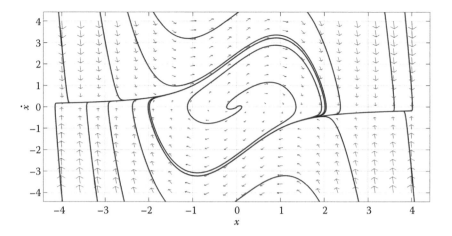

FIGURE 3.11   Phase portrait of the van der Pol oscillator.

$\gamma = 1$ is a bifurcation point and the bifurcation phenomenon in question is labeled as a *pitchfork bifurcation*.

Change in stability can be identified by tracking the eigenvalues in case of equilibrium states and the Floquet multipliers in case of periodic states with varying parameter. Passage of an eigenvalue from the left to the right half-plane or a Floquet multiplier going out of the unit circle would signify onset of instability in case of equilibrium or periodic states, respectively.

Bifurcations are commonly observed in a wide range of dynamical systems in varied fields such as chemical engineering, electronics, biology, economics, and, of course, in aircraft flight dynamics. Most dynamical systems encountered in practice, both natural and man-made, are nonlinear and multiparameter, and several bifurcation phenomena occur in these systems. Sometimes, a bifurcation phenomenon can lead to a more desirable or efficient operating state for a system, though more often than not it represents a loss of safety/control or a breakdown of the system's regular operation.

A bifurcation analysis maps all possible steady states of a dynamical system and their stability, and various occurrences of bifurcations as one or more parameters are varied. Such plots are often called *bifurcation diagrams*. Figure 3.4 is an example of a bifurcation diagram. If it is desired to operate the rotating hoop with bead system at the steady state $\theta = 0$, then the bifurcation diagram in Figure 3.4 reveals that the system must be operated at a value of γ less than 1. On the other hand, if the objective is to raise the bead to some angle θ on the hoop as it rotates, then the corresponding value of $\gamma > 1$ can be read from Figure 3.4 as well.

Depending on the nature of the bifurcation, the post-bifurcation behavior of the dynamical system can vary. Since each bifurcation event is accompanied by a change in stability, either of an equilibrium state or a periodic state, a catalog of different types of possible bifurcations can be built by observing the various ways in which the eigenvalues or Floquet multipliers can cause instability. In general, there are only two distinct ways for an eigenvalue to cross the imaginary axis in the complex plane to induce a bifurcation. Likewise, there are only three clearly different crossings of the unit circle by the Floquet multiplier. These are depicted in Figure 3.12.

At a stationary bifurcation, a steady state gives rise to one or more steady states of the same type (equilibrium or periodic, as the case may be) and there is an exchange of stability, or a pair of steady states annihilate each other. We shall examine these cases in more detail shortly. Stationary bifurcations are marked by an eigenvalue at the origin in case of equilibrium states or a Floquet multiplier at "1" in case of periodic states.

At a Hopf bifurcation, a pair of complex conjugate eigenvalues or Floquet multipliers, as the case may be, cross over as indicated in Figure 3.12. In case of equilibrium states, a Hopf bifurcation gives rise to a periodic state (limit cycle) whereas for a periodic state, it creates a quasi-periodic state. When viewed in the Poincaré plane, the periodic orbit of course appears as a *fixed point* and the quasi-periodic state is a closed orbit about that *fixed point*.

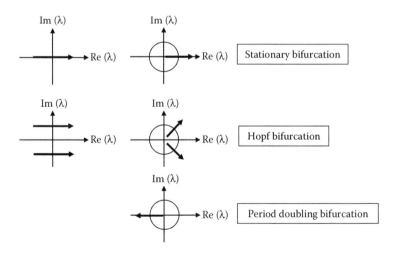

FIGURE 3.12   Types of bifurcations depending on the movement of eigenvalues (left) and Floquet multipliers (right).

In addition, for periodic states, a period-doubling bifurcation occurs when the Floquet multiplier crosses the unit circle at "−1." There is no equivalent of this bifurcation in case of equilibrium states. At a period-doubling bifurcation, typically, a periodic state of period T loses stability and a new stable periodic state of period $2T$ emerges. On the Poincaré plane, the original periodic state appears, as always, as a single *fixed point* and the new bifurcated period–$2T$ orbit alternates between two points.

3.3.1 Stationary Bifurcations of Equilibrium States

As seen in Figure 3.12, a stationary bifurcation of an equilibrium state is marked by a single real eigenvalue passing through the origin from the left to the right half complex plane. So, clearly, the equilibrium state in question transitions from stable to unstable. However, depending on other conditions (besides the single zero eigenvalue), the form and nature of the various equilibrium states, and hence the type of bifurcation, may be different. There are three main types of bifurcation of equilibrium states that arise under the condition of a single real eigenvalue crossing the origin in the complex plane. These are sketched in Figure 3.13 and discussed below.

3.3.1.1 Saddle-Node Bifurcation

Consider a single stable equilibrium state that with a varying parameter μ has a real eigenvalue that tends to the origin. At a critical value of the parameter μ, this eigenvalue crosses the origin and then moves into the right half-plane. This critical value corresponds to a saddle-node bifurcation and the shape of the curve of equilibrium states with parameter μ must appear as shown in Figure 3.13. It is remarkable that this curve is always locally quadratic around the critical point—it cannot take any other shape. Saddle-node bifurcations are observed in an amazingly large number of diverse dynamical systems and they appear identically every time.

There are two ways of understanding the changes that occur at a saddle-node bifurcation. One is to follow the curve of equilibrium states along the arc-length and observe the change in the sign of the eigenvalue, as we did above. The other is to vary the parameter μ and observe the change in the system dynamics. In case of the saddle-node bifurcation in Figure 3.13, for a negative value of μ, there are two equilibrium states—one stable and the other unstable. Perturbations lead away from the unstable state and toward the stable one, as marked in Figure 3.13. As μ tends to the critical value, the two states approach each other and coalesce—at the critical point, there is only one steady state with a zero

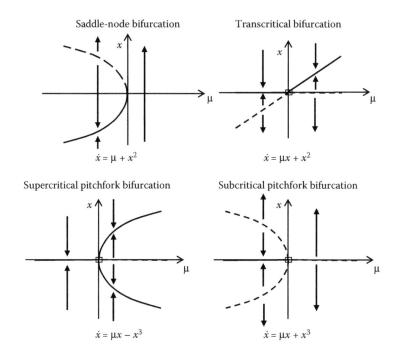

FIGURE 3.13   Three main types of stationary bifurcations of equilibrium states—stable ones are shown in full line, unstable ones are dashed.

eigenvalue that is neither clearly stable, nor clearly unstable. Trajectories approaching from negative values of the state variable tend to the (critical) steady state whereas for positive values of the state variable, trajectories tend to diverge from the steady state. Beyond the critical value of the parameter μ, there are no equilibrium states in the vicinity of the bifurcation point. Trajectories now pass unimpeded as indicated by the arrow in Figure 3.13 for negative μ. Thus, from this point of view, at a saddle-node bifurcation, a pair of equilibrium states collide and annihilate each other. Consequently, the dynamics on either side of the critical point are vastly different.

Imagine a dynamical system with a saddle-node bifurcation with its state on the stable branch in Figure 3.13. Let the parameter μ be varied slowly so that the state moves along the stable branch in a quasi-steady manner toward the critical point. Once the parameter goes beyond the critical point, there are no steady states in the vicinity and the system must then move along the upward arrow in Figure 3.13. It is as if the system stepped off a cliff and plunged downward to oblivion (except the arrow points upward in Figure 3.13).

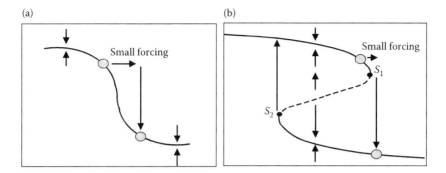

FIGURE 3.14 (a) Jump-like condition and (b) jump and hysteresis with a pair of saddle-node bifurcations.

In real-life physical systems though, there is usually another branch of steady states that the system can move to when it steps off a saddle-node bifurcation point. In that case, the system is said to "jump" from one branch of steady states to another. Jump phenomenon accompanied by hysteresis is fairly common in several dynamical systems. Figure 3.14b shows an example of a pair of saddle-node bifurcations, resulting in a jump phenomenon. Imagine moving along the upper stable branch until the critical point S_1 is reached. The system then jumps to the lower stable branch as marked by the downward arrow and proceeds further to the right in Figure 3.14b. However, from this state, as the parameter conditions change to the left, the system follows the lower stable branch until the other saddle-node point at S_2 where it jumps up, back to the upper stable branch. In this manner, a hysteretic response is created. Between the saddle-node points marked by S_1 and S_2, the system has multiple steady states (which by itself is not uncommon for nonlinear systems) and the branch the system occupies depends on the direction of change in the parameter conditions. Increasing parameter finds the system in one stable state and decreasing values of parameter in the other. Note that this multivalued behavior is not because the dynamical system function $f(x)$ is multivalued—it is not! For a contrast, Figure 3.14a shows a jump-like dynamics where the single stable branch of steady states falls steeply with changing parameter but does not fold over at a saddle-node bifurcation.

Homework Exercise: Jump phenomenon and hysteresis are known to occur in airplane flight dynamics as well, for example, in rolling maneuvers. Look up cases of jump in the roll rate with aileron deflection. We shall be studying this later in this text.

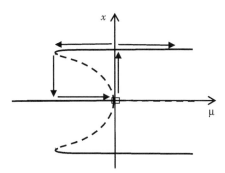

FIGURE 3.16   Jump phenomenon and hysteresis at a subcritical pitchfork bifurcation point.

by the arrows in Figure 3.16. Note that in this case too there is a range of values of the parameter where the system may find itself in one of three possible stable steady states. It is not possible to definitely predict which of these three steady states will be observed—it depends on the history and perturbation in the states, and changes in the parameter.

3.3.1.4 Perturbation to Transcritical and Pitchfork Bifurcations

Since all the stationary bifurcations in Figure 3.13 correspond to the same criterion—passage of a single real eigenvalue through the origin of the complex plane—one may wonder if one of them is more fundamental, in some sense, than the others. The answer is, yes. The saddle-node bifurcation is the most general form of stationary bifurcation in case of a zero real eigenvalue. The transcritical and pitchfork bifurcations are special cases that occur only when the system satisfies some additional constraint. Imagine a second parameter, in addition to the parameter μ in Figure 3.13. Only for a particular value of this second parameter would a transcritical or pitchfork bifurcation be formed. If the second parameter were perturbed (imagine an axis into/out of the plane of the paper in Figure 3.13), then the structure of the transcritical or pitchfork bifurcation may not be maintained. The typical forms of the transcritical and pitchfork bifurcations under perturbation of a second parameter are as sketched in Figure 3.17. In each case, the branches intersecting at the bifurcation point break apart to form either a saddle-node bifurcation or an unbifurcated branch of steady states, as indicated in Figure 3.17. Thus, for a general (or well-modeled) dynamical system, one would not expect to see a transcritical or pitchfork bifurcation unless it had an inherent constraint or symmetry that forced the second parameter to

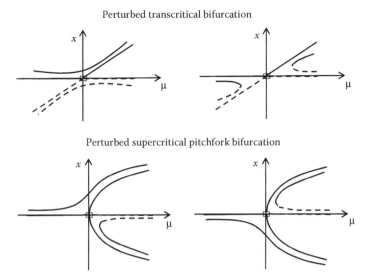

FIGURE 3.17 Form of transcritical and pitchfork bifurcation under perturbation of a second parameter.

have precisely that particular value for which one of these bifurcations would occur.

Homework Exercise: Review the bifurcation diagram in Figure 3.4 for the rotating hoop with a bead, which shows a *supercritical* pitchfork bifurcation. What is the constraint or symmetry that causes this particular bifurcation? Physically, how can this constraint be "broken" or the system perturbed to yield a perturbed bifurcation diagram as in Figure 3.17?

Homework Exercise: Sketch the perturbed bifurcation diagram for the *subcritical* pitchfork bifurcation in Figure 3.16. Identify the jump phenomenon and hysteresis loops. How is it different from the form of the bifurcation diagram for the jump in Figure 3.14b?

3.3.2 Hopf Bifurcation of Equilibrium States

As seen in Figure 3.12, a Hopf bifurcation occurs when a pair of complex conjugate eigenvalues cross the imaginary axis from left to right half-plane. At a Hopf bifurcation, the equilibrium state changes its stability and a family of periodic orbits emerges. Just as in the case of pitchfork bifurcations, Hopf bifurcation may be supercritical or subcritical. The supercritical Hopf bifurcation case is pictured in greater detail in Figure 3.18. When the eigenvalue pair is in the left half-plane, the equilibrium state is

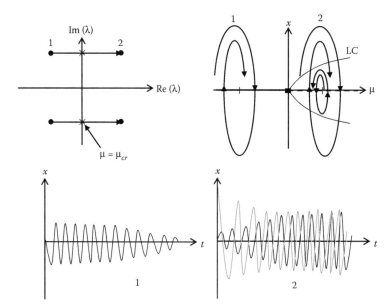

FIGURE 3.18 Emergence of stable limit cycle family at a supercritical Hopf bifurcation.

stable and perturbations spiral back into the equilibrium state in the phase plane. As the eigenvalues enter the right half-plane, the equilibrium state becomes unstable. Trajectories, perturbed from the equilibrium state, now spiral away from the unstable equilibrium state and approach the family of stable periodic orbits that emerges at the Hopf bifurcation. As seen in Figure 3.18, for a given value of the parameter μ, there is a single, isolated periodic orbit—a limit cycle. The upper and lower curves for the limit cycle (LC) denote its maximum and minimum amplitude. The periodic state cyclically oscillates between these extreme values.

Just as trajectories perturbed slightly off the equilibrium state spiral out to the limit cycle, trajectories beginning with large amplitudes (large x) spiral in to the limit cycle. Hence, the limit cycle in a supercritical Hopf bifurcation is a stable periodic state. It is remarkable that almost every limit cycling system observed in nature, from mechanical to biological to electronic systems, is created by the Hopf bifurcation mechanism and in some sense they are all very similar. In aircraft flight dynamics, *wing rock* is a well-known example of a limit cycling oscillation.

At a subcritical Hopf bifurcation point, a family of unstable equilibrium states changes its stability, and a family of unstable periodic orbits is created about the branch of stable equilibrium states.

Homework Exercise: Make a sketch similar to Figure 3.18 for a subcritical Hopf bifurcation.

Figure 3.19 compares the sense of the trajectories in the two cases of a stable and an unstable limit cycle in the plane. For a stable limit cycle, trajectories originating "within" the cycle spiral out toward it and those starting "outside" the cycle spiral in toward it. On the other hand, for an unstable limit cycle, any trajectory that starts "within" the cycle spirals in to the stable equilibrium state whereas trajectories perturbed "out" from the limit cycle spiral away, presumably to "infinity" or more reasonably to another stable state elsewhere. However, note that in case of general periodic states (closed orbits) in higher-dimensional space, there is no "inside" and "outside." Nevertheless, stable limit cycles are formed about an unstable equilibrium state, and unstable limit cycles are created about stable equilibria due to a Hopf bifurcation, but the boundary between trajectories that tend one way or another may not be so obvious.

3.3.3 Bifurcations of Periodic States

As we saw in Figure 3.12, periodic states can undergo three types of bifurcations depending on the manner in which the Floquet multipliers exit the unit circle in the complex plane. These three cases are reproduced in Figure 3.20 in greater detail showing the intersection of the periodic trajectories on the Poincaré plane.

In the first case (Figure 3.20a), one real Floquet multiplier exits the unit circle at +1. The bifurcation is called a *fold bifurcation* and is similar to the saddle-node bifurcation of equilibrium states. That is, a stable branch of periodic states folds over and turns unstable. The critical point, where it

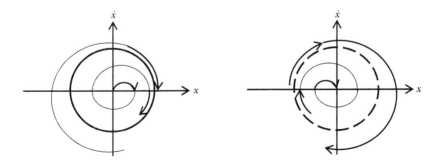

FIGURE 3.19 Trajectories spiraling toward a stable limit cycle (full line) and away from an unstable one (dashed line).

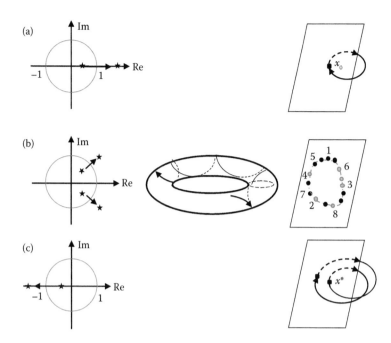

FIGURE 3.20 Bifurcations of limit cycles and Poincaré plane trajectories.

folds over, has one Floquet multiplier lying at +1. Just as in the case of a saddle-node bifurcation, jump phenomenon is likely at a *fold bifurcation*.

The second case deals with a pair of complex conjugate Floquet multipliers exiting the unit circle, as indicated in Figure 3.20b. This is called a *Torus bifurcation* or a *Neimark–Sacker bifurcation*. At this bifurcation, a stable limit cycle turns unstable and a quasi-periodic oscillation, as on the surface of a torus enveloping the limit-cycle-turned-unstable, is seen. Intercepts on the Poincaré plane appear, for example, as marked in Figure 3.20. No two intercepts coincide; otherwise, there would a closed orbit on the toroidal surface, hence a periodic state. A *Torus bifurcation* can also be supercritical or subcritical, as was the case for the *Hopf bifurcation*.

Homework Exercise: Try plotting the intercepts on the Poincaré plane for a trajectory on the torus where every intercept is at 150 deg clockwise relative to the previous one. Is the resulting trajectory periodic or quasi-periodic? If periodic, what is the ratio of periods in the longitudinal to the transverse (radial) sense?

The third possibility in Figure 3.20 has a single real Floquet multiplier leaving the unit circle at "−1." The resulting bifurcation is called a *period-doubling bifurcation* or a *flip bifurcation* (Figure 3.20c). In this instance, as the parameter in question is varied past the critical point, a periodic orbit of period T loses stability and is replaced with one of double the period, viz. $2T$. On the Poincaré plane, a single fixed point is replaced by a pair of points that the system cycles between. The bifurcated state of period $2T$ can undergo a further *flip bifurcation* resulting in a new cycle of period $4T$ and so on. This is usually called a *period-doubling cascade*.

Homework Exercise: Look up the possibility of a *period-halving bifurcation*.

These are the usual types of bifurcations that are frequently encountered. In addition, there are some bifurcations called *global bifurcations* that cannot be identified by tracking changes in eigenvalues or Floquet multipliers. In contrast, the bifurcations we have discussed are also called *local bifurcations* as they can be identified from *local* information, viz. eigenvalues or Floquet multipliers, which may be locally evaluated from a linearization about an equilibrium or a periodic state. Global bifurcations are sometimes encountered in systems of scientific and engineering interest, but we shall desist from talking about them any further here. Likewise, we shall also avoid going into the topic of chaos or chaotic states or chaotic attractors although such states have been reported in aircraft flight dynamics as well.

Finally, it may be instructive to compare the subcritical Hopf bifurcation with the subcritical pitchfork that we have already seen in Figure 3.13 in the context of jump and bifurcation. These two are pictured side by side in Figure 3.21. In either case, an equilibrium state changes stability across

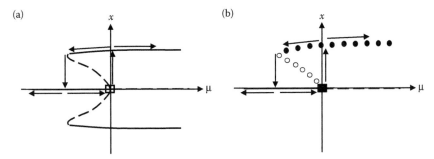

FIGURE 3.21 Subcritical (a) pitchfork and (b) Hopf bifurcations leading to jump in state and hysteresis (solid lines: stable equilibrium states; dashed lines: unstable equilibrium states; empty box: pitchfork bifurcation; filled box: Hopf bifurcation; filled circles: maximum amplitude of stable limit cycle state; empty circles: maximum amplitude of unstable limit cycle state).

the bifurcation point. In either case, a family of unstable steady states originates at the bifurcation point. For a Hopf bifurcation, this unstable family is periodic states, whereas in case of the pitchfork, it is unstable equilibrium states. However, note that in case of the pitchfork, two distinct, symmetric branches of unstable steady states are created and, in practice, the system can go one way or another, but not both simultaneously. Whereas for the Hopf bifurcation, only a single limit cycle branch is formed (indicated in Figure 3.21 by the maximum amplitude of the cycles). In either case, the unstable branch so created may fold over (as in a saddle-node or fold bifurcation) forming stable steady states—equilibria again in case of the pitchfork, and periodic states for the Hopf bifurcation. Thus, the jump phenomenon may be observed. As the parameter is increased past the bifurcation point, the primary steady state becomes unstable. A slight perturbation can then take the system to the large-amplitude stable state as marked by the up-arrow in Figure 3.21. From that point, for decreasing values of the parameter, the system jumps back to the primary steady state at the saddle-node (or fold) bifurcation point, thus forming a hysteresis loop.

3.4 CONTINUATION ALGORITHMS

In principle, bifurcation diagrams with a varying parameter, such as those presented earlier in this chapter, may be obtained by methodically solving for all the steady states at one value of the parameter, then stepping over to the next value of parameter and repeating the process, and so on. However, a far more efficient way of constructing a bifurcation diagram is to start at a steady state and "continue" along the branch. Thus, for example, in case of the saddle-node bifurcation diagram in Figure 3.13, instead of finding the two steady states at one value of parameter and doing so for every successive parameter value, one would start with a steady state and continue along that branch with increasing parameter, go around the fold (critical point), and track the other set of steady states with decreasing parameter. In this process, whenever bifurcations are encountered, those critical points are noted and the type of bifurcation identified. One then needs to go back to these critical points and further continue along bifurcated branches that originate from these points, if any. Algorithms that perform this kind of computation are called *continuation algorithms*.

While it is beyond the scope of this book to investigate continuation algorithms in depth, since we will be using them for our work and expect the reader to get familiar with one as well, some hints about how they operate may be worthwhile.

We would like the continuation algorithm to solve for steady states and periodic states of dynamical systems of the form Equation 3.1. Additionally, we would like it to compute the eigenvalues of the Jacobian matrix as in Equation 3.15 and Floquet multipliers of the Poincaré map as in Equation 3.21. And we would want it to identify the bifurcations in Figures 3.13 and 3.20 and correctly pick the newly bifurcated branches to be tracked.

Imagine that the continuation algorithm needs to track a branch of steady states $x^* = x^*(u)$ as sketched in Figure 3.22 starting from an equilibrium point (x^*, u^*). The steady states are solutions of the set of algebraic equations: $f(\underline{x}, u) = 0$. In general, these are a set of n equations in $(n + 1)$ unknowns—the n states and the single parameter. An obvious, though simplistic approach, would be to implement a predictor-corrector scheme. As pictured in Figure 3.22, for a fixed step Δu, one can find the "predicted" values of the states as follows:

$$f(\underline{x}, u) = 0 \Rightarrow \frac{\partial f}{\partial \underline{x}} \Delta \underline{x} + \frac{\partial f}{\partial u} \Delta u = 0 \Rightarrow \Delta \underline{x} = -\left(\frac{\partial f}{\partial \underline{x}}\right)^{-1} \cdot \left(\frac{\partial f}{\partial u}\right) \Delta u \quad (3.23)$$

In practice though, there is no need to always select the parameter u as the stepping variable; any one of the $(n + 1)$ variables can be used for this purpose, even variably at different points along a branch depending on convenience. Another option is to use an altogether different variable, such as arc-length along a branch, as the stepping variable. Several variants of this approach are possible; however, the predictor step basically uses "local" information at an equilibrium point (x^*, u^*) to advance in one direction along a branch, as depicted in Figure 3.22. Note that the Jacobian matrix

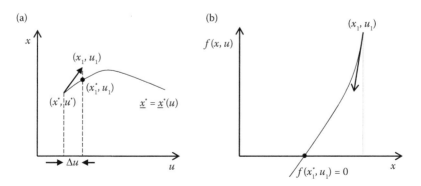

FIGURE 3.22 (a) Predictor step and (b) corrector step in a continuation.

$(\partial f / \partial \underline{x})_{(\underline{x}^*, \underline{u}^*)}$ used in Equation 3.23 is the same one that is needed to evaluate the stability of the equilibrium state $(\underline{x}^*, \underline{u}^*)$ anyway. At critical points where an eigenvalue may be zero, one needs to be careful in dealing with the Jacobian matrix. Here, one of the alternative formulations suggested above may be helpful. The predictor step yields an approximate state $(\underline{x}_1, \underline{u}_1)$, which must then be corrected to the "nearest" steady state on that branch. This may be done by a "corrector" step that may well be a standard Newton–Raphson algorithm. For further details about the continuation algorithm, one may refer to Reference 2.

EXAMPLE 3.1: NONLINEAR PHUGOID DYNAMICS

To illustrate the use of the continuation procedure on a dynamical system with one parameter, let us consider a simplified model of the airplane dynamics representing a nonlinear phugoid motion. Refer back to the equations of motion in Chapter 1 (Table 1.5). Assuming longitudinal motion and setting all lateral-directional variables to zero, from Equation 1.59, we have

$$m\dot{V} = T \cos \alpha - D - mg \sin \gamma$$
$$m(q - \dot{\alpha})V = T \sin \alpha + L - mg \cos \gamma \tag{3.24}$$

From Equation 1.47,

$$q - \dot{\alpha} = q_w \tag{3.25}$$

And further using Equation 1.28 we have

$$q_w = \dot{\gamma} \tag{3.26}$$

So, Equation 3.24 may be written as

$$\frac{\dot{V}}{g} = \frac{T}{mg} \cos \alpha - \frac{D}{mg} - \sin \gamma$$
$$\frac{\dot{\gamma}V}{g} = \frac{T}{mg} \sin \alpha + \frac{L}{mg} - \cos \gamma \tag{3.27}$$

Converting the velocity V to a nondimensional variable as $y = (V/V_0)$ where $V_0 = \sqrt{2mg/\rho S C_L}$, and transforming to a nondimensional time, $\tau = t/(V_0/g)$, Equation 3.27 may be represented as

$$
\begin{aligned}
\frac{dy}{d\tau} &= b - ay^2 - \sin\gamma = f_1(y, \gamma) \\
\frac{d\gamma}{d\tau} &= y - \frac{\cos\gamma}{y} = f_2(y, \gamma)
\end{aligned}
\tag{3.28}
$$

where small α has been assumed ($\sin\alpha \approx 0$, $\cos\alpha \approx 1$). The parameters in Equation 3.28 are: $b = (T/mg)$; $a = (C_D/C_L)$. Clearly, the dynamics in Equation 3.28 is nonlinear in y and γ. Note that the approximation in Equation 3.28 is used here only to illustrate the use of the continuation and bifurcation procedure. In general, in this text, we do not advocate the use of simplified or approximate models of the aircraft flight dynamics equations instead preferring to deal with the complete form of the equations as presented in Chapter 1.

Of the two parameters, let us select b as the continuation parameter. The first step of the analysis is to determine the equilibrium states. The equilibrium states of Equation 3.28 are given by

$$
\frac{dy}{d\tau} = 0 = b - ay^{*2} - \sin\gamma^*; \quad \frac{d\gamma}{d\tau} = 0 = y^* - \frac{\cos\gamma^*}{y^*}
\tag{3.29}
$$

where the $*$ represents a steady state. In terms of the parameters (a, b), one can arrive at the following analytical solution for equilibrium states:

$$
y^{*2} = \cos\gamma^* = \frac{ab \pm \sqrt{a^2 - b^2 + 1}}{(1 + a^2)}
\tag{3.30}
$$

Of course the analytical solution is not required since the continuation algorithm will determine the steady states numerically. But Equation 3.30 can be used to suggest a starting equilibrium state needed for the continuation method. For instance, setting $a = b$ in Equation 3.30 yields the steady state $\gamma^* = 0$, $y^* = 1$, which can be used to start the continuation algorithm.

The second step is to determine the stability of each of the computed equilibrium states. This is obtained by calculating the eigenvalues of the Jacobian matrix of the system in Equation 3.28. The Jacobian may be written as

$$
J = \begin{bmatrix} \dfrac{\partial f_1}{\partial y} & \dfrac{\partial f_1}{\partial \gamma} \\[2ex] \dfrac{\partial f_2}{\partial y} & \dfrac{\partial f_2}{\partial \gamma} \end{bmatrix}_{(y^*,\gamma^*)} = \begin{bmatrix} -2ay^* & -\cos\gamma^* \\[2ex] 1+\dfrac{\cos\gamma^*}{y^{*2}} & \dfrac{\sin\gamma^*}{y^*} \end{bmatrix}
\tag{3.31}
$$

Note that the Jacobian matrix does not directly depend on the parameter b. However, with varying b, the steady state (y^*, γ^*) changes, and the entries in the matrix of Equation 3.31 do depend on the values of (y^*, γ^*). Hence, one can expect a change in the steady states and their stability with varying parameter b. For instance, at the steady state $\gamma^* = 0$, $y^* = 1$,

$$
J = \begin{bmatrix} -2a & -1 \\ 2 & 0 \end{bmatrix}
\tag{3.32}
$$

Whose eigenvalues are $-a \pm \sqrt{a^2 - 2}$. Since a is usually of the order of 0.1, we may safely assume $a^2 \ll 2$; hence the eigenvalues are approximately $-a \pm i\sqrt{2}$. In other words, the damping of the phugoid motion depends primarily on the parameter a, which is the ratio of drag to lift, and the damped natural frequency is, to a first approximation, equal to $\sqrt{2}$, independent of any parameter.

Then again, we do not really need to make any approximation or even calculate the eigenvalues analytically—all the computations can be managed by the continuation algorithm. However, it is good practice to have an estimate of values that may be expected in the computation and to understand the nature of the changes in steady states and their stability as the continuation run is carried out.

We shall hold $a = 0.2$ for these computations and vary b between 0.2 and 0.8. Remember, $b = a = 0.2$ corresponds to a level, steady-state flight, and with increasing b, the thrust is being increased. Figure 3.23a shows the branch of steady states with varying parameter b. It can be noticed that beginning from $b = 0.2$, the equilibrium

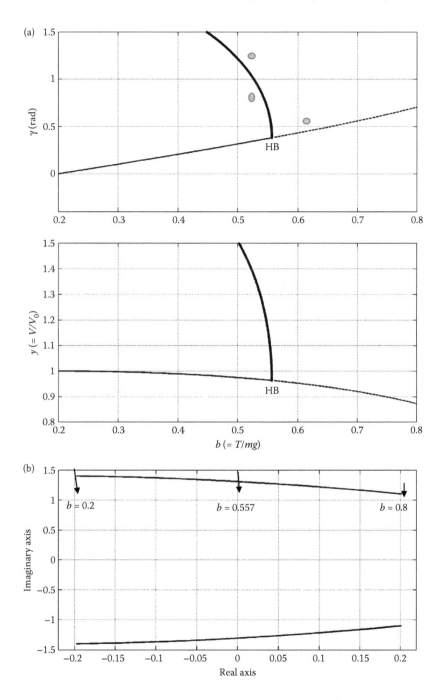

FIGURE 3.23   (a) Bifurcation diagrams of γ and y as function of parameter b (solid lines: stable states; dashed lines: unstable states; darker solid lines: unstable limit cycles; HB: Hopf bifurcation). (b) Plot of eigenvalues as function of parameter b.

states are stable and the steady-state climb angle increases with increasing thrust (parameter b), the velocity remaining nearly constant. The eigenvalues are a complex conjugate pair in the left half complex plane but are seen to move toward the imaginary axis with increasing b. At $b = 0.557$, the pair of complex eigenvalues cross the imaginary axis, indicating a Hopf bifurcation (HB)—the same is noted by the continuation algorithm and marked on the bifurcation diagram. As we have seen, a branch of limit cycles is expected to be created at the Hopf bifurcation point—either stable ones at a supercritical bifurcation or an unstable branch at a subcritical one. In this instance, the HB is subcritical and unstable limit cycles are formed—these are also computed and plotted by the continuation algorithm.

Beyond the HB point, the equilibrium branch has unstable states—the eigenvalue pair has moved into the right half complex plane as seen in Figure 3.23b. In this regime, according to the model (Equation 3.28), small perturbations off the steady state will diverge to infinity. In practice, alternative steady states and bifurcations may be expected when nonlinearities in the aerodynamic modeling, presently absent, are included. Nevertheless, the simple model in Equation 3.28 is of use to illustrate the use of the continuation and bifurcation method to a problem in flight dynamics.

To complement the bifurcation analysis, one can carry out time simulations for certain fixed values of the parameter b. Figure 3.24 shows the phase plane trajectories for four different values of b. It can be seen that for $b < 0.557$, trajectories within a certain bound spiral in to the stable steady state at $\gamma^* = 0$, $y^* = 1$. The boundary is determined by the unstable limit cycle seen in Figure 3.23a. As b tends to the critical value of 0.577, the extent of the limit cycle shrinks and the domain within the limit cycle wherefrom trajectories spiral in to the steady state reduces as well. For the case of $b = 0.7$ in Figure 3.24, the trajectories spiral out the unstable steady state and tend toward increasing values of γ. However, γ being a periodic variable with bounds $(-\pi, +\pi)$, one can imagine the trajectories to fold back along the X axis in Figure 3.24 and repeatedly pass from $-\pi$ to $+\pi$. Effectively, the aircraft appears to be pitching nose-over and looping repeatedly—an unrealistic solution.

Homework Exercise: Carry out a bifurcation analysis with parameter b held fixed and parameter a varied. Analyze your results.

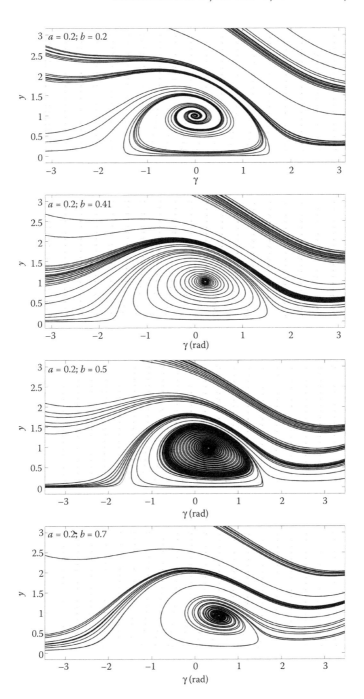

FIGURE 3.24 Phase portrait results for different sets of parameter values (fixed $a = 0.2$; different b).

3.5 CONTINUATION FRAMEWORK FOR MULTIPARAMETER SYSTEMS

In real life, most systems of interest have multiple parameters. Conventional aircraft flight dynamics itself has at least four parameters—the elevator, rudder, aileron, and throttle controls. Besides, depending on the analysis, there may be other parameters of interest, such as location of CG. The bifurcation analysis and continuation procedure that we have described so far is based on varying a single parameter at a time while all other parameters are held constant. Therefore, the parameters must be varied sequentially one by one if one wishes to explore the steady states and their bifurcations across the entire parameter space. There are two main issues with such an approach—First, collating all the different bifurcation analysis results and making sense of them collectively can be quite a task. Second, for most systems, it is not normal to operate them by varying one parameter at a time. In other words, it is natural to vary two or more parameters in some coordinated manner, as for example, the throttle and elevator in flight. So it would be useful if the bifurcation and continuation procedure could allow for multiple parameters, with perhaps the other parameters being varied as a function of a primary parameter.

In fact, for an operator of a dynamical system such as a pilot of an airplane, what is important is to hold certain values of the state variables or vary them in a particular manner. For example, a pilot may want to accelerate the airplane while maintaining level flight, or climb at a fixed speed. To realistically mirror the steady states and bifurcations encountered in flight, the continuation method should be able to handle constraints on the state variables. The variation of the parameters would be a consequence of satisfying the imposed constraints on the states. Such a continuation procedure has been developed by the authors and has been called the extended bifurcation analysis (EBA) method. In contrast, the traditional one-parameter continuation procedure is labeled as the standard bifurcation analysis (SBA) method. The EBA method has been widely used for a variety of problems of constrained flight maneuvers. In the following, we shall describe the EBA procedure with the nonlinear phugoid model as an example.

The phugoid model that we have seen in Equation 3.28 is a multiparameter dynamical system with two parameters (a, b). So far we have looked at bifurcations only in the parameter space constrained by $a = const$. Our aim now is to formulate the continuation algorithm such that it can handle arbitrary variation of parameters a and b, or impose certain constraints on the states variables, y and γ, and let the parameters a, b vary consequently.

For a general dynamical system of the form Equation 3.1, the vector of control parameters $\underline{U}$ can be split into a primary continuation parameter u, a set of parameters $\underline{p}_1(u)$ that are "scheduled" as a function of u, and another set of parameters that are held constant, $\underline{p}_2 = const$. In this manner, the parameters $\underline{p}_1(u)$ can be varied in conjunction with the primary continuation parameter u. The scheduling function $\underline{p}_1(u)$ can either be selected arbitrarily or selected such that the states satisfy a constraint condition of the form $\underline{g}(x) = 0$. Let us first illustrate this formulation with an example from aircraft longitudinal flight dynamics. We shall then present the computational procedure.

3.5.1 Scheduling the Parameters in a Multiparameter System

To explain some of the different ways in which the parameters may be scheduled in a multiparameter dynamical system, let us consider four different cases for the airplane flight dynamics problem:

Case 1: $\dot{\underline{x}} = \underline{f}(\underline{x}, T = \text{fixed}, \delta a = \delta r = 0, u = \delta e)$

Case 2: $\dot{\underline{x}} = \underline{f}(\underline{x}, u = T, \delta a = \delta r = 0, \delta e = \text{fixed})$

Case 3: $\dot{\underline{x}} = \underline{f}(\underline{x}, T(u), \delta a = \delta r = 0, u = \delta e)$ OR

$\dot{\underline{x}} = \underline{f}(\underline{x}, \delta e(u), \delta a = \delta r = 0, u = T)$

Case 4: $\dot{\underline{x}} = \underline{f}(\underline{x}, T, \delta a, \delta r, u = \delta e); \underline{g}(\underline{x}) = 0$

Case 1 is the SBA with elevator deflection δe as the continuation parameter and all the other parameters being held fixed. Especially since thrust is held fixed, almost every steady state would in general be a climbing/descending flight. Case 2 is again an SBA, this time with thrust as the continuation parameter and the elevator deflection being held fixed. In principle, one could run Cases 1 and 2 for several different values of the fixed parameter, creating a grid of solutions as it were, and then select the solutions of interest among this matrix of solutions. Alternatively, if one already knew the desired values of thrust as a function of elevator deflection or vice versa, then the bifurcation analysis could be formulated as Case 3. This is still an SBA procedure except that the solutions in the two-parameter $(T, \delta e)$ space lie along a predefined "cut" given by $(T, \delta e)$ or $\delta e\ (T)$. Case 4 is the EBA procedure—in this case, a "cut" in parameter space defined by $T(u)$, $\delta a(u)$, $\delta r(u)$, where u is the continuation parameter δe, is to be found such that the states satisfy the constraint given by $g(\underline{x}) = 0$. The key difference in Case 4 is that the "cut" in parameter space is not known *a priori* and must be computed as part of the EBA solution procedure.

For the general EBA formulation, the dynamical system in Equation 3.1 can be represented as follows along with a vector function $\underline{y} = \underline{g}(\underline{x}) = 0$.

$$\frac{dx}{dt} = \underline{f}(x, u, p); \quad \underline{y} = \underline{g}(\underline{x}) = 0 \tag{3.33}$$

where $u, p \in \underline{U}$, u is the scalar continuation parameter and p are the remaining $(m-1)$ parameters in $\underline{U}$, $y \in \Re^r$ is the vector of functions defining the constraints on states. In general, the number of independent constraints in y must be less than or equal to the number of free parameters p.

The EBA analysis of a system such as Equation 3.33 involves three steps:

1. Computing steady states $\underline{f}(x, u, p) = 0$ satisfying the constraints $\underline{y} = \underline{g}(\underline{x}) = 0$

2. Computing of stability of the constrained steady states

3. Identifying points of bifurcation and computing the bifurcation diagram, including bifurcated branches that may not satisfy the constraints $\underline{y} = g(\underline{x}) = 0$ (in the context of aircraft flight dynamics, these are called "departed" solutions, or "departures")

We shall examine each of these steps one by one.

Step 1: To find the "constrained" steady states, we must solve

$$\underline{f}(x, u, \; p) = 0; \; \underline{g}(\underline{x}) = 0 \tag{3.34}$$

Equation 3.34 is a set of $(n + r)$ equations (n from $\underline{f}$ and r from $\underline{g}$) to be solved in $(n + r + 1)$ unknowns (n of $\underline{x}$, r out of the $m-1$ elements of $\underline{p}$, and u). There is of course the question of whether this set of equations is well posed. To put it simply, consider one constraint $g_1(\underline{x}) = 0$ from the set $g(\underline{x}) = 0$ and let p_i be a parameter that can be "adjusted" to satisfy this constraint. Then, consider the next constraint $g_2(x) = 0$; perhaps, two parameters, p_i and p_j, can now be "adjusted" to satisfy both these constraints. And so on.

Obviously, two questions arise—First, how can we be sure that a particular constraint can indeed be met by one parameter or the other? What if none of the parameters selected influence the constraint function? Second, if there are multiple parameters in p that may be used to satisfy a particular constraint, should we make an "optimal" selection? Or does it not matter?

The way to answer these questions is to compute what is called the *Lie derivative* of g. For example, the derivative of the ith output equation will appear as follows:

$$\frac{dy_i}{dt} = \frac{dg_i(x)}{dt} = \frac{dg_i(x)}{dx_1} \cdot \frac{dx_1}{dt} + \frac{dg_i(x)}{dx_2} \cdot \frac{dx_2}{dt}$$

$$+ \cdots + \frac{dg_i(x)}{dx_n} \cdot \frac{dx_n}{dt} = \sum_{j=1,n} \frac{dg_i(x)}{dx_j} \cdot f_j \qquad (3.35)$$

where f_j is a function of (x, u, p). For a particular element of p to influence a constraint function $g_i(x)$, at least one of the derivatives (dg_i/dx_j) must be nonzero such that the corresponding f_j includes that particular element of p. If there are nonzero terms (dg_i/dx_j) but no element of p appears in the corresponding functions f_j, then higher *Lie derivatives* may be examined. The number of times an output y_i must be differentiated so that a parameter from p appears for the first time on the right-hand side as described above with reference to Equation 3.35 is known as the relative degree of the output y_i. Clearly, the relative degree of an output function cannot exceed the order of the system n.

Homework Exercise: Convince yourself of the last statement. If a function $y_i = g_i(x)$ has already been differentiated n times, as in Equation 3.35, without a particular element of the parameter vector p appearing in a nonzero manner, then that element can have no effect whatsoever on the function $y_i = g_i(x)$.

The other question is whether more than one parameter could be used to influence a given constraint $y_i = g_i(x) = 0$. This can also be determined from a similar exercise using the *Lie derivative*. In general, a parameter with a lower relative degree is the preferable one. In case, multiple parameters appear with the same relative degree, then the one with the greater influence, measured by $(dg_i/dx_j) \cdot (df_i/dp_k)$, is preferred. However, for most physical systems, and this is true of airplane flight dynamics as well, the parameters to be selected to meet a certain set of constraints are fairly obvious from the physics of the problem.

3.5.2 Influence of Aircraft Control Parameters on Constraints

Let us formally consider the issue of selecting the parameters to be used to achieve a certain constraint for aircraft flight. Let the constraint be a level flight condition where flight path angle is to be maintained zero, that is, $g = \gamma(\alpha,\beta,\phi,\theta) = 0$. The set of equations corresponding to Equation 3.34 in this case is

$$\underline{f}(V, \alpha, \beta, p, q, r, \phi, \theta, T, \delta e, \delta a, \delta r) = 0; \quad \gamma(\alpha, \beta, \phi, \theta) = 0 \tag{3.36}$$

where f are the vector functions defining the airplane dynamical system (as listed in Chapter 1), $(T, \delta e, \delta a, \delta r)$ are the control parameters, and the constraint equation for γ is the first of Equation 1.18. Of the parameters, let δe be the continuation parameter u. Then one of the parameters $(T, \delta a, \delta r)$ must be "freed" so that it may be varied appropriately to meet the constraint on flight path angle being zero for all steady states, as in Equation 3.36. Common sense suggests that the most suitable parameter for this purpose is the thrust, T. However, this conclusion can also be reached by carrying out the analysis using *Lie derivatives* as described above.

Homework Exercise: Evaluate the *Lie derivative* for the constraint equation $\gamma = (\alpha,\beta,\phi,\theta) = 0$. Specifically, evaluate (dg_i/dx_i) for the state α and confirm that it is nonzero. Then, examine the term $(dg_i(x)/dx_j).f_j$, where f_i is the state equation for α. Check whether the thrust T appears in this derivative term. In that case, what is the relative degree of $y = \gamma = 0$ with respect to T.

The solution of Equation 3.36 is just a continuation of $n + r$ variables with a single parameter. The result is the states x satisfying the constraint condition plus the variation of the "freed" parameter as a function of the continuation parameter, that is, $T(\delta e)$. A relationship of the form $p(u)$ is called a parameter schedule. Hence, $T(\delta e)$ is the schedule of the thrust as a function of elevator deflection required to maintain the airplane in a level flight trim state for every value of the elevator deflection.

Step 2: To determine the stability of the "constrained" steady state, we must solve

$$\dot{\underline{x}} = \underline{f}(\underline{x}, u, \underline{p}(u)) \tag{3.37}$$

where $\underline{p}(u)$ are the parameter schedules from the continuation solution in Step 1. The other parameters (that are not "freed") are held fixed, as always. Solution of Equation 3.37 is a SBA, which can solve for the steady states (same as those already found in Step 1) and also determine their stability, as we have seen earlier in this chapter.

Step 3: However, there is one additional issue to deal with in this SBA computation. While computing the "constrained" steady state and their stability, the SBA may come across bifurcation points. At these points, new steady states (equilibria as well as periodic solutions) may emerge. These bifurcated steady states will usually not satisfy the constraint conditions used in Step 1. However, they still obey the parameter schedule $\underline{p}(u)$. That is, they are solutions of Equation 3.37 with the parameter schedule $\underline{p}(u)$ other than those meeting the constraints in Equation 3.36. The SBA procedure can of course automatically compute these "unconstrained" steady states and their stability as well. These states are called *departures* from the desired "constrained" state solutions. For a complete solution of the airplane flight dynamics problem, it is essential to compute these "departed" states as well. For example, a departure from level flight trim may cause the airplane to bank and gradually spiral while losing altitude.

In the rest of this book, we shall present several examples of the use of bifurcation and continuation methods to analyze problems in airplane flight dynamics. For the moment though, we carry forward the nonlinear phugoid model from Example 3.1.

EXAMPLE 3.2: PHUGOID MOTION WITH FLIGHT PATH ANGLE CONSTRAINED

Recall the phugoid model in Equation 3.28. Of the two parameters, a, the inverse of L/D, had been held fixed at $a = 0.2$ and b, the thrust-to-weight ratio, had been used as the continuation parameter in Example 3.1. Now we shall allow for both parameters, a and b to vary and impose a constraint on the climb angle instead. The equations for the phugoid motion, in the manner of Equation 3.33, will now appear as

$$\frac{dy}{d\tau} = b - ay^{*2} - \sin\gamma^{*}; \quad \frac{d\gamma}{d\tau} = y^{*} - \frac{\cos\gamma^{*}}{y^{*}}; \quad \gamma^{*} = const. = k \quad (3.38)$$

where the flight path angle γ is now constrained to a particular value. The continuation parameter is still b; however, now a is simultaneously varied such that the constraint on γ is met at all steady states. Fortunately, in this case, Equation 3.38 can be solved analytically. For constant γ, the second of Equation 3.38 implies that y should be constant as well. Then the first of Equation 3.38 can be used to show that, in that case, a should vary linearly with b in order to maintain the constrained value of γ.

Homework Exercise: Solve for the steady states of Equation 3.38 and the schedule of parameter a as a function of continuation parameter b.

The constrained steady states can be computed numerically using a continuation algorithm following Steps 1–3 as outlined earlier. First, the parameter schedule $a = a(b)$ is found such that the steady states all satisfy the constraint condition. Two choices of the flight path angle constraint are made here: $\gamma = 0$ and $\gamma = 0.4$ rad.

Figure 3.25 shows the schedule of parameter a as a function of parameter b for two constrained values of $\gamma = 0$ and 0.4 rad. Then, in Step 2, an SBA analysis is carried out with the parameter schedule in Figure 3.25. The stability of the constrained steady states is determined and bifurcations are noted. The bifurcation diagrams for the same two constraint cases as in Figure 3.25 are shown in Figure 3.26 with γ as the variable. The plots for y are similar.

Clearly, in either case, the constrained value—$\gamma = 0$ or $\gamma = 0.4$ rad—is maintained. The steady states are all stable for $\gamma = 0$ within the prescribed limits of b while in case of $\gamma = 0.4$ rad, steady states are unstable for low values of b and they transition to stability via a Hopf bifurcation.

We can actually go one step further and track the location of the Hopf bifurcation in the two-parameter (a, b) space. As we have already seen, the occurrence of the Hopf bifurcation corresponds to a pair of pure imaginary eigenvalues of the Jacobian matrix. The Jacobian matrix for the phugoid dynamics in Equation 3.38 can be written as

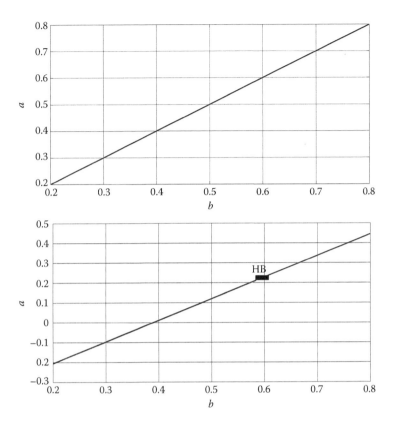

FIGURE 3.25 Parameter schedule $a = a(b)$ for the phugoid model in Equation 3.39 with the constraint, top: $\gamma = 0$ and bottom: $\gamma = 0.4$ rad (solid lines: stable solutions; dashed lines: unstable solutions; HB: Hopf bifurcation).

$$J = \begin{bmatrix} \dfrac{\partial f_1}{\partial y} & \dfrac{\partial f_1}{\partial \gamma} \\[2mm] \dfrac{\partial f_2}{\partial y} & \dfrac{\partial f_2}{\partial \gamma} \end{bmatrix}_{(y^*,\gamma^*)} = \begin{bmatrix} -2ay^* & -\cos\gamma^* \\[2mm] 1 + \dfrac{\cos\gamma^*}{y^{*2}} & \dfrac{\sin\gamma^*}{y^*} \end{bmatrix} \tag{3.39}$$

The trace and determinant of the matrix are given by

$$\mathrm{trace}(J) = -2ay^* + \frac{\sin\gamma^*}{y^*};$$

$$\det(J) = -2a\sin\gamma^* + \cos\gamma^* + \frac{\cos^2\gamma^*}{y^{*2}} \tag{3.40}$$

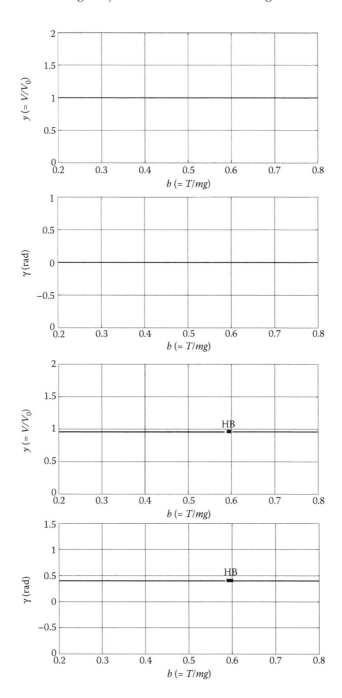

FIGURE 3.26 Bifurcation diagram for the phugoid model in Equation 3.39, variables γ and y, with the constraints, top plots: γ = 0 and bottom plots: γ = 0.4 rad (solid lines: stable solutions; dashed lines: unstable solutions; HB: Hopf bifurcation).

Homework Exercise: Use the condition $trace(J) = 0$ for the occurrence of a Hopf bifurcation to determine the critical values of parameters a, b, which correspond to a Hopf point as follows:

$$a_{cr} = \frac{\sin \gamma^*}{2 \cos \gamma^*} = \frac{\tan \gamma^*}{2}, \quad b_{cr} = \frac{3}{2} \sin \gamma^*$$

Hence, plot the track of Hopf bifurcations in (a, b) parameter space. Physically, what does it mean in terms of loss of damping of the phugoid mode as a function of thrust-to-weight ratio and (inverse of) lift-to-drag ratio?

Step 3 is the tracking of the limit cycles arising at the Hopf bifurcation point in Figure 3.26; that is left as a Homework Exercise.

EXERCISE PROBLEMS

1. Determine analytically equilibrium states/fixed points and their stability of the following first-order equations:

   a. $\dot{x} = x^2 - 1$

   b. $\dot{x} = 1 - x^2$

   c. $\dot{x} = x^3$

   d. $\dot{x} = x(1 - x^2)$

   e. $\dot{x} = x(1 - x)$

   f. $\dot{x} = 1 - 2 \cos x$

   g. $\dot{x} = e^{-x} \sin x$

   h. $\dot{x} = 1 - 3x(1 - x)$

   i. $\dot{x} = \sin x$

2. Determine analytically equilibrium states/fixed points and their stability of the following second-order systems:

   a. $\dot{x} = x(y - x); \quad \dot{y} = y(2x - y)$

   b. $\dot{x} = y; \quad \dot{y} = y - x(1 + xy)$

   c. $\ddot{x} - \dot{x}(1 - x) + 3(x^2 - 1) = 0$

d. $\ddot{x} - \dot{x} + 1 - e^x = 0$

e. $\dot{x} = \sin y; \quad \dot{y} = x^2 - 9$

f. $\dot{x} = 1 + y - e^{-x}; \quad \dot{y} = x^3 - y$

g. $\dot{x} = -y - x^3; \quad \dot{y} = x$

h. $\dot{x} = xy - 1; \quad \dot{y} = x - y^3$

i. $\dot{x} = x(3 - 2x - y); \quad \dot{y} = y(2 - x - y)$

j. $\ddot{x} = x^3 - x$

3. Sketch phase portraits of the system equations given in Exercise 2 and confirm if the conclusions made earlier from analysis is correct or not. [*Hint:* Use pplane8.m, available in public domain, for generating phase portraits.].

4. Determine periodic orbits and their stability of the following difference (map) equations:

a. $x_{n+1} = \sin x_n$

b. $x_{n+1} = \cos x_n$

c. $x_{n+1} = \sqrt{x_n}$

d. $x_{n+1} = x_n - x_n^2$

e. $x_{n+1} = 3(x_n - x_n^2)$

f. $x_{n+1} = x_n^2$

g. $x_{n+1} = x_n - x_n^3$

h. $x_{n+1} = 1 + \frac{1}{2}\sin x_n$

5. Write a code in MATLAB to simulate longitudinal motions using all four longitudinal dynamics equations of aircraft using ODE45 subroutine. Choose values of parameters (*a*, *b*) and appropriate initial conditions to examine phugoid behavior in regions around Hopf bifurcation. Discuss the influence of other two equations on the phugoid dynamics.

6. Determine numerically or otherwise steady states and stability of the following parameterized systems at different values of parameters: First-order systems:

i. $\dot{x} = 0.4x\left(1 - \dfrac{x}{k}\right) - \dfrac{x^2}{1+x^2}$; $\quad k = 2, 8, 10$

ii. $\dot{x} = 1 + rx + x^2$; $\quad r = -1, 0, 2$

iii. $\dot{x} = x - rx(1-x)$; $\quad r = -1, 0, 1$

iv. $\dot{x} = x + \dfrac{rx}{1+x^2}$; $\quad r = -2, 0, 3$

v. $\dot{x} = rx + x^3 - x^5$; $\quad r = -0.5, -0.25, 0, 0.5$

Second-order systems:

i. $\dot{x} = u - x^2$, $\dot{y} = -y$; $\quad u = -2, 0, 2$

ii. $\dot{x} = -x + y$, $\dot{y} = \dfrac{x^2}{1+x^2} - uy$; $\quad u = 0, 0.5, 2$

iii. $\dot{x} = y - 2x$, $\dot{y} = u + x^2 - y$; $\quad u = -1, 0, 1$

iv. $\dot{x} = x[x(1-x) - y]$, $\dot{y} = y(x-u)$; $\quad u = -2, 0, 3$

v. $\dot{x} = y + ux$, $\dot{y} = -x + uy - x^2 y$; $\quad u = -1, 1, 2$

7. Sketch phase portraits of second-order system equations in Exercise 6 by hand and by using the pplane7.m code in MATLAB.

8. Usually, it is more practical to work with reduced-order models of systems to study a particular phenomenon represented by specific bifurcations. Look for the simplest models (also known as "normal forms" given in Figure 3.13 for static bifurcations) of each of the bifurcations discussed in this chapter.

9. Origination and bifurcations of limit cycle states are best studied by rewriting the dynamical systems equations in polar (r, θ) coordinates, where r represents radial distance of the trajectory from the origin and θ represent anticlockwise winding of the trajectory. Convert the following equations in xy-coordinates to polar coordinates and discuss about fixed points and stability of the system as function of the parameter a.

$$\dot{x} = -y + ax(x^2 + y^2)$$
$$\dot{y} = x + ay(x^2 + y^2)$$

10. Analyze fixed points/stability and bifurcations of the two-dimensional system with respect to the parameter μ.

$$\dot{r} = \mu r + r^3 - r^5$$
$$\dot{\theta} = \omega + b r^2$$

11. Using the low-angle-of-attack aircraft models given in Chapter 2, examine best suitable controls for specific constraints.

12. Discuss various types of equality/inequality constraints a physical systems may be expected to operate under with or without active control systems.

13. AUTO continuation code for bifurcation analysis of dynamical systems, and MATCONT, a continuation code on MATLAB platform, are available in the public domain. Use these tools for studying different physical systems described by ordinary differential equations.

14. Use ODE45 in MATLAB to set up simulation code for Phugoid dynamics treated in Example 3.1. Verify predictions made out based on bifurcation analysis results of the model in Equation 3.28.

15. Carry out phase portrait analysis of constrained Phugoid dynamics in Example 3.2.

REFERENCES

1. Lyapunov, A. M., *The General Problem of the Stability of Motion*, A. T. Fuller (trans), Taylor & Francis, London, ISBN 978-0-7484-0062-1, 1992.
2. Nayfeh, A. H., and Balachandran, B., *Applied Nonlinear Dynamics: Analytical, Computational, and Experimental Methods*, Wiley-Interscience, New York, 1995.

CHAPTER 4

Longitudinal Flight Dynamics

Longitudinal flight dynamics deals with flight restricted to the longitudinal plane—that is, the body X^B–Z^B plane, which is usually a plane of symmetry for most airplanes. A significant percentage of flight time is actually spent in longitudinal flight. Besides, a study of longitudinal flight dynamics permits us to explore various aspects of airplane trim, stability, and performance with a greater ease due to the presence of fewer variables. Hence, it makes sense to begin with an analysis of longitudinal flight. Of course, an airplane in longitudinal flight can be affected by disturbances that perturb its motion *out of the longitudinal plane*. However, for the present, we shall restrict ourselves to disturbances only in the longitudinal variables; hence, the airplane motion remains restricted to the longitudinal plane.

In the preceding chapters, we have written out the equations for the dynamics of a rigid airplane in flight in various forms. Then we have presented the aerodynamic model—when combined with the equations of motion, we could simulate the dynamics of the airplane in flight to different control inputs. In the previous chapter, we have presented a methodology to analyze the steady states of a dynamical system and their stability in a systematic manner, and thus understand the totality of its dynamic behavior. It is time now to apply the continuation and bifurcation analysis method to the dynamic model of an airplane in flight—first restricted to flight in the longitudinal plane.

The equations of motion of an airplane in the longitudinal plane may be read from Table 1.8 (but the velocity V is not assumed to be constant here):

$$\dot{V} = \frac{1}{m}(-mg\sin\gamma - D + T\cos\alpha)$$

$$\dot{\alpha} = \frac{1}{mV}(mg\cos\gamma - L - T\sin\alpha) + q$$

$$\dot{q} = \frac{1}{I_{yy}}M \qquad (4.1)$$

$$\dot{\theta} = q$$

where $\gamma = \theta - \alpha$.

To complete Equation 4.1, it is necessary to introduce the aerodynamic model from Equations 2.1 through 2.3, as below:

$$D = \bar{q}SC_D = \bar{q}S\left[\begin{array}{c} C_{Dsta}(Ma,\alpha,\delta e) + C_{Ddyn}\left(\dfrac{(q_b - q_w)c}{2V}\right) \\[2mm] + C_{Dflo}\left(\dfrac{q_w c}{2V}\right) + C_{Ddow}(\alpha(t-\tau)) \end{array}\right]$$

$$L = \bar{q}SC_L = \bar{q}S\left[\begin{array}{c} C_{Lsta}(Ma,\alpha,\delta e) + C_{Ldyn}\left(\dfrac{(q_b - q_w)c}{2V}\right) \\[2mm] + C_{Lflo}\left(\dfrac{q_w c}{2V}\right) + C_{Ldow}(\alpha(t-\tau)) \end{array}\right] \qquad (4.2)$$

$$M = \bar{q}ScC_m = \bar{q}Sc\left[\begin{array}{c} C_{msta}(Ma,\alpha,\delta e) + C_{mdyn}\left(\dfrac{(q_b - q_w)c}{2V}\right) \\[2mm] + C_{mflo}\left(\dfrac{q_w c}{2V}\right) + C_{mdow}(\alpha(t-\tau)) \end{array}\right]$$

Further, the dynamic terms can be written in terms of the dynamic derivatives as in Equation 2.10, the downwash terms may be modeled as described in Chapter 2, and the flow curvature terms may be ignored as discussed earlier, giving

$$D = \bar{q}SC_D = \bar{q}S\left[C_{Dsta}(Ma,\alpha,\delta e) + C_{D_{q1}}\frac{(\dot{\alpha})c}{2V} + C_{D\dot{\alpha}}\frac{(\dot{\alpha})c}{2V}\right]$$

$$L = \bar{q}SC_L = \bar{q}S\left[C_{Lsta}(Ma,\alpha,\delta e) + C_{L_{q1}}\frac{(\dot{\alpha})c}{2V} + C_{L\dot{\alpha}}\frac{(\dot{\alpha})c}{2V}\right] \quad (4.3)$$

$$M = \bar{q}ScC_m = \bar{q}Sc\left[C_{msta}(Ma,\alpha,\delta e) + C_{m_{q1}}\frac{(\dot{\alpha})c}{2V} + C_{m\dot{\alpha}}\frac{(\dot{\alpha})c}{2V}\right]$$

Inserting the aerodynamic model from Equation 4.3 into the longitudinal dynamics Equation 4.1 and rearranging terms, we have

$$\dot{V} = -g\sin\gamma - \frac{\bar{q}S}{m}\left[C_{Dsta}(Ma,\alpha,\delta e) + \left(C_{D_{q1}} + C_{D\dot{\alpha}}\right)\frac{(\dot{\alpha})c}{2V}\right] + \frac{T}{m}\cos\alpha$$

$$\dot{\alpha} = \frac{g\cos\gamma}{V} - \frac{\bar{q}S}{mV}\left[C_{Lsta}(Ma,\alpha,\delta e) + \left(C_{L_{q1}} + C_{L\dot{\alpha}}\right)\frac{(\dot{\alpha})c}{2V}\right] - \frac{T\sin\alpha}{mV} + q \quad (4.4)$$

$$\dot{q} = \frac{\bar{q}Sc}{I_{yy}}\left[C_{msta}(Ma,\alpha,\delta e) + \left(C_{m_{q1}} + C_{m\dot{\alpha}}\right)\frac{(\dot{\alpha})c}{2V}\right]$$

$$\dot{\theta} = q$$

It may be instructive to rewrite Equation 4.4 in a slightly different, but totally equivalent, form as below:

$$\frac{\dot{V}}{V} = \frac{g}{V}\left\{-\sin\gamma - \frac{\bar{q}S}{W}\left[C_{Dsta}(Ma,\alpha,\delta e) + \left(C_{D_{q1}} + C_{D\dot{\alpha}}\right)\frac{(\dot{\alpha})c}{2V}\right] + \frac{T}{W}\cos\alpha\right\}$$

$$\dot{\gamma} = \frac{g}{V}\left\{-\cos\gamma + \frac{\bar{q}S}{W}\left[C_{Lsta}(Ma,\alpha,\delta e) + \left(C_{L_{q1}} + C_{L\dot{\alpha}}\right)\frac{(\dot{\alpha})c}{2V}\right] + \frac{T}{W}\sin\alpha\right\} \quad (4.5)$$

$$\ddot{\theta} = \frac{\bar{q}Sc}{I_{yy}}\left[C_{msta}(Ma,\alpha,\delta e) + \left(C_{m_{q1}} + C_{m\dot{\alpha}}\right)\frac{(\dot{\alpha})c}{2V}\right]$$

The division by V in the first of Equation 4.5 is helpful numerically as it brings all the variables to a common range of magnitude. It also brings out the fact that the natural (as against forced due to application of control inputs) variation of V with time occurs at the timescale g/V, which for most airplanes is of the order of 10 s. The second of Equation 4.5 is more usefully recast in terms of the flight path angle γ. In contrast, α, in terms of which the second equation was originally written in Equation 4.4, is not as useful flight dynamically though its aerodynamic significance is undisputed. Significant flight conditions such as maximum endurance, best range, optimal cruise, and stall are usually specified in terms of flight speed V. And maneuvers such as climb/descent and pull-up/push-down are described in terms of γ or its rate of change $\dot{\gamma}$. Note that the natural variation of γ also takes place at the timescale g/V that is of the order of 10 s. The third and fourth equations in Equation 4.4 have been combined into a single equation for $\ddot{\theta}$ in Equation 4.5, thus eliminating the pitch rate q as a variable. The variation of the body-axis pitch (Euler) angle is of interest from the point of view of stability of the airplane in flight. Changes in θ with time occur at the timescale $\sqrt{I_{yy}/\bar{q}Sc}$, which for airplanes is usually of the order of 1 s.

Homework Exercise: Estimate the timescale factor $\sqrt{I_{yy}/\bar{q}Sc}$ for the F-18 aero model in Table 2.1. Assume sea level flight at Mach 0.3. (It works out to 0.51 s, which is of the correct order.)

Thus, Equation 4.5 represents changes in the airplane trajectory (variables V, γ) and the body attitude (variable θ). There are three aerodynamic parameters—Ma, α, $\dot{\alpha}$—however, these may be calculated from the available values of the three variables (V, γ, θ) at every instant. Then there are the two control inputs—the elevator deflection δe and the thrust T, which is usually specified in terms of the throttle setting η and is a function of the aerodynamic parameters as well, as we shall see shortly.

4.1 LONGITUDINAL STEADY STATES (TRIMS)

Steady states (also called trims) of Equation 4.5 correspond to constant values (not changing with time) of the three variables (V, γ, θ). They may therefore be obtained by solving Equation 4.5 after setting the time derivatives of (V, γ, θ) to zero. Note that α being equal to $\theta - \gamma$, if the latter two variables are constant, then so is α; hence $\dot{\alpha} = 0$ as well. Thus, we have the following trim equations:

$$0 = \frac{g}{V}\left\{-\sin\gamma - \frac{\bar{q}S}{W}\left[C_{Dsta}(Ma,\alpha,\delta e)\right] + \frac{T}{W}\cos\alpha\right\}$$

$$0 = \frac{g}{V}\left\{-\cos\gamma + \frac{\bar{q}S}{W}\left[C_{Lsta}(Ma,\alpha,\delta e)\right] + \frac{T}{W}\sin\alpha\right\} \qquad (4.6)$$

$$0 = \frac{\bar{q}Sc}{I_{yy}}\left[C_{msta}(Ma,\alpha,\delta e)\right]$$

Homework Exercise: Aircraft trimming routines may be used to solve for steady-state solutions of Equation 4.6. Check out some of the standard trimming routines and figure out how they work. For example, some of them use an optimizer—somewhat different from the methodology we shall follow.

4.1.1 Modeling Engine Thrust

Modern airplane engines are largely turbofans though turboprops and piston-props are used for specific roles. The plain turbojet is rarely used but since the principle of modeling engine thrust is fairly similar in all cases, we shall illustrate it here with an example of a turbojet.

The thrust obtained from an engine in flight depends chiefly on four factors—the flight Mach number, the altitude, the throttle (or engine rpm) setting, and the losses (due to installation, atmospheric conditions, flight conditions, etc.). Therefore, the engine thrust can be most generally modeled as

$$T = (1 - \varepsilon_T) \cdot \eta \cdot \sigma \cdot T_{SL}^{100}(Ma) \tag{4.7}$$

where ε_T is the thrust loss coefficient (usually of the order of 3%–4%), η is the throttle (or engine rpm) setting, σ is a factor that accounts for the effect of altitude away from sea level (SL), and T_{SL}^{100} is the specified engine thrust at 100% throttle setting under SL conditions. Clearly, $\eta = 1$ at 100% throttle setting and $\sigma = 1$ at the SL condition. Typical variation of the thrust of a turbojet engine as a function of Mach number for different throttle settings and at two different altitudes is shown in Figure 4.1. Note the lower thrust at the higher altitude and for lower throttle settings.

It is not uncommon to find the thrust modeled in terms of a thrust coefficient C_T as $T = \bar{q}SC_T$ similar to the modeling of aerodynamic drag; however, there is no rational basis for such a model nor is it really useful. Instead, if needed, a thrust coefficient that captures the effect of flight Mach number on the engine thrust may be modeled as below:

$$C_T = \frac{T_{SL}^{100}(Ma)}{T_{SL}^{100}(Ma \to 0)} \tag{4.8}$$

With the definition in Equation 4.8 incorporated, Equation 4.7 appears as

$$T = \underbrace{(1-\varepsilon_T)}_{\substack{\text{after deducting} \\ \text{losses}}} \cdot \underbrace{\eta}_{\substack{\text{engine} \\ \text{rpm}}} \cdot \underbrace{\sigma}_{\text{altitude}} \cdot \underbrace{C_T(Ma)}_{\substack{\text{Mach} \\ \text{number}}} \cdot T_{SL}^{100}(Ma \to 0) \tag{4.9}$$

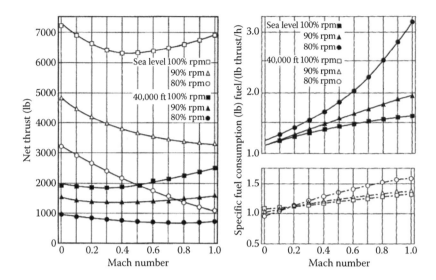

FIGURE 4.1 Typical jet engine thrust and thrust specific fuel consumption characteristics. (From http://people.clarkson.edu/~pmarzocc/AE429/AE-429-6.pdf.)

where the effects captured by the different terms have been named below them.

The other engine figure of merit of interest is the (thrust) specific fuel consumption (SFC), which is the mass of fuel consumed per unit thrust per unit time. The SFC may be modeled similar to the thrust as follows:

$$sfc = \underbrace{(1+\varepsilon_{sfc})}_{\substack{\text{addition} \\ \text{due to losses}}} \cdot \underbrace{\eta}_{\substack{\text{engine} \\ \text{rpm}}} \cdot \underbrace{\sigma}_{\text{altitude}} \cdot \underbrace{C_{sfc}(Ma)}_{\substack{\text{Mach} \\ \text{number}}} \cdot sfc_{SL}^{100}(Ma \to 0) \tag{4.10}$$

where the terms are similar to those in the thrust model in Equation 4.9 and ε_{sfc} is the SFC loss coefficient (usually of the order of 2%). Typical variation of the SFC of a turbojet engine as a function of Mach number for different throttle settings and at two different altitudes is shown in Figure 4.1.

The other parameter that steady-state solutions of Equation 4.5 given by Equation 4.6 depend on is the elevator deflection angle δe. Typical variations of aircraft force and moment coefficients with elevator deflection are shown in Figure 2.12 via the derivatives $C_{L\delta e}$, $C_{D\delta e}$, and $C_{m\delta e}$ as functions of the angle of attack. To effect any change in longitudinal motion of an airplane, both elevators on the right and on the left sides of an airplane are together deflected up or down; thus, the total effect can be summed up as

$$C_{(.)\delta e}\delta e = \frac{C_{(.)\delta e,l}\delta e_l + C_{(.)\delta e,r}\delta e_r}{2}; (.) = L, D, m.$$

4.2 LONGITUDINAL TRIM AND STABILITY ANALYSIS

Equations 4.6 for the airplane longitudinal trim may be solved by using a bifurcation and continuation algorithm as described in Chapter 3. There are three variables of interest—V, γ, and θ—and additionally the angle of attack α, and two control parameters—the elevator deflection δe and the throttle setting η. The SBA as explained in Chapter 3 allows only one control parameter to be varied at a time while the others are to be held fixed. Therefore, longitudinal trims first with varying elevator deflection alone and then with varying throttle setting alone will be considered below. The SBA method also computes the stability of each trim state; α is useful here because the onset of instability often correlates with certain changes in some of the aerodynamic coefficients that are predominantly a function of α.

4.2.1 Longitudinal Trim and Stability with Varying Angle of Attack

Let us select the F-18/HARV high-angle-of-attack airplane data set in Chapter 2 for these computations and set the throttle to a fixed value: $\eta = 0.346$. The precise value is not very important—the results may be expected to be qualitatively similar for a range of throttle settings. We shall pick a Mach number of $M = 0.3$—low speed typical of landing approach. For a given throttle setting and flight speed, one may expect the airplane to trim over a wide range of angles of attack with different elevator settings. One may choose any convenient combination of α and δe, and set off a continuation run from that initial steady state. The SBA method should expectedly capture all other possible trim states with varying $(\alpha, \delta e)$.

Let us select an angle of attack of nearly 6.5 deg, corresponding to 0.113 rad. The initial trim state then turns out to be defined as follows:

$$[M, \alpha, q, \theta] = [0.3, \; 0.113 \, \text{rad}, \; 0, \; 0.113 \, \text{rad}], \quad [\eta, \delta e] = [0.346, -0.01 \, \text{rad}]$$

It may be noticed that since $\alpha = \theta$ at this trim state, the flight path angle γ is actually zero; that is, it is a level flight trim. However, as we mentioned earlier, it is not required to select this specific initial trim state.

Let this initial trim state be labeled "1." Then, SBA computations with varying elevator deflection as the continuation parameter are presented in Figure 4.2. Both increasing and decreasing elevator deflection cases are combined into a single plot in Figure 4.2 and some of the trim states are labeled "2," "3," "4," and so on. The initial trim state is marked by a triangle

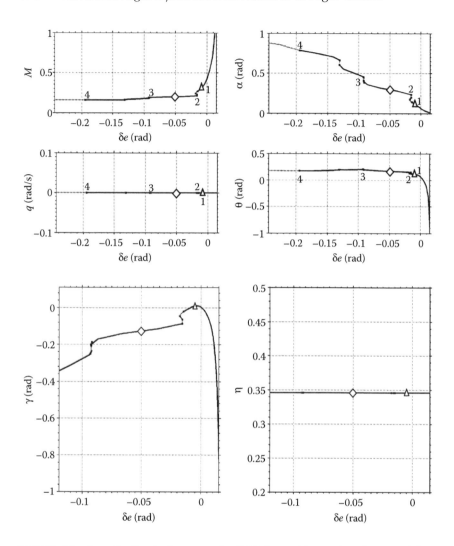

FIGURE 4.2 Longitudinal trim states for F-18 data with varying elevator deflection (throttle setting held fixed at $\eta = 0.346$). Full line: stable; dashed line: unstable.

symbol and a diamond symbol is used for the steady state with trim angle of attack 0.3 rad. The steady state with the diamond symbol can be compared with another trim state at $\alpha = 0.3$ rad ≈ 17.2 deg that was used in the simulation cases of Chapter 2 (see Trim State 1). In that case, the state variables were: $[M, \alpha, q, \theta] = [0.206, 0.3 \text{ rad}, 0, 0.3 \text{ rad}]$, $[\eta, \delta e] = [0.54, -0.0522 \text{ rad}]$.

Homework Exercise: Compare the above trim state with the one marked with the diamond symbol in Figure 4.2. Although they both trim at the same angle of attack and elevator deflection, the thrust setting is different and hence the flight Mach number differs between the two cases. Also, the steady state in Figure 4.2 at the diamond symbol is not a level flight unlike the trim state listed above.

Examining the plot of α versus δe in Figure 4.2, it is clear that larger negative (up) elevator deflection produces trims at higher angles of attack, as expected. However, the $\alpha - \delta e$ trim curve is not entirely smooth. There are disruptions—short stretches of unstable trims between points marked "1" and "2," in the region marked "3," and so on. Also, beyond the point marked "4," the steady states are unstable—the point "4" corresponds to an angle of attack of 0.8 rad, approximately 45 deg.

Figure 4.2 shows that $q = 0$ for all the trims, as it should be for steady-state solutions. Also, Mach number and pitch angle θ vary little between the trims marked "1" to "4"—consequently, the flight path angle is negative, indicating that the trims correspond to descending flight. Sometimes, this makes it difficult to draw reasonable and useful conclusions from SBA plots since not all the trim states are similar in some sense (level flight, for instance).

Homework Exercise: Can you think of a way to adjust the thrust such that each of these trim states becomes a level flight state? Can you anticipate whether that will cause any change in the stability of the trim states?

The transition to instability at trim point label "4" in Figure 4.2 is interesting. Referring back to the aerodynamic data in Figure 2.12, it can be seen that the derivative C_{mq} changes sign at precisely that value of angle of attack. Since C_{mq} predominantly affects the short-period damping, one can conclude that the instability at label "4" is the undamping of the short-period mode leading to oscillations in pitch called *pitch bucking*. A scan of the modal eigenvalues output by the SBA method confirms this deduction. Other instabilities such as the ones marked at label "3" and label "2" are marginal—either slow divergence or over a very brief stretch of angle of attack—and need not cause concern.

Thus, the SBA can be used to provide a snapshot of all likely trims and their stability over a range of varying parameter.

4.2.2 Longitudinal Trim and Stability with Varying Throttle Setting

Now, let us take the case of throttle as the varying parameter while the elevator deflection is held fixed. As always, any initial trim state may be selected—the following state is chosen for this computation:

$$[M, \alpha, q, \theta] = [0.27, 0.138 \text{ rad}, 0, 0.157 \text{ rad}], \quad [\eta, \delta e] = [0.43, -0.014 \text{ rad}]$$

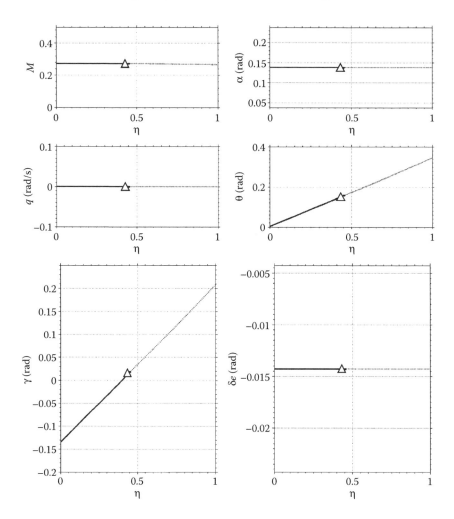

FIGURE 4.3 Longitudinal trim states for F-18 data with varying throttle setting (elevator deflection held fixed at $\delta e = -0.014$ rad) (full line: stable, dashed line: unstable).

Note that the initial steady state marked by the triangle in Figure 4.3 is not a level trim. The elevator deflection is held fixed and instead the throttle setting is varied over the stretch 0–1 in this case. The resulting trim states as computed by the SBA method are shown in Figure 4.3. As expected from the sample simulations of Chapter 2, increased throttle puts the airplane into climbing trims, whereas lowered throttle values yield descending trims. The airplane nose pitches up or down almost in sync with the climb/descent; changes in angle of attack are therefore very marginal as seen in the plots of Figure 4.3.

4.3 LEVEL FLIGHT TRIM AND STABILITY ANALYSIS

The SBA analysis in the previous section, unfortunately, does not reflect the airplane flight realistically. In practice, multiple inputs are used simultaneously to fly a certain maneuver and to steer the airplane, whereas the SBA analysis is restricted to changing only one control input at a time. While the SBA analysis is still useful for predicting the onset of instabilities, especially at moderate-to-high values of angle of attack, for flight dynamic analysis, it is better to study specific maneuvers such as level flight or climbing/descending flight. This can be carried out by the EBA method presented in Chapter 3.

Here, the EBA analysis is first carried out for level flight trims. For this, an additional constraint equation $\gamma = 0$ is included to the set of trim Equation 4.6 and the throttle setting η is appropriately calculated for every value of elevator deflection parameter such that $\gamma = 0$ is maintained. The resulting variation of the throttle setting η and the flight path angle (constrained to zero) are shown in Figure 4.4. The need to satisfy the level trim constraint dictates the throttle setting in Figure 4.4.

Computations with both increasing and decreasing values of the elevator deflection parameter are halted when the throttle setting reaches the maximum thrust limit of 1.0. The corresponding limits on the elevator deflection are found to be around −0.09 rad (up-elevator) and 0.01 rad

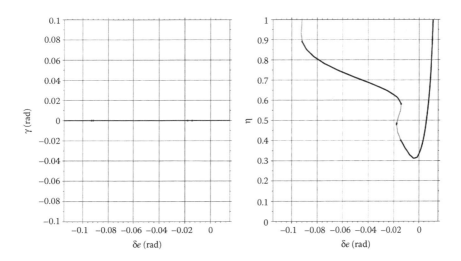

FIGURE 4.4   Throttle schedule for F-18 data with varying elevator deflection to satisfy level flight constraint.

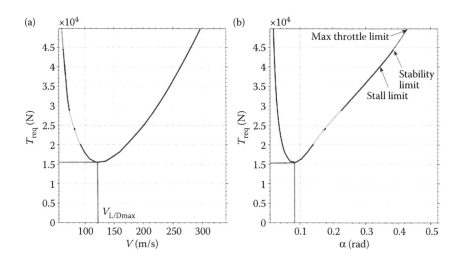

FIGURE 4.5   Trim thrust for F-18 data with varying elevator deflection to satisfy level flight constraint plotted as a function of (a) trim velocity and (b) trim angle of attack (full line: stable, dashed line: unstable).

(down-elevator). These correspond to an angle of attack of 0.42 rad (maximum) and 0.02 rad (minimum), respectively, as seen in Figure 4.5.

Since thrust equal to drag is being maintained at all values of elevator deflection in Figure 4.4, the variation of throttle reflects the nearly quadratic functionality of drag versus elevator deflection (equivalently, angle of attack). This is more explicitly seen in the plot of thrust versus angle of attack or thrust versus trim velocity in Figure 4.5, which will be useful for further analysis as well. For instance, the airplane's maximum velocity is constrained by the upper limit of throttle setting. From Figure 4.5, this reads to be a little under 300 m/s at this particular flight condition (this number is to be taken with a pinch of salt as Mach-number-dependent effects have likely not been included in this aerodynamic model).

The balance of forces in level flight requires that (T/W) be equal to $1/(L/D)$; that is, the minimum thrust level corresponds to the maximum (L/D) flight. The velocity at which the maximum (L/D) flight occurs may therefore be marked in Figure 4.5 at the point of minimum thrust $(\eta = 0.31)$. This value turns out to be around 120 m/s. Since maximum endurance flight implies minimum fuel consumption at every instant, and from Equation 4.10 that is seen to correspond to the lowest permissible throttle setting to maintain level flight, the best endurance for this airplane data and flight condition is obtained at the maximum (L/D)

condition at a velocity of 120 m/s. The EBA analysis can thus be used to analyze the aircraft performance.

The EBA analysis also provides information about the stability of each trim state. For instance, Figure 4.5 suggests that all trims on the "front side" of the power curve (as the T vs. V plot is often called) are indeed stable. (Of course, this analysis does not presently include stability to disturbances out of the longitudinal plane.) Interestingly, on the "back side" of the power curve, there are multiple stable trim branches. This is in contrast to the popular statement often seen in *aircraft performance* texts that trims on the "back side" of the power curve are unstable by considering a limited notion of stability in terms of thrust and drag variation with velocity alone. In fact, as marked in Figure 4.5, the lower velocity (higher α) limit of trimmed level flight can arise due to one of three reasons:

- The first reason could be the maximum throttle limit beyond which thrust can no longer balance drag and level flight cannot be maintained; the airplane then enters a descending trim. In the present case, this limit is ruled out as the EBA analysis in Figure 4.5 marks the low-velocity $\eta = 1$ trim around $\alpha = 0.42$ rad as unstable.

- The second reason could be a limit due to loss of stability, for instance, as seen in Figure 4.5 for $\alpha = 0.38$ rad. The nature of the instability depends on the aerodynamic details of the airplane and the flight condition.

- The third cause may be aerodynamic stall, which, if indicated by the peak of the lift coefficient versus angle of attack curve in Figure 2.12, occurs near $\alpha = 0.35$ rad. Often, one is led by textbooks to believe that after stall there is a sharp drop in lift coefficient and stable, level trimmed flight can no longer be maintained. Figures 2.12 and 4.5 show that this is not necessarily so; in this instance, the drop in lift coefficient post-stall is gradual and stable level flight is possible at stall and slightly beyond.

If the instability at $\alpha = 0.38$ rad is critical, then it marks the limit of low-velocity level trimmed flight. Otherwise, the limit is due to the throttle setting reaching $\eta = 1$, where the trim is technically unstable but the instability may be effectively tolerable.

The other flight variables from the constrained level flight EBA analysis are shown in Figure 4.6. With increasing negative (up-) elevator, the

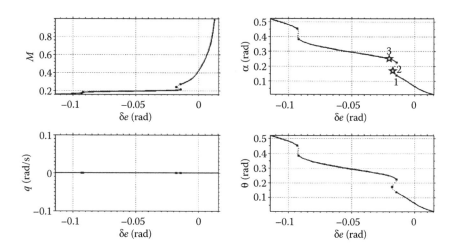

FIGURE 4.6 Flight variables in constrained level flight using the throttle schedule for F-18 data with varying elevator deflection in Figure 4.4 (full line: stable, dashed line: unstable).

airplane trims at higher angles of attack, as expected. There are three segments seen in the plots of Figure 4.6—a low-α segment with a steeper slope, and mid-α and high-α segments with a shallower slope. As can be found from Section 4.6.2 of Reference 1, the slope of the trim angle of attack versus trim elevator deflection curve is related to the airplane stability in pitch. In general, a steeper slope means a lower degree of stability and a larger sensitivity of trim α to elevator deflection.

Homework Exercise: Can you correlate the slope of the various segments on α—δe plot with the appropriate aerodynamic derivatives in Chapter 2?

The gap between the stable segments in Figure 4.6 needs to be analyzed. Referring back to the aerodynamic data in Figure 2.12, there is no obvious discontinuity in the plots of C_m versus α or C_m versus elevator deflection that may suggest the sudden onset of instability in Figure 4.6 in the angle of attack range of these gaps. Hence, the instability is probably not connected to pitch stability and the short-period dynamics, in which case it must be due to the phugoid mode. The EBA analysis may also be used to output the modal eigenvalues; hence, these can be examined to ascertain the cause of these unstable gap regions. Table 4.1 reports the longitudinal mode eigenvalues at three points (labels "1," "2," "3" in Figure 4.6) around one such unstable gap region. It is seen that the short-period mode is stable in all cases. The phugoid mode goes unstable at a Hopf bifurcation point

TABLE 4.1   Longitudinal Eigenvalues for Labels Marked in Figure 4.6

Label in Figure 4.6	Nature of Stability	Phugoid Mode Eigenvalue	Short-Period Mode Eigenvalue
1	Hopf bifurcation	$0.0 \pm j0.1260$	$-0.5586 \pm j0.6896$
2	Unstable trim	$0.0047 \pm j0.1195$	$-0.5136 \pm j0.4437$
3	Stable trim	$-0.0206 \pm j0.1908$	$-0.2570 \pm j0.9580$

(label "1"), where the eigenvalues lie on the imaginary axis. At the point labeled "2," the phugoid mode is marginally unstable, and at the point with label "3," it has regained stability. Thus, the small detours of the phugoid mode eigenvalues across the imaginary axis are responsible for the limited regions of instability in the gaps between the stable segments in Figure 4.6. Note that these unstable gap regions happen to be on the "back side" of the power curve in Figure 4.5 where an unstable phugoid mode is understandable.

Homework Exercise: Traditionally, the instability on the "back side" of the power curve is attributed to loss of "speed stability," that is, by considering the dynamics in V alone due to changes in thrust and drag. On the other hand, the phugoid mode predominantly involves variation in V and γ, the flight path angle, as well. Are these two—unstable phugoid and "speed instability"—related?

4.3.1 Level Flight Airplane Performance

As seen above, the EBA analysis can also serve as a tool to analyze the performance of the aircraft in addition to its trim and stability. Some of the performance parameters of interest in level flight are briefly discussed below with reference to the EBA results in Figure 4.5—a deeper analysis will be available in any text on *aircraft performance*.

1. *Maximum speed:* As seen previously, the maximum speed in level trim usually corresponds to the throttle limit of $\eta = 1$, subject to that trim state being stable.

2. *Minimum speed:* As discussed above with reference to Figure 4.5, the minimum speed is traditionally attributed to the stall limit; however, depending on the stalling behavior of the airplane and the resultant stability of the level flight trim states, it may be limited by the onset of phugoid instability (on the "back side" of the power curve) or by the maximum throttle limit.

3. *Speed for best endurance:* The best endurance or the longest duration of flight (for a given amount of fuel) is obtained when the fuel is consumed as slowly as possible. That is, when the SFC (Equation 4.10) is the least, which happens when the throttle setting is the minimum for level trim flight to be possible. The condition for *endurance* can be written formally as follows:

$$\dot{W}_f = \frac{dW_f}{dt} = -\text{sfc}T \Rightarrow dt = -\frac{dW_f}{\text{sfc}T} \Rightarrow \int_0^t dt = -\int_{W_f}^0 \frac{dW_f}{\text{sfc}T} \quad (4.11)$$

Thus, the best endurance is possible at a flight speed corresponding to minimum trim thrust, which happens to be the same as maximum (L/D) as marked in Figure 4.5.

4. *Speed for best range (or optimum cruise condition):* Clearly, for a commercial airliner, it is not flying at top speed alone or consuming fuel at the lowest rate (remaining aloft for longest time) that matter. Instead, a combination of high speed (high flight Mach number) and low fuel consumption (high aerodynamic efficiency, L/D) is desired. Therefore, it is common to optimize a parameter such as $V(L/D)$ (or $M(L/D)$) to get the ideal cruise condition. This can be seen more formally by writing an expression for the *range* below:

$$\frac{dx}{dW_f} = \frac{V}{-\text{sfc}T} \Rightarrow dx = -\frac{V}{\text{sfc}(D/L)} \cdot \frac{dW_f}{W} \Rightarrow x = \int_0^x dx$$

$$= \frac{1}{\text{sfc}} V\left(\frac{L}{D}\right) \int_{W-W_f}^W \frac{dW}{W}$$

$$(4.12)$$

where the factor $V(L/D)$ is evident. To examine this condition, we plot the variable T/V versus velocity and angle of attack in Figure 4.7. The minima in Figure 4.7 corresponds to flight at the condition $(VL/D)_{\text{max}}$ that gives best range (optimal cruise condition). In this instance, the best-range velocity and angle of attack may be read off Figure 4.7 to be approximately 150 m/s and 3 deg, respectively.

Note that the best-range velocity is a compromise between the maximum velocity and the speed for minimum fuel consumption.

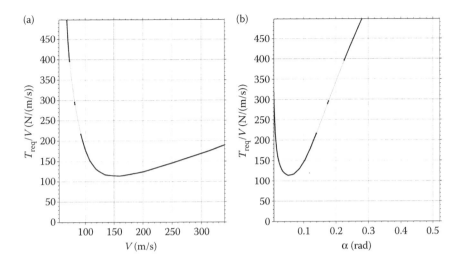

FIGURE 4.7   Trim thrust divided by velocity for F-18 data with varying elevator deflection to satisfy level flight constraint plotted as a function of (a) trim velocity and (b) trim angle of attack (full line: stable, dashed line: unstable).

For many airplanes, the typical value of best-range angle of attack is 4 ± 1 deg (as is the case in Figure 4.7 as well).

Homework Exercise: Assuming a parabolic drag polar, obtain analytical relations for the best-endurance speed and the best-range speed. Evaluate these speeds and match them against the values from the EBA analysis in the plots.

5. *Speed for minimum power:* Just as the case of speed for minimum thrust in Figure 4.5, one can also find the speed at which the airplane consumes the minimum power, where $P = TV$. However, this is more a figure of merit for propeller-powered airplanes and not so useful for jet-powered ones. Nevertheless, such a plot can be produced from the EBA analysis as shown in Figure 4.8. In this case, we plot the variable $T.V$ versus velocity and angle of attack. The velocity and angle of attack for minimum power flight may be read off Figure 4.8 to be approximately 100 m/s and 5.5 deg, respectively.

4.4 CLIMBING/DESCENDING FLIGHT TRIM AND STABILITY ANALYSIS

Two other figures of merit for airplanes are the steepest climb angle and the fastest climb rate. As is well known, any excess of thrust over drag may

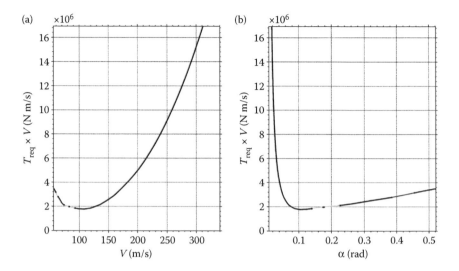

FIGURE 4.8   Trim power (thrust times velocity) for F-18 data with varying eleva-tor deflection to satisfy level flight constraint plotted as a function of (a) trim velocity and (b) trim angle of attack (full line: stable, dashed line: unstable).

be used to either accelerate the airplane (gain kinetic energy) or to increase its altitude (gain potential energy). Presently, we shall look at steady climb, that is, with no acceleration. Then the difference between thrust and drag is entirely used to gain altitude and clearly this gain in altitude will be maximum when the thrust is a maximum. Therefore, analysis of the best climb flights should be carried out with maximum throttle, that is, $\eta = 1$.

However, it still remains to determine the ideal velocity at which to carry out a climb maneuver so that either the climb angle is maximized (steepest climb) or the climb rate is maximized (fastest climb). To calcu-late this, an SBA analysis may be carried out with fixed throttle $\eta = 1$ and elevator deflection varied as the parameter. Results from such an analysis are shown in Figure 4.9. The trim angle of attack and the correspond-ing steady-state climb/descent angle are plotted along with the stability of each trim state. More positive elevator deflections create a nose-down moment that push the trim angle of attack to lower values, hence drag is lower and the thrust–drag difference puts the airplane into a climbing trim ($\gamma > 0$). On the other hand, the larger negative (up-) elevator deflec-tions push the airplane nose up into higher angle of attack trims where the higher drag overwhelms the thrust, which is limited by the throttle at $\eta = 1$. The airplane then steadily descends ($\gamma < 0$) as it would, for instance, during the landing approach. As in the case of level flight (Figure 4.6),

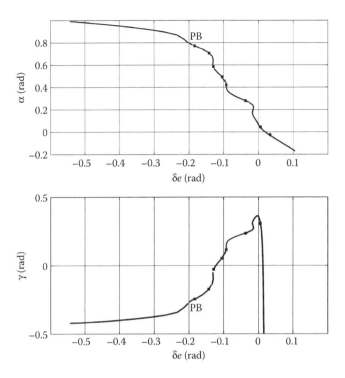

FIGURE 4.9 Analysis of climbing flights by SBA, throttle $\eta = 1$, and elevator deflection varied as the parameter (full line: stable, dashed line: unstable).

there are multiple stable trim branches with brief interludes of instability, largely due to the phugoid mode as seen previously. With increasing α, the final transition to instability at the point labeled "PB" is due to loss of short-period damping leading to pitch bucking—same as in Figure 4.2.

To deduce the climb figures of merit referred to earlier, it is preferable to plot the climb angle γ versus the velocity (or Mach number) for steepest climb and the climb rate $V \sin \gamma$ in case of the fastest climb. These are shown versus the Mach number in Figure 4.10.

From the upper panel of Figure 4.10, the steepest climb angle is approximately 20 deg and this occurs at a Mach number of around 0.35. These numbers are quite representative of aircraft of this class. The velocity at this Mach number is pretty close to the $V_{(L/D)\text{max}}$ value of 120 m/s in Figure 4.5, which is fairly obvious since that is the point of minimum drag and with thrust set to maximum at $\eta = 1$, the difference between thrust and drag is the most when drag is the least.

The lower panel of Figure 4.10 yields the best climb rate, which works out to nearly 60 m/s (3.6 km/min) when multiplied by the speed of sound.

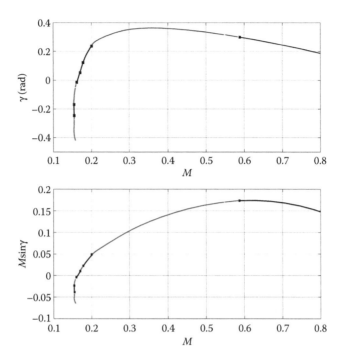

FIGURE 4.10   Climbing flight figures of merit by running SBA with throttle $\eta = 1$ and elevator deflection varied as the parameter plotted as a function of trim Mach number (full line: stable, dashed line: unstable).

The corresponding Mach number for the fastest climb is around 0.62, which corresponds to a velocity of 210 m/s. Note that this is the sustained (steady) climb rate. At this Mach number, the climb angle would be around 17 deg, marginally lower than the steepest climb angle of 20 deg. While the steepest climb angle may be of interest in clearing an obstacle after take-off, it is usually the fastest climb (maximum climb rate) that is more useful as a performance figure of merit.

In Figure 4.10, the SBA analysis reveals that the fastest climb trim is stable, whereas the steepest climb trim occurs at a point of phugoid instability. This is not an unusual occurrence and some airplanes are known to have a mildly unstable phugoid dynamics during climb following take-off. Pilots can usually handle this instability manually. (See Exercise 2 for an interesting case study.)

4.5 PULL-UP AND PUSH-DOWN MANEUVERS

Besides the steady straight-line flight paths that we have considered so far, airplanes also fly along curvilinear flight paths in the longitudinal plane.

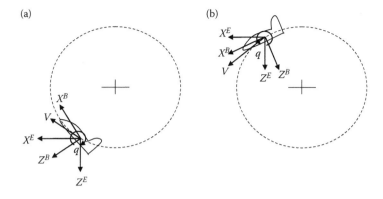

FIGURE 4.11 Sketch of (a) pull-up and (b) push-down maneuver in longitudinal plane (airplane is at the bottom/top of the curved flight path in each case).

These maneuvers are called pull-up or push-down and are pictured in Figure 4.11.

Unfortunately, these maneuvers are necessarily accelerated ones, that is, they do not satisfy all the equilibrium (trim) conditions in Equation 4.6. Hence, they cannot be directly analyzed per se by using the bifurcation and continuation method applied to the dynamic Equation 4.5 or the trim Equation 4.6. However, pull-up/push-down flight can still be analyzed using what is called a quasi-steady-state approximation.

Let us begin with the longitudinal dynamics equations in the form given by Equation 4.5 (reproduced below).

$$\frac{\dot{V}}{V} = \frac{g}{V}\left\{-\sin\gamma - \frac{\bar{q}S}{W}\left[C_{Dsta}(Ma,\alpha,\delta e) + \left(C_{D_{q1}} + C_{D\dot{\alpha}}\right)\frac{(\dot{\alpha})c}{2V}\right] + \frac{T}{W}\cos\alpha\right\}$$

$$\dot{\gamma} = \frac{g}{V}\left\{-\cos\gamma + \frac{\bar{q}S}{W}\left[C_{Lsta}(Ma,\alpha,\delta e) + \left(C_{L_{q1}} + C_{L\dot{\alpha}}\right)\frac{(\dot{\alpha})c}{2V}\right] + \frac{T}{W}\sin\alpha\right\}$$

$$\ddot{\theta} = \frac{\bar{q}Sc}{I_{yy}}\left[C_{msta}(Ma,\alpha,\delta e) + \left(C_{m_{q1}} + C_{m\dot{\alpha}}\right)\frac{(\dot{\alpha})c}{2V}\right]$$

(4.5)

Consider a pull-up maneuver (a push-down is very similar). Begin with the airplane in a steady trim flight state ($\gamma = 0$ and known values of δe, η, V and all the other flight variables) at the bottom of the curved flight path. Neglect changes in velocity; hence, the first of Equation 4.5 is to be ignored.

Now imagine an upward (negative) deflection of the elevator. The consequent nose-up moment will pitch the airplane up to a higher angle of attack, hence a larger lift (in general, pre-stall). The additional lift will produce a centripetal acceleration, which curves the flight path up as shown by the pull-up sketch in Figure 4.11a. From the second of Equation 4.5, still considering the flight state at the bottom of the curve where $\gamma = 0$, we have

$$\dot{\gamma} = \frac{g}{V}\{-1+n_z\} \qquad (4.13)$$

where,

$$n_z = \frac{\bar{q}S}{W}C_{Lsta}(Ma,\alpha,\delta e) + \frac{T}{W}\sin\alpha \qquad (4.14)$$

is called the load factor and the rate derivatives with respect to C_L in the second of Equation 4.5 have been dropped. Clearly, $n_z = 1$ corresponds to level flight with $\dot{\gamma} = 0$, as suggested by Equation 4.13. With reference to Figure 4.11, $\dot{\gamma}$ is the angular velocity of the airplane velocity vector in the pull-up maneuver. Note that $\dot{\gamma}$ is not necessarily equal to $\dot{\theta}$—the angular velocity of the airplane's body-fixed axis—at all times; the difference between the two depends on the rate of change of angle of attack. However, at a quasi-steady state, $\dot{\alpha} = 0$ and then $\dot{\theta}$ and $\dot{\gamma}$ are the same, that is, the body-axis and the velocity vector follow the curved flight path in sync.

The stability of the quasi-steady state is related to perturbations in the angle of attack and must be deduced from the last of Equation 4.5. For a fixed value of velocity and throttle setting, for a point at the bottom of the curved flight path ($\gamma = 0$), the quasi-steady state and its stability can be obtained by solving

$$\ddot{\theta} = \frac{\bar{q}Sc}{I_{yy}}\left\{ C_{mstab}(Ma,\alpha,\delta e) + (C_{mq1}+C_{m\dot{\alpha}})\frac{\dot{\alpha}c}{2V} \right\} \qquad (4.15)$$

Effectively, the perturbations in angle of attack represent short-period dynamics about a curved flight path (in contrast to the straight-line flight path considered previously). Each value of the elevator deflection parameter decides the load factor as given by Equation 4.14, which in turn sets the angular velocity of the curved flight path in Equation 4.13. These different quasi-steady states, distinguished by their load factor and angular velocity, can have different stability properties since the aerodynamic coefficient/derivatives in Equation 4.15 are themselves functions of the trim angle of attack and elevator deflection.

The quantum of elevator deflection required to change from one quasi-steady state to another (or from a level flight steady state to a curved flight quasi-steady state) can be estimated from Equations 4.14 and 4.15. In terms of perturbations in elevator deflection, angle of attack, and load factor between two quasi-steady states,

$$(\Delta n_z - 1)\frac{W}{\bar{q}S} = C_{L\alpha}\Delta\alpha + C_{L\delta e}\Delta\delta e \tag{4.16}$$

where the thrust term has been ignored in addition to the $(C_{Lq1} + C_{L\dot{\alpha}})$ term that had been dropped previously. Similarly,

$$\Delta\ddot{\theta} = \frac{\bar{q}Sc}{I_{yy}}\left\{(C_{m\alpha}\Delta\alpha + C_{m\delta e}\Delta\delta e) + (C_{mq1} + C_{m\dot{\alpha}})\frac{\Delta\dot{\alpha}c}{2V}\right\} \tag{4.17}$$

However, since $\Delta\ddot{\theta}$ and $\Delta\dot{\alpha}$ are individually zero at each (quasi-) steady state, Equations 4.16 and 4.17 form a pair of equations in terms of $\Delta\alpha$, $\Delta\delta e$ and $(\Delta n_z - 1)$. Eliminating $\Delta\alpha$ yields a result for the elevator deflection required for a certain change in load factor, popularly called "elevator deflection per g" as below:

$$\frac{\Delta\delta e}{(\Delta n_z - 1)} = \frac{C_{m\alpha}(W/\bar{q}S)}{C_{m\alpha}C_{L\delta e} - C_{L\alpha}C_{m\delta e}} \tag{4.18}$$

Equation 4.18 offers an alternative interpretation of stability—onset of instability corresponds to the point where a vanishingly small elevator deflection can induce a unit change in load factor ("g"). This, according to Equation 4.18, matches the condition $C_{m\alpha} = 0$, which leads to the standard definition of *neutral point* and *static margin* (see Reference 1). This would suggest that there is no fundamental difference in pitch stability between a straight-line flight path and a curved one, a conclusion that is reaffirmed

---

BOX 4.1 STABILITY IN PULL-UP MANEUVERS—THE TRADITIONAL VIEWPOINT

Traditionally, the pitch damping term in Equation 4.15 has been modeled as $C_{mq1}(q_1c/2V)$, where q_1 is the body-axis angular velocity. In that case, expressing q_1 in terms of $\dot{\alpha}$ and $\dot{\gamma}$, there appears an additional term involving C_{mq1} and $\dot{\gamma}$ in the equation for the short-period dynamics about a curved flight path. The effect of this term on the short-period dynamics is to provide an additional source of stability in case of a curved flight path over and above the short-period stability in level flight. In terms of "elevator angle per g," Equation. 4.18 in the traditional version would appear as

$$\frac{\Delta\delta e}{(\Delta n_z - 1)} = \frac{(C_{m\alpha}(W/\bar{q}S) + C_{mq1}(c/2V)C_{L\alpha}(g/V))}{(C_{m\alpha}C_{L\delta e} - C_{L\alpha}C_{m\delta e})}$$

with the additional term $C_{mq1}(c/2V)C_{L\alpha}(g/v)$ in the numerator. As a result, it has been traditionally believed that an airplane in a curved flight path with load factor different from 1 somehow acquires additional stability in pitch, leading to the definition of *maneuver point* and *maneuver margin* for curved flight paths, which replace the *neutral point* and *static margin* in case of straight-line 1–g trim flight.

---

by the numerical example that follows shortly. For a traditional view of "elevator angle per g" and the concepts of the so-called *maneuver point* and the *maneuver margin*, see Box 4.1.

The standard phugoid dynamics is no longer possible in this quasi-steady-state analysis since the velocity is being held fixed. However, note that perturbations in the angle of attack will also affect the lift coefficient in Equation 4.15, thus perturbing the load factor n_z and thereby the curvature of the flight path (Equation 4.13). As long as the α dynamics in Equation 4.15 is stable, the perturbations in Equation 4.13 should also die down and the quasi-steady state with its angular velocity $\dot{\gamma}$ is regained.

An initial level flight trim state at $Ma = 0.4$, $\eta = 0.38$ is used as the starting point for a numerical study of the pull-up maneuver. Velocity and throttle setting are held unchanged at these values. The second and third of Equation 4.5 are solved in the form below with elevator deflection varying as the continuation parameter. The point at the bottom of the vertical loop is considered.

$$0 = -\dot{\gamma}_{ss} + \frac{g}{V}\left\{-1 + \frac{\bar{q}S}{W}\left(C_{L\alpha}\alpha + C_{L\delta e}\delta e + C_{L\dot{\alpha}}\frac{\dot{\alpha}c}{2V}\right)\right\}$$

$$\ddot{\theta} = 0 = \frac{\bar{q}Sc}{I_{yy}}\left\{C_{m\alpha}\alpha + C_{m\delta e}\delta e + C_{m\dot{\alpha}}\frac{\dot{\alpha}c}{2V}\right\} \qquad (4.19)$$

$$\dot{\theta} = 0 = q$$

The quasi-steady rate of change of flight path angle $\dot{\gamma}_{ss}$ is obtained from the first of Equation 4.19. The results of the continuation run are plotted in Figure 4.12.

From first of the equation in Equation 4.19, it is apparent that $\dot{\gamma}_{ss}$ is directly proportional to the lift-coefficient C_L at constant speed; hence, the $\dot{\gamma}_{ss}$ profile in Figure 4.12 closely mirrors the variation of C_L with angle of attack in Figure 2.1. The maximum value of $\dot{\gamma}_{ss}$, approximately 14.5 deg/s, therefore occurs at C_{Lmax} (refer Figure 2.1) or α_{stall}. The second

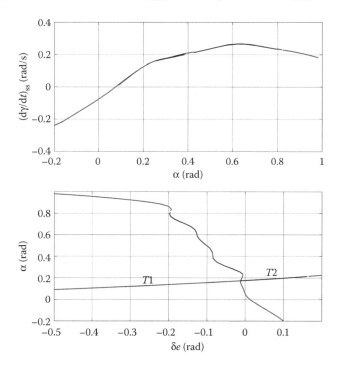

FIGURE 4.12 Plots of quasi-steady rate of change of flight path angle $\dot{\gamma}_{ss}$ (in rad/s) as a function of angle of attack (in rad) and angle of attack (in rad) as a function of elevator deflection (in rad) for fixed Mach number $Ma = 0.4$, and throttle, $\eta = 0.38$.

of Equation 4.19 at fixed trim speed results in $\alpha - \delta e$ variation by solving $C_m = 0$. The quasi-steady variation of α against δe is shown in Figure 4.12 as well. For both steady, level flight and quasi-steady pull-up maneuvers, a certain value of elevator setting corresponds to a particular angle of attack as given by the second of Equation 4.19. However, in case of steady, level flight, for each value of elevator setting, the velocity is varied such that the level flight condition is maintained, whereas in case of the pull-up maneuver, since the velocity is held fixed at the initial trim value, depending on whether the lift at that value of elevator setting is more or less than the weight, the airplane will either pull up or push down—the rate of pull-up or push-down is found from the first of Equation 4.19.

Of the equations in Equation 4.19, the first is a faux dynamic equation, arranged in this manner to calculate the quasi-steady pull-up/push-down rate. Hence, the single eigenvalue representing this equation must be ignored as it has no physical relevance. The transverse branch of trim angle of attack values in Figure 4.12 (T1, T2) is a consequence of a faux bifurcation of this first equation of Equation 4.19 and must be ignored as well. The stability of the pull-up or push-down quasi-steady state is obtained by the eigenvalues representing the other two equations in Equation 4.19. The stability characteristics for the quasi-steady pull-up maneuver in Figure 4.12 are very similar to those of the steady, level flight and the climbing flight trims in Figures 4.2 and 4.9.

4.6 WIND EFFECTS ON LONGITUDINAL DYNAMIC MODES

In Figure 4.13, an airplane in straight-line level flight trim condition encounters a wind-shear. This linear wind-shear is defined as $\partial W_x / \partial h$.

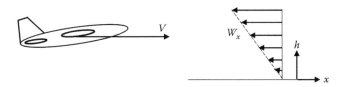

FIGURE 4.13 Airplane in straight and level flight state with a linear wind gradient ahead.

In order to analyze airplane response to a wind gradient in longitudinal plane, Equation 4.5 can be used with an additional equation to model the variation in altitude as follows:

$$\dot{h} = V \sin \gamma \qquad (4.20)$$

The effect of the wind shear is introduced by writing the dynamic pressure $\bar{q}$ as:

$$\bar{q} = \frac{1}{2} \rho \left(V + \frac{\partial W_x}{\partial h} h \right)^2 \qquad (4.21)$$

Assuming the airplane to be flying a level flight trim condition $V^* = \text{const}$, $\alpha^* = \text{const}$, $q^* = 0$, $\theta^* = \alpha^*$ at constant altitude $h^* = \text{const}$, an analysis is carried out to study the effect of a linear wind gradient on longitudinal modes of airplane. The magnitude of the wind gradient term $\partial W_x / \partial h$ is varied in the typical range $0-0.2/\text{s}$ and analysis is carried out for F-18/HARV model. The eigenvalue plots in Figure 4.14 show significant variations both in the short-period and phugoid mode behavior of airplane. While the short-period mode remains stable with some loss of damping as the magnitude of wind gradient increases, the phugoid mode actually becomes unstable in this range of wind gradient. Additional eigenvalue on the real axis corresponds to the exponential altitude variation.

EXERCISE PROBLEMS

1. Study the literal approximations to the short-period and phugoid modes from any elementary textbook on flight mechanics (e.g., Reference 1). Make a list of geometric/inertial (e.g., wing taper ratio, flap deflection) and flight parameters (air density, etc.) that influence the stability (frequency and damping) of the longitudinal modes.

2. Turboprop airplanes such as the C-130 Hercules are known to have a mildly unstable phugoid oscillation under low-speed flight conditions. While pilots can usually handle an airplane with a slightly unstable phugoid mode, the airplane's flight in case the pilots are incapacitated is a matter of concern. Some accidents have been

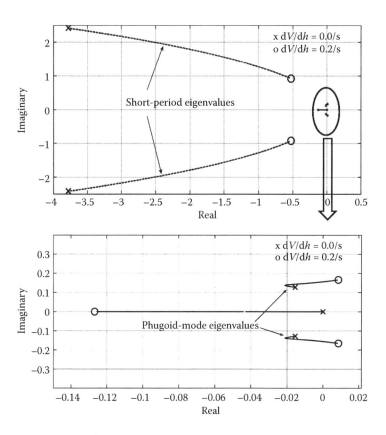

FIGURE 4.14 Variation of longitudinal mode eigenvalues, including wind gradient effects.

attributed to such an instance.[*] Attempt an analysis such as that in Figure 4.5 for a turboprop airplane like the C-130 and record your observations. Are your results and conclusions any different from that for a turbofan airplane in Figure 4.5?

3. Show explicitly that the best-range speed may be obtained from a plot of T/V versus the trim velocity as in Figure 4.7.

4. Why is the minimum thrust speed of interest to jet-powered airplanes and the minimum power speed in case of propeller-powered ones? If the minimum thrust speed corresponds to $(L/D)_{max}$, then what does the minimum power speed correspond to and what is its physical significance?

_____

[*] www.edwardjayepstein.com/archived/zia_print.htm.

5. The climb performance metrics in Figure 4.10 do not relate to efficiency. Is there any way of incorporating an additional requirement for maximizing aerodynamic efficiency or minimizing fuel consumption along with the climb performance criterion to derive an "optimal" climb trim (such as was done for optimal cruise by maximizing a combination of V and (L/D)? If so, then carry out the analysis and produce a plot like that in Figure 4.10 and make your deductions.

6. Oftentimes, for commercial airplanes, the throttle setting permitted at take-off and climb is limited due to noise level restrictions at the airport. Find a realistic value of the throttle setting used during the early climb phase and estimate the loss in climb rate due to this restriction. You can use the same data that was employed in producing Figure 4.10 (even though it is not for a commercial airliner) or use another data set of your own.

7. Derive the equation for the traditional version of "elevator angle per g" as written in Box 4.1. Try to interpret the additional term $C_{mq1}(c/2V)\,C_{L\alpha}(g/V)$ physically.

8. Carry out numerical simulation (integration) of Equation 4.5 with Equation 4.20 to study the effects of wind on the modal dynamics of airplanes. (*Hint:* Modal dynamics can be better captured in numerical simulation of the linearized equations around the equilibrium state. Refer to Chapter 3 for the linearization technique to develop one on your own for the longitudinal dynamics model in Equation 4.5 with Equation 4.20).

REFERENCE

1. Sinha, N. K., and Ananthkrishnan, N., *Elementary Flight Dynamics with an Introduction to Bifurcation and Continuation Methods*, CRC Press, Boca Raton, Florida, 2013.

CHAPTER 5

Longitudinal Feedback Control

T HE PREVIOUS CHAPTER EXPLORED airplane trim and stability in longitudinal flight maneuvers—level flight, climbing/descending flight, and pull-up/push-down maneuvers. Under different conditions, either of the modes, short-period or phugoid, was noticed to become unstable. One option is to curtail the flight envelop to avoid regions of instability. Otherwise, it can be left to the pilot to manually handle the airplane despite the instability. But, most often, a flight control system (FCS) is used as an interface between the pilot and the airframe. The FCS augments the airframe's inherent stability and helps reduce pilot workload in many ways. For instance, limiters can be automatically enforced and a more uniform response of the airframe to pilot inputs under varied airframe and flight conditions can be presented. In this chapter, we present a broad framework for a generic FCS with emphasis on flight in the longitudinal plane. We then analyze the airplane's closed-loop dynamics using the tools described in Chapter 3 and used in Chapter 4 to study the open-loop longitudinal flight dynamics.

5.1 GENERIC FLIGHT CONTROL SYSTEM

The key blocks in a generic FCS are presented in Figure 5.1. The pilot command in terms of stick, pedal, or throttle lever is first interpreted by the FCS through the block labeled "feedforward control law" to meet a

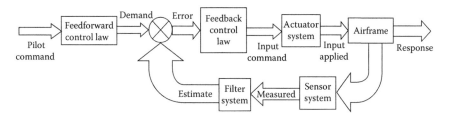

FIGURE 5.1   Layout of a generic FCS.

certain demand on a particular physical quantity such as pitch rate or load factor, for instance. Limiters can also be implemented here and the control can be "shaped" as in the proportionality factor between the pilot command and the physical demand being adjusted depending on some criteria.

The demanded variable is then compared against its presently esti-mated value obtained from the "filter system," which is a part of the feed-back loop. The error between the demand and the estimate forms the input to the "feedback control law" whose task is to broadly nullify this error. If successful, the eventual estimate of the variable is the same as its demanded value and the airplane is flying exactly as commanded by the pilot and interpreted by the feedforward control law.

The feedback control law computes the control input required and passes a command to the "actuator system." The actuator system then, subject to its limitations, applies the control input to the "airframe." The response of the airframe to control inputs has already been explored in the previous chapters. The airframe's response is picked up and measured by the "sensor system" but these signals may be noisy and contain errors due to various sources; also not every physical variable required for feedback can usually be measured. The filter system is supposed to filter out the noise and errors from the measured signals and provide estimates of the unmeasured variables.

5.2 AIRFRAME, SENSOR, FILTER, ACTUATOR

In case of longitudinal flight, the response of the airframe is usually given in terms of its velocity, flight path angle, angle of attack (AOA), pitch angle, pitch rate, etc. For a rigid-body airframe, the model equations for flight in the longitudinal plane are available from Chapter 4:

$$\frac{\dot{V}}{V} = \frac{g}{V}\left\{-\sin\gamma - \frac{\overline{q}S}{W}\left[C_{Dsta}(Ma,\alpha,\delta e) + \left(C_{D_{q1}} + C_{D\dot{\alpha}}\right)\frac{(\dot{\alpha})c}{2V}\right] + \frac{T}{W}\cos\alpha\right\}$$

$$\dot{\gamma} = \frac{g}{V}\left\{-\cos\gamma + \frac{\overline{q}S}{W}\left[C_{Lsta}(Ma,\alpha,\delta e) + \left(C_{L_{q1}} + C_{L\dot{\alpha}}\right)\frac{(\dot{\alpha})c}{2V}\right] + \frac{T}{W}\sin\alpha\right\}$$

$$\ddot{\theta} = \frac{\overline{q}Sc}{I_{yy}}\left[C_{msta}(Ma,\alpha,\delta e) + \left(C_{m_{q1}} + C_{m\dot{\alpha}}\right)\frac{(\dot{\alpha})c}{2V}\right]$$

(4.5)

They are augmented by the navigation equations in longitudinal dynamics:

$$\dot{x}^E = V\cos\gamma$$
$$\dot{z}^E = -V\sin\gamma$$

(5.1)

Given a set of control inputs (here, thrust and elevator deflection), the solution of the model equation yields the model response in terms of the flight variables. However, as pictured in Figure 5.2, there may be errors and uncertainties in the airframe model due, for instance, to atmospheric conditions or aerodynamic modeling issues. Additionally, there may be physical effects not modeled in Equation 4.5 such as wing flexure or fuel slosh that may also respond to the control inputs. The real response of the

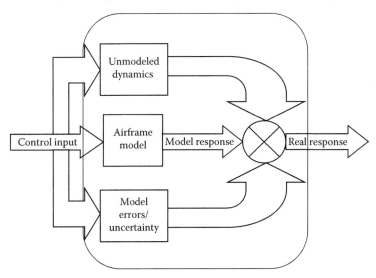

FIGURE 5.2 Typical airframe model and sources of error.

airframe therefore may well be somewhat different from that of the airframe model. Also note that the "airframe model" itself may actually be a stack of different models for different CG or fuel state or store condition, for landing gear or flaps deployed or not, etc.

The large block in Figure 5.2 covering all the smaller blocks represents the "real" airplane. In practice, this is the "hardware" out there that exhibits the real response to any control input. Then why do we need the models, one may wonder—that is because we need a model of this hardware to design the FCS. Unfortunately, the FCS cannot be designed with the hardware itself at hand. So, the airframe must be represented by a model with the attendant issues raised above—different models for different states (CG, fuel, landing gear, etc.), errors, uncertainties, and unmodeled dynamics. The FCS must be designed such that it meets a certain standard for all the models in the airframe model stack. And when implemented on the hardware (the real airplane), it must allow the airplane to be flown and to perform satisfactorily notwithstanding the presence of errors and uncertainties in the model and the unmodeled dynamics.

More about this can possibly be found in a book on *flight control systems*. For the present textbook, we focus on the fundamental "flight dynamic" issues of designing a feedback control law for a given airplane model such as the one in Equation 4.5. However, it may be worthwhile to underline that in practice several such control laws would be designed to cover all possible flight states and these laws would be stitched together by some form of "scheduling." Further, these laws would be tested with generic or specific models of errors, uncertainty, and unmodeled dynamics before being cleared to be loaded on the airframe for flight testing.

Likewise, a discussion of which sensors may be placed where on the airframe to measure which variables, and the kind of errors that may be expected in the measurement, would be too lengthy to find a place in this book. For our purposes, all variables required for feedback may be assumed to be available, either measured or estimated, with no noise/error. In practice, however, sensor models must be incorporated during the FCS design/evaluation process. And since the sensor measurements will inevitably be noisy and erroneous, a "noise/error model" should form part of the sensor system. Downstream, a filter system is needed to correct the noisy/erroneous measurement. The filter will itself use a model of the airframe and thereby try to estimate variables that are not directly measured by the sensors. So, that is another block where the airframe model is needed, warts and all (errors/uncertainty, unmodeled dynamics).

Actuators used to deflect the control surfaces are electromechanical systems, whereas in the case of thrust, the "actuator" is the engine itself. The input actually applied by the actuator system may be different from that commanded by the feedback control law due to time lags, saturation, and rate limits. At the very least, a low-order actuator model incorporating these effects is needed during preliminary design of the FCS. At a later stage, a more detailed "actuator model" may be used and eventually the FCS may be tested with hardware-in-the-loop, that is, with the actuator in place. Actuators may come with a control system of their own, such as an engine control system (ECS) for the thrust. In fact, the "engine model" with its ECS may even be more complex and involved than the airframe model with its FCS. By and large, the engine controller and the FCS design take place independently though there is interest in integrated flight-engine controller design on and off. The engine and flight control systems get tested together when the closed-loop airframe model is mated with the closed-loop engine model for simulations. Further discussion on modeling and controller design for the engine, or any other actuator system, is beyond the scope of this textbook. For our purposes, simple lag models with saturation and rate limits will capture the essential closed-loop dynamics of the actuator systems.

5.3 GENERIC LONGITUDINAL FCS STRUCTURE

At first glance, the longitudinal Equations 4.5 and 5.1 reproduced earlier in this chapter appear to feature about half a dozen variables and only two control inputs; so it may not be clear how the control problem may be solved.

In fact, the airframe model equations may be dealt with sequentially as shown in Figure 5.3—an approach that is sometimes referred to as "backstepping." In this approach, the entire control problem is divided into three sequential control subproblems, represented by the blocks in Figure 5.3 loosely labeled "flight path," "attitude," and "rate."

Let us say that an input from the pilot can be interpreted as a command to either accelerate or climb or both or neither (maintain steady, level flight). Each of these commands can be expressed in terms of a certain velocity and flight path angle profile. For example,

- Accelerate in level flight: V profile, $\gamma = 0$
- Steady climb: fixed V, fixed γ

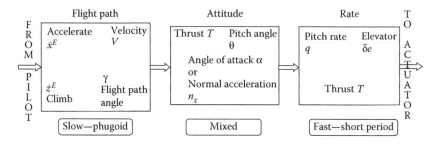

FIGURE 5.3   Generic structure of longitudinal FCS—sequential, "back-stepping" type.

- Accelerate and climb: such as, pull-up with fixed V, varying γ

- Steady, level flight: fixed V, $\gamma = 0$

That is, if the appropriate variation of V and γ is possible, then the maneuver as commanded by the pilot has been achieved. Thus, the flight path control problem can be solved by using V and γ as intermediate variables or "surrogate control."

The output of the flight path block—the desired flight path profile—is passed on to the next block in the sequence, the attitude block. The attitude block computes the variation of AOA/normal acceleration and thrust required to meet the V, γ profile demanded by the flight path block. To take the same example cases as previously:

- Accelerate in level flight: V profile, $\gamma = 0$—thrust required to accelerate; change in AOA to compensate for the increase in V so that lift is unchanged and level flight is held

- Steady climb: fixed V, fixed γ—thrust required to hold steady climb angle; AOA for lift to balance out weight component at fixed V

- Accelerate and climb: such as, pull-up with fixed V, varying γ—normal acceleration for the pull-up; thrust required to hold velocity fixed

- Steady, level flight: fixed V, $\gamma = 0$—AOA for lift to balance weight at fixed V; thrust to balance drag

In general, the attitude block determines the thrust and AOA/normal acceleration variation necessary to fly a certain flight path profile, and hence the desired pitch angle. Thus, AOA/normal acceleration act as "surrogate variables" at this point of the sequence.

Variations on this theme are possible—for instance, in case of a pull-up maneuver, it may be easier to directly demand a normal acceleration than have it calculated through a γ profile, but that is more a matter of detail. The basic principle of sequencing or back-stepping is as captured in Figure 5.3.

Next in sequence is the rate block. Given a desired change in pitch angle, the body-axis pitch rate required should be met by an appropriate deflection of the elevator. And the thrust required should be provided by a suitable setting of the engine parameters. The pitch rate is the surrogate variable in this block.

Figure 5.4 is a graphic that further explains the definition and use of surrogate variables in the FCS structure of Figure 5.3. At "Level 1" is the actual command, which is realized by the variables (V, γ) in surrogate control–1. At "Level 2," these very variables now form the surrogate command–2; they in turn are realized by surrogate control–2 consisting of the variables $(\alpha/n_z, T)$. Of these, (α/n_z) is the surrogate command–3 at "Level 3" where the surrogate control–3 being solved for is (q). Finally, (T)

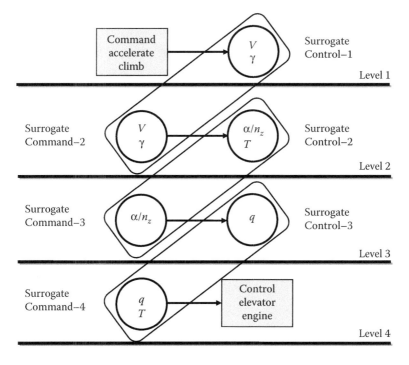

FIGURE 5.4 Different levels of the sequential, back-stepping-type longitudinal FCS structure with the surrogate variables marked.

from surrogate control–2 and (q) from surrogate control–3 come together at "Level 4" to form the surrogate command–4 and at this level, the actual control variables are solved for. Thus, even though the number of variables in the problem and the number of independent controls do not match, by this structure, at every level, the commanded and controlled variables are commensurate.

Another manner of distinguishing between the various blocks in Figure 5.3 is in terms of the natural time scales of the variables in that block. For instance, in the flight path block, the variables V, γ are known to vary at the slower, phugoid time scale, whereas in the rate block, the pitch rate variable q varies at the faster, short-period time scale. The pitch angle in the attitude block varies due to both short-period and phugoid dynamics; it can therefore be classified as "mixed." Note, however, that this distinction is appropriate only in case the commanded maneuver is not an accelerated one; otherwise, the time scale/frequency of the input (forcing function) will pervade through all the blocks.

As an example, consider a pilot command requiring the airplane to accelerate while maintaining level flight. This requires the flight path angle γ to be held at zero, or in other words, zero normal acceleration. The axial acceleration profile determines the required increase in velocity and hence the desired increase in thrust. The increase in velocity requires a simultaneous decrease in AOA to hold the flight path angle unchanged (maintain zero normal acceleration), which can be achieved by pitching the airplane nose down. That means a negative pitch rate, which requires an up-elevator deflection. Simultaneously, the increase in thrust requires the throttle to be opened. The precise control inputs at each instant are to be calculated by the control law but the time scale depends on the axial acceleration profile subject to actuator rate limits.

5.4 LONGITUDINAL FLIGHT CONTROL MODES

A longitudinal FCS may have several modes any of which may be selected at a given point of time in the flight. More generally, how the pilot command is to be interpreted by the feedforward control law block in Figure 5.1 is a matter of the FCS design. Some examples are given in Figure 5.5.

The "altitude hold" mode, as the name implies, requires the airplane to maintain a steady altitude. This is standard for commercial airliners, which are assigned fixed altitudes (FL, flight level) for cruise. Imagine an airplane in such a flight encountering a headwind. The relative flight

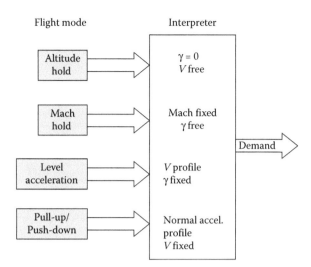

FIGURE 5.5 Examples of longitudinal FCS flight modes with the interpreted demands to the feedback control law.

velocity and the aerodynamic Mach number will therefore increase, resulting in the airplane pulling up and gaining altitude if unchecked. The altitude hold mode demands that γ is maintained at zero; the flight control law will therefore adjust the AOA by inducing a pitch rate using elevator deflection. The drag will also change—increase due to increase in relative velocity/aerodynamic Mach number but decrease due to the adjustment in the AOA. The velocity is left free to settle down to a level depending on the relative values of thrust and drag.

A second example in Figure 5.5 is the "Mach hold" mode. As fuel is consumed during cruise, the airplane becomes lighter and tends to accelerate. If the Mach hold mode is engaged, the elevator is used to increase the AOA so as to bring down the trim velocity. In the process, the airplane is usually put into a gradual climb, usually called a cruise-climb flight.

The level acceleration or "throttle slam" was discussed at the end of the previous section. Similarly, for a pull-up maneuver, the "stick" is pulled back, resulting in a rapid up-elevator deflection. The nose is pulled up; the pitching motion increases the AOA, which raises the lift and produces a normal acceleration. The throttle is adjusted to maintain the velocity. These flight modes are more common to combat airplanes.

Effectively, as depicted in Figure 5.5, a desired velocity and flight path profile is input as the demand to the flight control law depending on the selected flight mode.

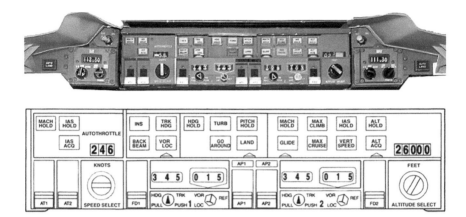

FIGURE 5.6 Layout of the flight mode buttons in the Concorde cockpit. (From www.concordesst.com/autopilot.html.)

A snapshot of the Concorde cockpit with the various flight mode buttons is shown in Figure 5.6. A brief description of the auto-throttle and longitudinal (vertical) auto-pilot modes as implemented in the Concorde FCS is presented in Box 5.1.

5.5 LONGITUDINAL FEEDBACK CONTROL LAW

Following the examples in Figure 5.5, we consider the input to the feedback control law to be a demanded profile of velocity V and flight path angle γ. The control inputs to the actuators/engine/airframe are the elevator deflection and the thrust (throttle setting). Between the (V, γ) demand and the control inputs is the feedback control law as depicted in Figure 5.1. The feedback control law is structured in the form of three back-stepping loops along the lines of Figure 5.3. A block diagram of such a three-loop back-stepping longitudinal feedback control law appears in Figure 5.7.

Estimated values of the states are assumed to be available from the sensor/filter. The outermost of the three loops involves the flight path variables, (V, γ). The error between the demanded and estimated values of (V, γ) is input to the "outer (flight path) loop" controller whose job it is to calculate the AOA and thrust required to satisfy the $(V, \gamma)_{\text{demand}}$. The error between the AOA so demanded and its estimate forms the input to the "middle (attitude) loop" controller. To achieve the desired AOA, it must demand a pitch rate. Then, the "inner (rate) loop" uses the error between

BOX 5.1   CONCORDE AUTO-THROTTLE AND AUTO-PILOT LONGITUDINAL (VERTICAL) FLIGHT MODE FUNCTIONALITIES

Auto-throttle: There are three main auto-throttle modes:

- *MACH HOLD*: This will ensure the engines are automatically throttled to maintain the current Mach number being followed when this mode was engaged. This is used during cruise as it will maintain the correct Mach number no matter what the outside conditions are that effect the Mach number, such as temperature.
- *IAS HOLD*: This the mode that the throttle will hold the indicated air speed (IAS) that the aircraft was flying at when they were engaged.
- *IAS ACQ*: The third mode will allow the aircraft to fly or acquire to a set air speed, for example, on approach, the aircraft could be flying at 250 knots, with the next speed being 210 knots. This can be keyed in and then, when requested to 210 knots by ATC, the model selected. When 210 knots is achieved, the mode will revert to IAS HOLD.

Auto-pilot vertical modes:

- *PITCH HOLD*: This is the basic mode of the auto-pilot and will hold the existing aircraft pitch when engaged. It comes on as default when the auto-pilot is engaged.
- *MACH HOLD*: This function will hold the current Mach number by pitch changes and not throttles changes. If the auto-throttles are engaged, they will take precedence and the auto-pilot will default to PITCH HOLD.
- *MAX CLIMB*: This is selected at or near V_{mo} (maximum operating speed) and will hold the air speed to a figure around V_{mo}. As the speeds approached V_{mo} at the top of the climb it will disengage and hold the speed with pitch changes.
- *MAX CRUISE*: This engages shortly after Mach2 and is an extension of MAX CLIMB. It is normally used in conjunction with the auto-throttles primed in MACH HOLD to keep the aircraft flying at Mach2.0. If the aircraft begins to overspeed, due to temperature changes, the auto-throttles will slow the aircraft down. Once back at the correct speed, MACH HOLD will disengage and MAX CRUISE will re-engage. It also prevents the aircraft exceeding the maximum operating temperature (T_{mo}) of 127°C on the tip of the nose.
- *IAS HOLD*: Holds the current-indicated air speed by means of pitch changes.

- *ALT HOLD*: Holds the aircraft's existing altitude.
- *VERT SPEED*: Sets up the aircraft to hold a vertical speed as set up on the vertical rate of climb indicator.
- *ALT ACQ*: Similar to IAS ACQ on the auto-throttles. A preset altitude to fly to can be programmed in on the selector, and when the ALT ACQ button is pressed, the aircraft will fly to that altitude. The prime light will light during the operation and the button will itself light when the speed is acquired.
- *TURB*: Turbulence mode, only used in moderate or severe turbulence. It holds the existing pitch attitude and heading; it reduces the trim rate of the electric trim system to smoothen the ride.
- *LAND*: Automatic landing mode. When this is pressed, the prime triangle will light. It causes the aircraft to capture the glideslope and track to localizer that has been selected. When the glideslope has been captured, the button will light and the small triangle prime light under the button will go out. The VOR/LOC button will light when the localizer has been captured. During the capture process, the prime light on VOR LOC will light. After the LAND mode is selected, the second auto-pilot can be engaged for redundancy.
- *GO AROUND*: Indicates an automatic go around has been initiated. This is carried out when more than two of the throttle levers are moved fully forward in LAND or GLIDE modes. It will pitch the aircraft up at 15 degrees and hold the wings level until the next command is made by the crew.
- *GLIDE*: This is used when it is expected that the pilot will not carry out an automatic landing and simply wants the aircraft to automatically fly the path down to the runway, where he will take over for a manual landing.

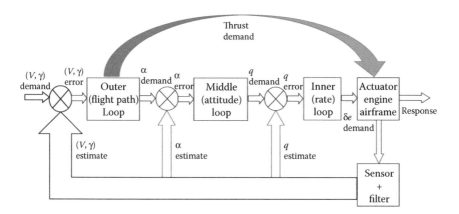

FIGURE 5.7 Typical three-loop longitudinal feedback control law.

demanded and estimated pitch rate to command an elevator deflection. The thrust demand from the outer (flight path) loop and the elevator demand from the inner (rate) loop form the inputs to the airframe block (with actuator, engine).

The precise logic used within each of these controller blocks to calculate the output demand for a given input error is a matter of personal choice. By and large, the standard practice is to use a version of the proportional–integral–differential (PID) law as below where the output demand is the sum of three terms—one proportional to the error, one related to the integral of the error, and the third a function of the differential of the error:

$$demand = k_p(error) + k_i \int error + k_d \frac{d}{dt}(error) \qquad (5.2)$$

The constants k_p, k_i, k_d, also called the "gains," may be computed by your favorite linear control design methodology—there are several. Typically, the frequency-domain control design methods are preferred because they naturally provide information that is required for the assessment and certification of the flight control law.

Our interest in this text is the airplane dynamics in flight, both with and without a flight control law in the loop. Since linear control design is not within the scope of this book, we would prefer not to play favorites and rather let the reader experiment with their pet technique.

5.5.1 Gain Scheduling

The gains in Equation 5.1 are usually calculated for one flight condition and one airframe state. While the same values of the gains can be used for other flight conditions/airframe states, typically, the airplane dynamics at different conditions will be different; so a single "static" set of gains will give differing closed-loop airplane responses under different conditions. It is preferable for the closed-loop airframe system to present the pilot with a uniform response over as wide a range of flight conditions as possible as this makes the task of handling the aircraft more predictable and reduces pilot workload. Therefore, the gains in Equation 5.2 are invariably re-calculated for each flight condition/airframe state giving an array of values. These must be selected/interpolated depending on the flight condition prevailing during the flight.

EXAMPLE: Gain scheduling for stability augmentation of short-period dynamics of F-18 model in level flight.

Consider Equation 4.5 for the longitudinal dynamics of an airplane. We have seen in Section 4.3 how to augment Equation 4.5 with constraint equations that will enforce a level flight condition on all steady states. Note that the airplane dynamics as a whole is not constrained to level flight—just the trim states. In perturbations about these level flight trims, the airplane is allowed to be disturbed arbitrarily in all the state variables. In a similar manner, we can also impose constraints on the eigenvalues of one or more modes. The parameters that can be "scheduled" to enforce the eigenvalue constraints are the *gains* in the flight control law—hence, these are called the "gain schedules."

In order to constrain the eigenvalues through the control law, we must first consider the closed-loop airplane dynamics. Let us go back to the longitudinal dynamics model in Equation 4.5 and represent it as follows:

$$\dot{\underline{x}} = \underline{f}(\underline{x}, \underline{U}) \tag{5.3}$$

where $\underline{x}$ are the states $[V, \alpha, q, \theta]$ and $\underline{U} = [\delta e, \eta]$ is the vector of longitudinal control parameters. Steady states of Equation 5.3 are given by $\underline{f}(\underline{x}^*, \underline{U}^*) = 0$, where the "asterisk" denotes a trim condition. First, let us insist on all the steady states satisfying a level flight constraint. This can be enforced by introducing a constraint equation $\gamma^* = 0$. Next, consider a feedback control law of the following form:

$$\Delta \underline{U} = -K \Delta \underline{x} \tag{5.4}$$

so that the closed-loop dynamics of the airplane appears as

$$\dot{\underline{x}} = \underline{f}(\underline{x}, \underline{U}^* + \Delta \underline{U}) \tag{5.5}$$

with the control law as given in Equation 5.4. Clearly, by changing (some of) the elements of the gain matrix K, the airplane response to a perturbation $\Delta \underline{x}$ about a trim state $(\underline{x}^*, \underline{U}^*)$ can be altered.

Homework Exercise: Work out the closed-loop dynamics of the airplane explicitly by considering the linearized dynamics about a selected trim state. Thus, write the linearized longitudinal model in the form, $\Delta\dot{x} = A\Delta\underline{x} + B\Delta\underline{U}$, and incorporating the control law as in Equation 5.4, the closed-loop linearized model as, $\Delta\dot{x} = (A - BK)\Delta\underline{x}$. Then work out the relationship between a certain element of the feedback gain matrix K and the modal eigenvalues.

We shall carry out this *gain scheduling* exercise numerically using the EBA method introduced in Chapter 3 and applied to several constrained longitudinal flights in Chapter 4. Let us again select the F-18 HARV model presented in Chapter 2 and constrain the short-period dynamics. To that end, we shall additionally require the short-period eigenvalues to be constrained as below at all the level flight trim states:

$$\lambda_{SP} = \xi_r \pm i\xi_i = -1.5 \pm i1.33$$

In the EBA framework, the constrained trim states must satisfy the following set of equations:

$$\begin{pmatrix} \underline{f}(x, \eta, \delta e^*, k_1, k_2, \delta a = 0, \delta r = 0) \\ \gamma \\ \lambda_r^{SP} - \xi_r \\ \lambda_i^{SP} - \xi_i \end{pmatrix} = 0 \qquad (5.6)$$

The first of Equation 5.6 is the set of aircraft trim equations where the aileron and rudder deflections are held at zero each. The elevator deflection δe is the continuation parameter. The throttle parameter η is freed so that the level flight constraint—the second of Equation 5.6—can be met. The feedback control law, Equation 5.4, is implemented as follows:

$$\Delta\delta e = -k_1\Delta\alpha - k_2\Delta q, \quad \text{that is, } \delta e = \delta e^* - k_1\Delta\alpha - k_2\Delta \qquad (5.7)$$

The feedback gains, k_1 and k_2, from Equation 5.7, appear as parameters in the first of Equation 5.6—they are to be tuned so that the short-period eigenvalue is constrained as specified by the third and fourth of Equation 5.6. Thus, there are three "freed" parameters—η, k_1, and k_2—and three corresponding constraints in Equation 5.6.

First, let us observe the open-loop eigenvalues by applying the level flight constraint but not enforcing the short-period eigenvalue constraint. The variation of the short-period and phugoid eigenvalues in this case is shown in Figure 5.8. The level flight trims cover a range of AOA from 2 to 26 deg. Both the short-period and phugoid eigenvalues move toward the imaginary axis with increasing AOA; however, only the phugoid mode eventually becomes unstable at an AOA of around 22 deg. Nevertheless, the change in the short-period dynamics due to the movement of the eigenvalues is significant and maintaining a relatively unchanged short-period behavior would be of interest.

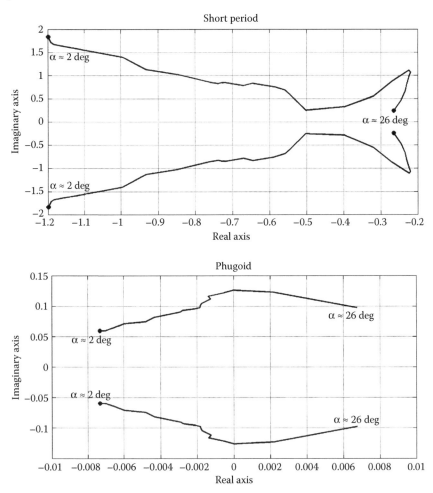

FIGURE 5.8 Open loop short-period and phugoid mode eigenvalues for F-18 model under consideration.

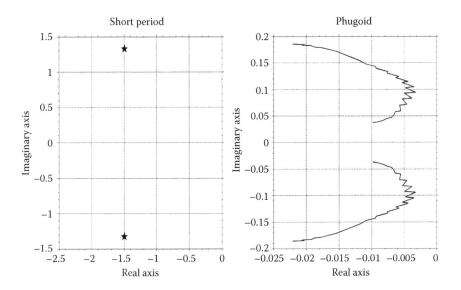

FIGURE 5.9 Closed-loop short-period and phugoid mode eigenvalues with short-period eigenvalues constrained.

Next, the level trims are recomputed, this time with the short-period eigenvalues constrained as in Equation 5.6. The modal eigenvalues over the same range of trim AOAs is plotted in Figure 5.9. Notice that the short-period eigenvalues are now located at the same point for all the trim solutions. The phugoid mode eigenvalues, which have not been constrained, show a shift with AOA, but the phugoid mode no longer goes unstable.

The gain schedules, $k_1(\alpha)$ and $k_2(\alpha)$, are shown in Figure 5.10. Note that these schedules are computed automatically as part of the EBA procedure. Traditionally, the aircraft dynamics would be linearized at a selected set of operating points and the gains computed for each of them. The gains designed at these operating points would then be interpolated or "stitched together" to create a *gain schedule*. The advantage of formulating this problem as in Equation 5.6 and using the EBA method for solution is that the choice of operating points (including constraints on the trim states) and the computation of the gains is automated. Variations of the formulation in Equation 5.6 are possible—for instance, one need not select a fixed value for the short-period eigenvalues; instead the constrained eigenvalue location can itself be made a function of AOA. The solution procedure using EBA is equally adept under these variations.

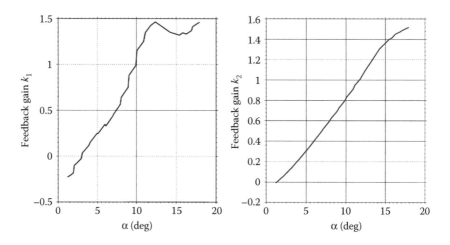

FIGURE 5.10   Closed-loop feedback gains between AOA and pitch rate to elevator.

5.6 DYNAMIC INVERSION CONTROL LAW

For academic purposes, a convenient way to study the airplane closed-loop behavior is to use what is called a "dynamic inversion" controller for each of the blocks in Figure 5.7. Dynamic inversion basically "inverts" the equation concerned with the dynamics in a controller block in Figure 5.7 to compute the output demand given an input error. Consider the dynamic equation associated with a controller block to be symbolically of the form:

$$\frac{d}{dt}(input\ error) = f(states,\ output\ demand) \tag{5.8}$$

Inverting Equation 5.8 yields

$$output\ demand = f^{-1}\left(states,\ \frac{d}{dt}(input\ error)\right) \tag{5.9}$$

The function f^{-1} is effectively the symbolic dynamic inversion control law.
To clarify matters, let us demonstrate the dynamic inversion controller with an example of the inner (rate) loop in Figure 5.7. This loop is reproduced in Figure 5.11 within an inset box. The equation in question is the one for the pitching moment relating the pitch rate to the elevator deflection:

$$\dot{q} = \frac{\bar{q}Sc}{I_{yy}}\left[C_{msta}(Ma,\alpha,\delta e) + (C_{mq1} + C_{m\dot{\alpha}})\frac{\dot{\alpha}c}{2V}\right] \tag{5.10}$$

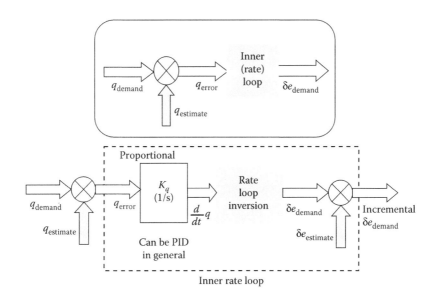

FIGURE 5.11   Example of dynamic inversion feedback controller for rate loop.

The error in the pitch rate is the input available:

$$e_{(q)} = q_{demand} - q_{estimate} \tag{5.11}$$

The dynamic inversion law first "models" $\dot{q}$ as a first-order dynamical system:

$$\dot{q} = K_q e_{(q)} \tag{5.12}$$

where K_q has units of 1/s and is called the "bandwidth" of the rate loop. Note that Equation 5.12 is merely a model; there is nothing sacrosanct about it. Just as K_q is a proportional term, one can think of an integral and a differential term as well, converting Equation 5.12 into a PID form. However, for our purposes, the proportional form of Equation 5.12 suffices.

Then the model in Equation 5.12 is inserted into the LHS of Equation 5.10; thus

$$K_q e_{(q)} = \frac{\overline{q}Sc}{I_{yy}} \left[C_{msta}(Ma, \alpha, \delta e) + (C_{mq1} + C_{m\dot{\alpha}}) \frac{\dot{\alpha}c}{2V} \right] \tag{5.13}$$

and Equation 5.13 is inverted to solve for δe assuming the current values of all the states to be known from the filter/sensor. The difference between the demanded and current values of δe is the incremental change in δe that will act to change the current values of the aircraft states.

Similarly, each of the loops in Figure 5.7 can be implemented as a "dynamic inversion" controller, by "inverting" the appropriate equation, each with its own "bandwidth." This is depicted schematically in Figure 5.12. In addition to the natural flight dynamic time scales and the time scale (frequency) of the command (forcing function) discussed earlier, the bandwidths of the dynamic inversion controller introduce yet another set of time scales. The effect of each bandwidth is to control the response rate in its loop. In case of the back-stepping structure in Figure 5.12, the bandwidths are selected such that there is a step-up in each successive loop from the outermost to the innermost.

For example, the ratio of bandwidths (Hz or 1/s) in Figure 5.12 may be selected as

$$K_{(V,\gamma)} : K_\alpha : K_q = 1:5:25$$

The increasing bandwidths imply decreasing time constants τ for each loop, such as, for example,

$$\tau_{(V,\gamma)} : \tau_\alpha : \tau_q = 10\,\text{s} : 2\,\text{s} : 0.4\,\text{s}$$

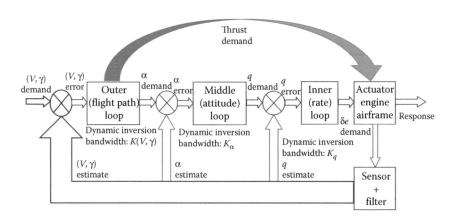

FIGURE 5.12 Back-stepping dynamic inversion longitudinal flight control with the individual loop bandwidths.

This ensures that each loop responds 5 times faster than its immediate outer counterpart.

Note that besides the bandwidths (one per loop), there are no other parameters or "gains" to be selected/tuned in case of dynamic inversion control. Also, these bandwidths are usually chosen to be the same under all flight conditions/airframe states. Thus, no explicit "gain scheduling" is required in case of a dynamic inversion controller.

The bandwidths literally ensure the stability of each loop and ensure that each "demanded" variable responds in a first-order dynamic fashion, as in Equation 5.12, for example, with a time constant fixed by the value of the bandwidth. By selecting the bandwidths of adjacent loops to be at least 5 times apart, one can minimize the interactions between the loops. Nevertheless, when successive dynamic inversion loops are "backstepped" in the manner of Figure 5.12, there is a certain degree of interaction between the loops; hence, the effective time constants of the entire, coupled system may be somewhat different from that prescribed by the bandwidths.

5.7 CLOSED-LOOP STABILITY ANALYSIS

In this section, we demonstrate the use of the analysis tools in Chapter 3 to analyze the dynamics of a closed-loop airframe dynamics (airframe + controller). The actuators are modeled either by rate and saturation limits or by low-order lag models. The structure of the closed-loop dynamical system being studied is represented by the block diagram of Figure 5.13.

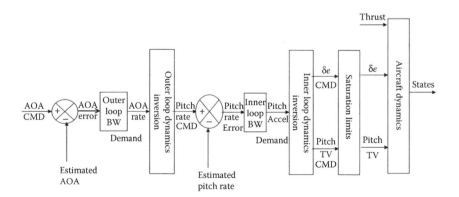

FIGURE 5.13 Back-stepping dynamic inversion longitudinal flight control with additional thrust vectoring (TV) control and saturation limits.

The controller in Figure 5.13 has two loops—the outer one is an AOA loop, where AOA is the commanded variable. The error between the commanded and estimated values of AOA is converted via the outer loop bandwidth to an AOA rate demand. The outer loop inversion block inverts the AOA dynamics to command the pitch rate. For the inner loop, the pitch rate is the commanded variable, which is compared against the measured pitch rate. The pitch rate error is passed through the inner loop bandwidth to produce a pitch acceleration demand. The inner loop inversion inverts the pitch rate dynamics to find the elevator command necessary to satisfy the demanded pitch acceleration. The elevator actuator system is modeled in Figure 5.13 by only a saturation block. If the command is below the saturation limit, it is passed unchanged; otherwise, it is limited by the saturation value. The positive and negative saturation limits and the aircraft dynamics model are the same as that for the F-18 aircraft presented in Chapter 2. There is an option for an additional pitch control in the form of thrust vectoring (TV), but that is not activated for now. The inner and outer loop bandwidths are set at 10 and 2 rad/s, respectively, obeying the 5:1 rule of thumb. The throttle is the other input to the aircraft dynamics block. In practice, the throttle input will differ from the commanded throttle, in the simplest model, due to lags in the engine dynamics. For this particular simulation, the engine model is ignored.

The closed-loop airplane dynamics now consists of

- The aircraft dynamics model, in this case, the longitudinal dynamics equations from Chapter 4

- The two dynamic inversion relations as in Equation 5.13

- The dynamics defined by the two bandwidths as in Equation 5.12

The input or the control variable, also the primary continuation parameter, in case of the closed-loop model in Figure 5.13 is the commanded AOA. The other parameter is the throttle, which may be used to impose a constraint, such as level flight trim. The outputs are the states. The estimated AOA and pitch rate are fed back, but note that feedback of other state variables may be needed to evaluate the two dynamic inversion blocks.

Homework Exercise: Based on the listing above, write out explicitly the set of equations that represent the closed-loop dynamics of Figure 5.13 that would need to be solved.

Let us compute the EBA solution for the F-18 airplane model with the controller in the form as in Figure 5.13 and saturation limits placed on the elevator deflection and the throttle setting. Additionally, we also require the steady states to be level trim flights. Figure 5.14 shows the trim AOA and the trim flight path angle and the stability of the steady states with commanded AOA as the continuation parameter. As expected, the commanded AOA and the actual trim AOA are the same, but only up to a point. Steady level flight is approximately satisfied until a commanded AOA of about 0.6 rad. Beyond that point, the trim states are descending

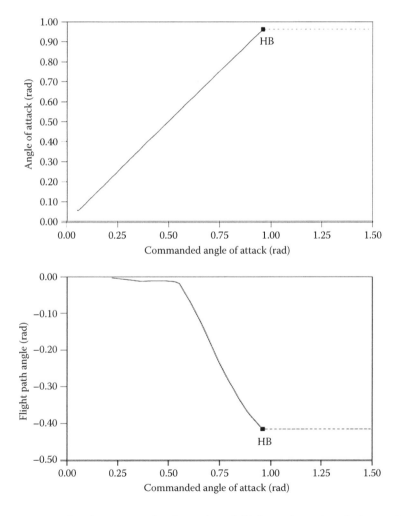

FIGURE 5.14 Steady-state angle of attack and flight path angle solutions with commanded AOA as parameter.

flights with increasingly negative values of trim flight path angle as the commanded AOA is increased. The loss of level flight trim solutions can be correlated with the plot of throttle input shown in Figure 5.15. Once the throttle saturates at $\eta = 1$ around a commanded AOA value of 0.6 rad, the throttle can no further be used to enforce the level flight constraint. Thus, once the throttle saturates, the trim states are no longer level flight solutions. However, elevator deflection is still available to meet the commanded elevator; hence the commanded AOA is still achieved at this point.

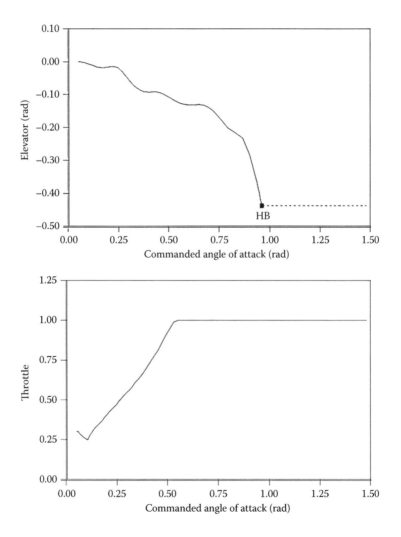

FIGURE 5.15   Elevator and throttle inputs with saturation for the steady states in Figure 5.14 with commanded AOA as parameter.

As the commanded AOA is increased further from this point, the trim AOA is limited around 0.95 rad. With reference to Figure 5.15, it can be seen that this limit is due to the elevator having hit its upper (negative) saturation level. Beyond this point, even as the commanded AOA is increased, there is no further physical elevator deflection possible to pitch the airplane up to a higher AOA.

Thus, two kinds of limits are seen in the EBA solutions in Figure 5.14. The point where the throttle saturates is the limit of level trim flight. Beyond that point, as AOA is increased, it is no longer possible to maintain level trims—the steady states must necessarily be descending flights. Interestingly, the limit of steady *level* flight occurs at an AOA very close to the "stall" AOA; that is, the AOA where the lift coefficient reaches a peak value. Beyond this stall value of AOA, the airplane cannot be held in level flight trim and will enter into descending steady flights. It may be mistakenly assumed that this is due to lack of lift caused by the fall in the lift coefficient beyond the stall AOA. In fact, as seen in Figure 5.15, this is due to the lack of thrust since the throttle has reached its maximum level of 100%. The second limit is the point where the elevator deflection saturates; then it is no longer possible to trim at higher AOAs. This point marks the maximum trim AOA for the airplane. Even though aerodynamically at higher AOAs there may still be enough lift coefficient for flight to be theoretically possible, it is not possible to pitch the airplane up to those AOAs for lack of elevator deflection capability.

In terms of stability, the trims prior to the point where the elevator saturates are all stable. This is practically ensured by the controller in Figure 5.13 by the choice of the two bandwidth values. However, once the elevator saturates, the loop in Figure 5.13 is effectively "open" since the link between the commanded elevator and actual elevator deflection is broken. In other words, no matter what value of elevator is commanded, the actual elevator value is fixed at the saturation value. Therefore, the variables fed back have no influence on the control input and hence on the airplane dynamics. Once the closed-loop has snapped at a point in the loop, the effective airplane dynamics reverts to the open-loop case. This is what is observed in Figure 5.14 at the saturation point where an HB is marked. However, unlike a regular Hopf bifurcation, the movement of the eigenvalues in this case is not smooth across the imaginary axis on the complex plane. As seen in Figure 5.16, prior to the HB point, the short-period mode is stable with a complex pair of eigenvalues in the negative real part. Just beyond the HB point, this pair instantaneously crosses over

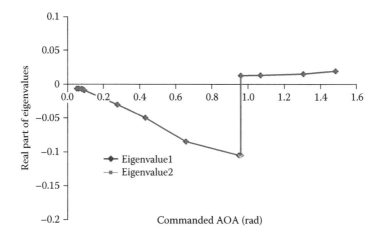

FIGURE 5.16 Movement of real part of short-period eigenvalue with commanded AOA as parameter showing the jump at onset of saturation.

to the right half-plane taking up a position corresponding to the open-loop short-period dynamics, which in this case happens to be unstable. Therefore, the onset of saturation in this case also corresponds to the onset of instability in the steady states. However, this is clearly not expected to always be so. If the airplane open-loop dynamics happens to be stable at the point where saturation occurs, then the dynamics will simply revert to the stable open-loop condition and there will be no onset of instability. There may be a change in the dynamics (frequency and damping of the modes) as the eigenvalues, while remaining stable, reposition themselves from their closed-loop locations dictated by the controller gains to their open-loop positions. Thus, onset of instability coincident with saturation coming into effect is unrelated to the saturation mechanism itself.

5.7.1 With Thrust Vectoring Control Included

How do the results of the previous section change when pitch TV is used to supplement the aerodynamic control due to elevator? Once the elevator saturates, we shall allow the pitch TV control to provide pitching moment and hence permit the airplane to trim at higher AOAs. Other conditions are unchanged from the previous computation—a level flight trim constraint is imposed, which is satisfied as long as the throttle is not saturated, beyond which the trims are descending flights. The revised results with additional pitch TV control are plotted in Figure 5.17.

It can be seen from Figure 5.17 that the pitch TV control cuts in where the elevator saturates and supplies the requisite pitching moment to allow

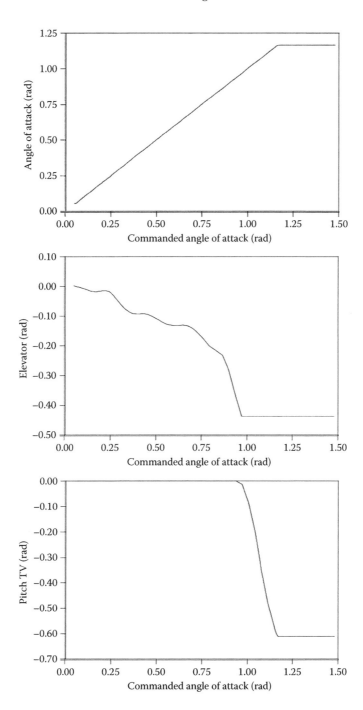

FIGURE 5.17   Steady-state angle of attack with commanded AOA as parameter and corresponding elevator and pitch TV inputs.

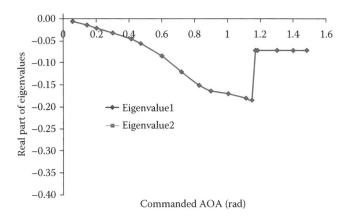

FIGURE 5.18 Movement of real part of short-period eigenvalue with commanded AOA as parameter at onset of pitch TV control saturation.

the airplane to trim at higher AOAs. Thus, the trim AOA curve continues uninterrupted with increasing commanded AOA until the pitch TV control also saturates around an AOA of 1.15 rad. Beyond that point, there is no more nose-up pitch control moment available and hence the trim AOA solutions are limited to a maximum of around 1.15 rad. However, interestingly, these limited AOA solutions are actually stable. As discussed previously, at saturation, the closed-loop system in Figure 5.13 reverts to its open-loop dynamical behavior, which happens to be stable. Figure 5.18 shows the jump in the short-period eigenvalues as the system switches from closed-loop to open-loop instantaneously at onset of pitch TV saturation. In this case, the open-loop eigenvalues remain in the left-half complex plane. Thus, the onset of instability or otherwise at the point of control saturation is not related to the saturation process itself but on the inherent stability of the dynamical system with the saturated loops being snapped open.

EXERCISE PROBLEMS

1. Research into flight modes besides the examples in Figure 5.5 that may be incorporated into a commercial airlines FCS. See Box 5.1 for one example.

2. Construct a linear (state-space) model of low alpha F-18 data trimmed at $\alpha = 20$ deg in level flight. Design a feedback law for augmenting

longitudinal stability of aircraft at this trim condition as per flying and handling qualities specifications (see References 1 and 2).

3. Short-period dynamics is governed mainly by the pitch dynamic equations:

$$\ddot{\alpha} = \frac{\overline{q}Sc}{I_{yy}} \left[C_{m\alpha}\alpha + C_{mq}\frac{\dot{\alpha}c}{2V} + C_{m\delta e}\delta e \right]$$

Use the following data of F-18 model to design a stability augmentation system (feedback law) to improve short-period damping by 15% and damped time period by 10%.

$$m = 15{,}119.28 \text{ kg}, S = 37.16 \text{ m}^2, I_{yy} = 239720.76 \text{ kg-m}^2, c = 3.511 \text{ m}$$

$$C_{m\alpha} = -0.00279/\deg; C_{mq} = -0.0842/\deg; C_{m\delta e} = -0.0075/\deg$$

Trim state: $(V^*, \alpha^*, \delta e^*) = (151.74 \text{ m/s}, 3.28 \text{ deg}, 0 \text{ deg})$

4. Use the F-18 model data in Chapter 2 and linear model developed in exercise 2 to study the effect of the stability augmentation system designed in exercise 3 on the phugoid dynamics at the selected trim state.

5. Using AOA and pitch rate feedback, and setting short-period mode specifications carry out gain scheduling over the complete range of level flight trim conditions for F-18/HARV.

6. Gain scheduling for controlling eigenvalues and further for modifying eigenvectors (eigenstructure assignment) is usually a tedious task. A straightforward method probably is the one presented in this chapter. Read up about various numerical techniques used for gain scheduling [3].

REFERENCES

1. O'Hara, F., Handling criteria, *Journal of Royal Aeronautical Society*, 71(676), 271–291.
2. MIL-F-8785C, *Flying Qualities of Piloted Airplanes*, U.S. Department of Defense Military Specifications, 1980.
3. Nelson, R. C., *Flight Stability and Automatic Control*, Second Edition, McGraw Hill Publication, New York, USA, 2007.

CHAPTER 6

Lateral-Directional Flight Dynamics and Control

S O FAR WE HAVE STUDIED longitudinal airplane flight—both trim states and perturbations from trim were limited to the longitudinal plane— and control. Indeed, a large proportion of airplane flight time is restricted to the longitudinal plane and many of the performance parameters and maneuvers of interest pertain to longitudinal flight. We have explored much of this in Chapter 4 and then been introduced to control of longitudinal flight dynamics in Chapter 5. However, motions out of the longitudinal plane are also of significance. Even for a longitudinal trim state, disturbances may take the airplane away from the longitudinal flight—either a yawing (directional) motion out of the longitudinal plane or a rolling (lateral) motion. Thus, lateral-directional stability is of interest even if the trim state is a longitudinal one. Besides, there are a couple of trim states of interest that are not limited to the longitudinal plane. The more significant one is a level turn maneuver, which is of great importance to combat airplanes, but is also required for other aircraft. The other is a straight line trim flight with sideslip where the direction of the flight (the velocity vector) is not aligned with the airplane nose (body X^B axis)—this is of interest during landing in crosswind. So, in this chapter, we shall look at flight dynamics and control out of the longitudinal plane. However, in many maneuvers of interest, the lateral-directional dynamics is coupled with the longitudinal dynamics; hence a true six degree of freedom coupled lateral-longitudinal analysis is required—that will be addressed in the next chapter.

6.1 LATERAL-DIRECTIONAL MODES IN STRAIGHT AND LEVEL LONGITUDINAL FLIGHT

Straight and level flight trims have already been demonstrated in Chapter 4, Figure 4.6, for the F-18 data set. There, it was observed that the branch of trim states could become unstable due to loss of stability of either the short period mode or the phugoid mode. These modes were limited to disturbances in the longitudinal plane. At the same time, one can also consider disturbances out of the longitudinal plane. These are usually represented by the following variables—body-axis roll rate ($\dot{\phi}$ or p), body-axis roll angle ϕ, and sideslip angle β. Unlike the longitudinal modes where there is a relatively clean separation between the variables involved in the *short period* and *phugoid* modes, the lateral (roll) and directional (sideslip) motions in the lateral-directional dynamics are usually coupled. There are usually three lateral-directional modes—*roll (subsidence)*, *Dutch roll*, and *spiral*. To a first approximation, one can imagine the lateral-directional variables to be associated with these modes as follows: *roll* (roll rate), *Dutch roll* (sideslip with some roll angle coupled in), and *spiral* (roll angle with some sideslip coupled in). However, this is to be used only as a guideline to understand the lateral-directional motions physically. Below, we shall evaluate the lateral-directional modes and their stability numerically.

Considering the same initial trim condition as in Chapter 4, Figure 4.6, and imposing a level flight constraint as before, that is, flight path angle, sideslip, and bank angle constrained to zero at trim, level flight trim states are computed by the EBA method. The variation of throttle with Mach number required to maintain level flight is plotted in Figure 6.1. The shape of the throttle schedule in Figure 6.1 is identical to that in Figure 4.6; the only difference is the additional instability at the point marked A in Figure 6.1. The other points of instability—B, C, D, and E—are common to both Figures 4.6 and 6.1. The instability at point A is that of a lateral-directional mode, in this case, the spiral mode. Hence, it was not visible in Figure 4.6 where only stability of the longitudinal modes was considered.

Figure 6.1 also shows the flight path angle solutions obtained using this particular throttle schedule. As expected, the level flight trims between points A and D are unstable. At point A, the flight path angle plot shows a branch of stable states A–C, which are *departures* from the level flight trims. Since γ is negative, these are descending trims. The stable *departed*

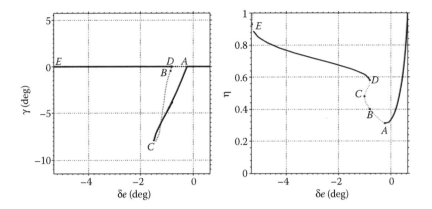

FIGURE 6.1   Bifurcation diagram of flight path angle and throttle variation as a function of elevator deflection in level flight trim.

trims themselves lose stability at point C and the branch C–B–D seen in the flight path angle plot is unstable. Thus, between the values of elevator deflection marked by points D and C, there are two possible stable trim states—one is the primary level flight trim and the other is the *departed* descending trim solution.

To better understand the nature of the *departed* trims in Figure 6.1, we shall examine the results for the other variables from this EBA run—the longitudinal variables are plotted in Figure 6.2 and the lateral-directional variables in Figure 6.3.

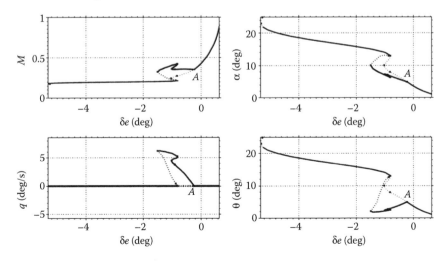

FIGURE 6.2   Bifurcation plot of aircraft longitudinal variables in level flight as a function of the elevator deflection (full line: stable, dashed line: unstable).

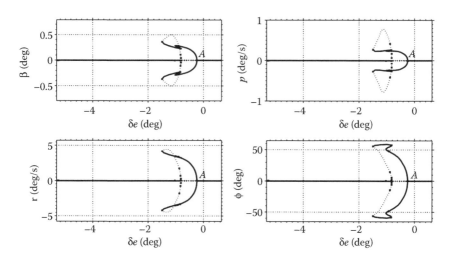

FIGURE 6.3 Bifurcation diagram of lateral-directional variables in level flight as a function of the elevator deflection (full line: stable, dashed line: unstable).

As expected, the primary level flight trim states in Figure 6.3 have zero values of the lateral-directional variables. However, at the bifurcation at point A, which appears to be a *pitchfork bifurcation*, symmetric stable branches emerge with nonzero values of the lateral-directional variables, the primary variable being the roll angle—the values of the other variables are relatively small. The instability at point A may therefore be identified as a *spiral mode instability*. Beyond point A, the airplane *departs* from stable, level, longitudinal flight and enters a gradual banked turn with a little sideslip. The left- and right-turn solutions in Figure 6.3 are symmetric; it is impossible to predict from the bifurcation diagram of Figure 6.3 which way the airplane will bank—that depends on the disturbances at the moment and the perturbations they create in the state variables. For both the left- and right-bank solutions, the longitudinal variables in Figure 6.2 are identical, that is, either way the airplane descends. The AOA in the descent is a little lower than that for the corresponding level flight trims; the pitch angle θ is considerably lower, suggesting that the nose has pitched down relative to its attitude in level flight. The solution in Figure 6.2 may be compared with that in Figure 4.6. The positive pitch rate for these *departed* trims will be explained shortly.

The banked turning motion implies an angular velocity vector approximately pointing downward, along the earth Z^E axis. However, since the airplane is significantly banked, the airplane angular velocity vector will largely be resolved along the body Y^B and Z^B axes, which implies a body pitch rate q and yaw rate r, respectively, as is evident from Figures 6.2 and 6.3.

How significant is this *spiral mode instability* and what implication does it have for the airplane level flight performance as analyzed in Chapter 4? To address this, we should examine the eigenvalues at these trim states— they are also output as part of the EBA analysis. The eigenvalues obtained from the computation are plotted in Figure 6.4a and b in the form of a root locus with AOA as the parameter. From Figure 6.4a and b, the individual modes can be studied for their stability characteristics as below:

- *Short period mode*: This mode is represented by a pair of complex conjugate eigenvalues that remain in the left-half plane over this range of angles of attack; hence, the short period mode is stable. At higher angles of attack, this pair of eigenvalues may transgress into the right-half plane when pitch oscillations referred to as *pitch bucking* earlier may occur.

- *Roll mode*: A single real eigenvalue that shows the roll mode to be always stable. In fact, an unstable roll mode is very unlikely for most airplanes.

- *Dutch roll mode*: A complex conjugate eigenvalue pair, usually better damped than the *phugoid mode* but less damped as compared to the *short period mode*. The *Dutch roll* mode is stable in Figure 6.4 but in many cases the *Dutch roll* pair may cross over to the right-half plane with increasing AOA leading to the onset of oscillations called *wing rock*.

- *Phugoid mode*: A pair of complex conjugate eigenvalues very close to the imaginary axis indicating a lightly damped oscillatory mode. The phugoid eigenvalues cross and recross the imaginary axis, leading to the onset and recovery of stability seen on the "back-side of the power curve."

- *Spiral mode*: A single real eigenvalue that is marginally unstable for small values of AOA, then moves into the left-half plane, indicating stability, as the AOA is increased. This is the instability seen at point A in Figure 6.1. Usually, the spiral mode diverges in bank so slowly that the pilot has adequate opportunity to manually correct the departure. Therefore, oftentimes, the spiral instability is ignored and the bifurcation diagram is drawn as in Figure 4.8, which assures that all trims on the front side of the power curve are effectively stable.

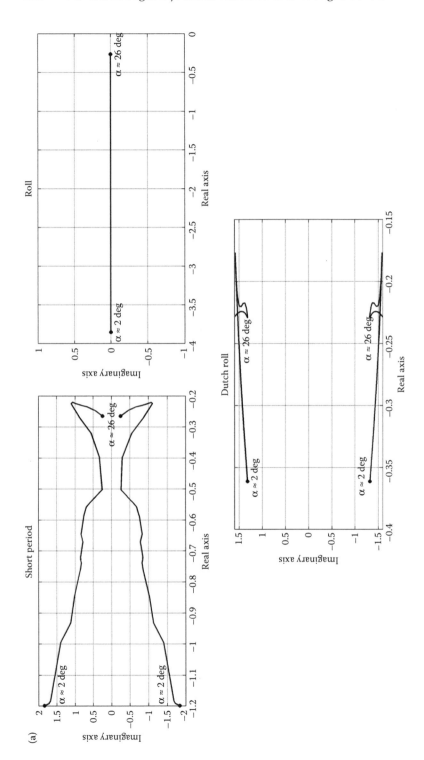

FIGURE 6.4   (a) Root locus plot showing fast mode eigenvalues—short period, roll, and Dutch roll.   (*Continued*)

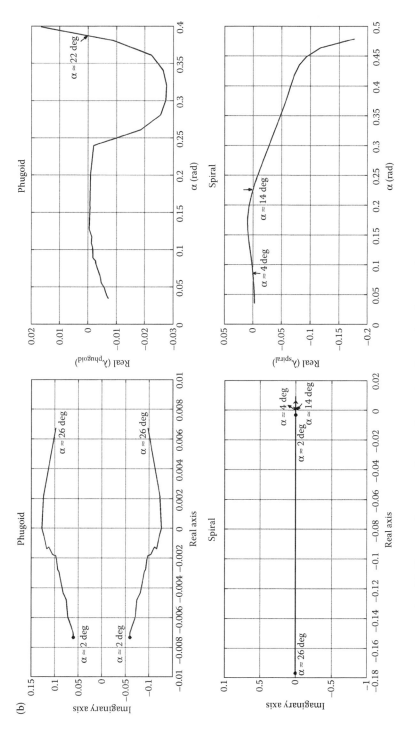

FIGURE 6.4 (Continued) (b) Slow mode eigenvalues—phugoid and spiral—along with their real parts plotted against trim angle of attack showing zero crossing.

Thus, two of the three lateral-directional modes are likely to show instability for a typical airplane—*spiral mode* is likely to lose stability at lower values of AOA and *Dutch roll* instability is likely at moderate-to-high values of AOA. The first is usually not a major concern as long as the time to double amplitude is large; then it can be manually handled by the pilot. *Dutch roll* instability leading to *wing rock* on the other hand may be an issue for several airplanes placing an upper limit on allowable AOA.

Homework Exercise: It helps to look at several examples of bifurcation diagrams for airplane flight dynamics to gain experience in interpreting the results. Every aircraft will have a fairly unique bifurcation diagram with regard to the number and type of instabilities, especially at higher AOA. For example, take a look at the bifurcation diagram in Figure 6.5 for the F-15 airplane [1]. Try to identify the various bifurcation points and the likely instability induced at each. For example, the Hopf bifurcation around 20 deg AOA signals an unstable Dutch roll mode and onset of wing rock oscillations. Note that the bifurcation is subcritical and the unstable limit cycle branch then folds over to create the stable wing rock oscillations.

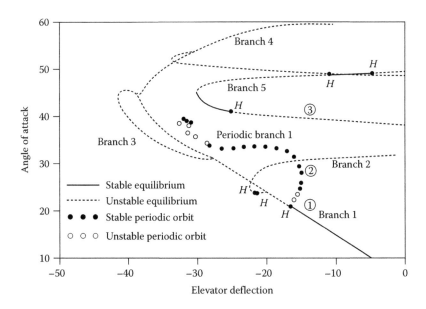

FIGURE 6.5   Bifurcation diagram for the F-15 aircraft. (From Nolan, R.C., Wing rock prediction method for a high performance fighter aircraft, Thesis, Air Force Institute of Technology, AFIT/GAE/ENY/92J-02, 1992.)

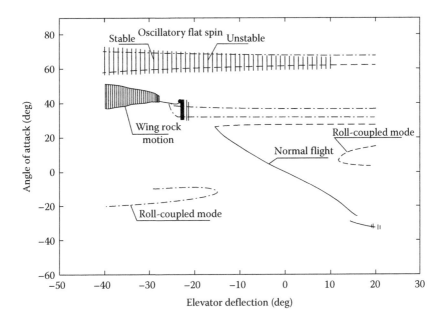

FIGURE 6.6 Bifurcation diagram of HHIRM model. (From https://www.slideshare.net/Project_KRIT/goman-modeling-and-analysis-of-aircraft-spin)

Homework Exercise: Another example of a bifurcation diagram is presented in Figure 6.6 [2]. Use this example to identify and understand the various instabilities and corresponding nonlinear dynamical phenomena in airplane flight.

6.2 HORIZONTAL LEVEL TURN TRIMS

Banked level turn trims are of importance to most airplanes. For commercial airliners and civil aircraft, they represent a possible steady state. For example, while loitering over an airport, an airplane would fly alternating segments of banked turn and level flight as it carries out a horizontal loop. For combat aircraft, turn rate and turn radius are accepted parameters used to characterize airplane performance. For superior air combat capability, the airplane should have the maximum possible turn rate and the minimum possible turn radius. The maximum turn rate achieved is typically constrained by the maximum coefficient of lift that can be obtained (C_{Lmax}) and the maximum load factor (n_{max}) that can be sustained by the airplane. These constitute aerodynamic and structural/human constraints, respectively, with maximum thrust further limiting the turn rate for sustained, level maneuvers. Most texts consider these criteria while obtaining closed-form expressions for maximum turn rate, leaving stability out of the picture. Knowledge of stability is important for deciding the envelope of operations and designing control laws that can stabilize the airplane and provide adequate handling qualities. Further, instantaneous turn maneuvers where the velocity cannot be maintained due to thrust–drag imbalance are also significant. In such cases, a steady equilibrium value of velocity is not possible.

In the following, we shall use the EBA method to analyze horizontal turn performance and stability together as was done for longitudinal flight in Chapter 4. We shall also obtain turn performance metrics in this manner in a similar fashion as the climb metrics in Chapter 4.

6.2.1 Formulation and Constraints

The analysis of turn maneuvers requires the complete set of eight equations for the airplane dynamics as in Table 1.6. A typical choice of states $\underline{x}$, primary control u, and other parameters $\underline{p}$ is as below:

$$\underline{x} = [V, \alpha, \beta, p, q, r, \phi, \theta], \quad u = \delta e, \quad \underline{p} = [\eta, \delta a, \delta r] \qquad (6.1)$$

Constraints are to be placed on the states to ensure that the trims correspond to a level turning flight. Two of the constraints are common to all conditions—these are level ($\gamma = 0$) and zero sideslip ($\beta = 0$). Additionally, the turn flight can be constrained in one of three ways—lift-limited, thrust-limited, or load factor-limited. The selected constraints for each of these three cases are indicated in Table 6.1. In case of lift limiting, either the AOA or the lift coefficient is to be constrained, usually close to the maximum (stall) value. For the thrust-limited case, the throttle value can be set—this is not really a constraint in the EBA sense. If the load factor is the limiting constraint, either the bank angle or the load factor itself can be set to a fixed value.

TABLE 6.1 Summary of Turn Trim State Constraints

Label	Common Constraints	Additional Constraint	Freed Parameters in p Vector	Fixed Parameters
Thrust limited, A	$\gamma = 0, \beta = 0$	–	$\delta a, \delta r$	$\eta = 1$
Lift limited, B	$\gamma = 0, \beta = 0$	$\alpha_{CL\max} = 0.6283$ rad	$\eta, \delta a, \delta r$	–
Load factor limited, C	$\gamma = 0, \beta = 0$	$\mu = -1.3845$ rad OR $n = 5.4g$	$\eta, \delta a, \delta r$	–

The high-angle-of-attack F-18 HARV airplane data are used for the calculations. For the lift-limited turn, an AOA of 0.6283 rad (36 deg) is selected—this is very close to the stall value of AOA of F-18 HARV. Thrust-limited turns are computed by setting the throttle to a maximum of 1.0. For the load factor limited turns, a load factor of 5.4g is set—this may be implemented by setting the corresponding value of bank angle (1.3845 rad = 79.326 deg) as a constraint.

6.2.2 Parameter Schedules

The EBA method schedules the freed parameters in Table 6.1 such that the constraints are met at all trim states. The computed parameter schedules are plotted in Figure 6.7. The range of elevator deflections is roughly

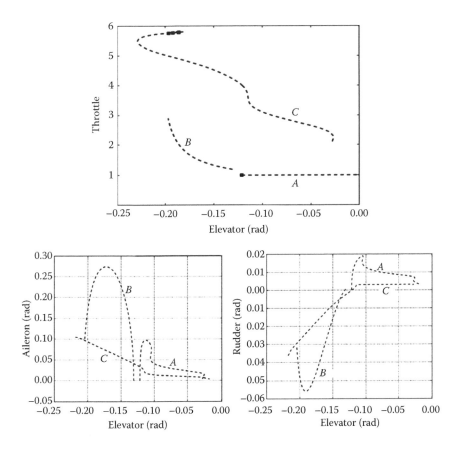

FIGURE 6.7 Parameter schedules computed to satisfy the level turn constraints in Table 6.1.

between 0 and −14 deg (up elevator). This is required to increase the AOA, thereby to create additional lift, which is needed to produce a load factor ($n = L/W$) as desired.

For Case A, the throttle is held fixed at $\eta = 1$; for the other cases, the computed η variation is plotted. It may be noticed that the throttle required to maintain level flight is greater than the maximum allowed ($\eta > 1$) for most of the trims. In other words, the drag is too large to be balanced by the thrust possibly due to the high induced drag caused at these trim states with large C_L. Therefore, the velocity will continuously bleed and cannot be maintained at a given value. Thus, if an optimum turn performance metric requires a certain value of velocity, that metric can be optimum only momentarily. The velocity required to optimize that metric cannot be held; the turns with $\eta > 1$ are therefore called "instantaneous" as against the "sustained" turns with $\eta = 1$, which can be held level, with a fixed value of velocity. So, in Figure 6.7, the trims on the branch "A" are *sustained turns*, branches labeled "B" and "C" correspond to *instantaneous turns*.

Relatively large values of aileron deflection are seen in Figure 6.7 because of asymmetric aerodynamics between the right and left elevator for this airplane model. This creates a rolling moment due to elevator deflection, which is nullified by the use of aileron. A slight yaw is produced in the process, which is canceled by the use of rudder, as shown in Figure 6.7.

6.2.3 Turn Performance

To create the zero sideslip, level turn performance plots, we need to solve the EBA problem thrice, one for each of the cases in Table 6.1, and merge the results. Let us first look at the bifurcation diagrams produced by the three individual runs. Remember that the throttle, elevator, and rudder are varied in sync with the elevator as in Figure 6.7. Essentially, the rudder maintains zero sideslip, the throttle maintains level flight even with nonphysical values of $\eta > 1$ in Cases B and C, and the aileron creates and holds the bank angle while the elevator sets the AOA that produces the load factor required for the turn.

Figure 6.8 shows the variation of trim AOA in the three turn cases. For Case B (bottom left), of course, the trim AOA is held constant, as seen. The trims are largely unstable, most probably due to an unstable phugoid mode at that particular value of AOA, as we have seen in Figure 4.2. The other cases do not require as high a value of AOA as the lift-limited Case B. In fact, the variation of AOA in Case A is very similar to that over a

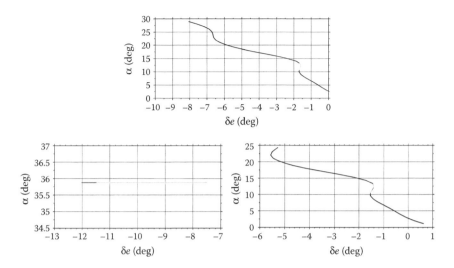

FIGURE 6.8 Variation of trim angle of attack with elevator deflection as continuation parameter (top—Case C; bottom left—Case B; bottom right—Case A).

similar range of elevator values in Case C. Higher AOA trims in Case A become impossible due to lack of thrust to trim out in level flight.

Figure 6.9 shows the trim bank angle for the three turn cases in Table 6.1. Remember that the centripetal acceleration in the turn depends on the bank angle and the lift (hence the lift coefficient and the velocity). Hence,

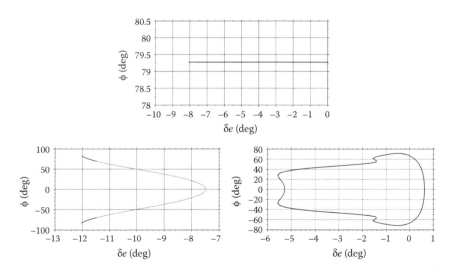

FIGURE 6.9 Variation of trim bank angle with elevator deflection as continuation parameter (top—Case C; bottom left—Case B; bottom right—Case A).

combinations of large AOA and large bank angle should suggest regions of high turn rate (as we shall see when we explicitly plot the turn rate shortly). In Figure 6.9, Case C corresponds to a fixed bank angle, hence the turn rate will be closely correlated with the AOA—larger AOA with more up (negative) elevator will imply larger turn rates. On the other hand, Case B is a fixed AOA case (fixed C_L), so a steeper bank angle in general will lead to a higher turn rate. However, since the AOA is always so high, there is a large drag and hence a need for high thrust as seen in Figure 6.7. Therefore, this turn cannot be sustained. For the third case, Case A, the bank angle is nearly flat over a large stretch somewhat mimicking the constant bank angle of Case C. The AOA in this case is limited by the thrust available to balance out the drag.

By combining the EBA results for the three cases in Table 6.1, we can create a single, composite plot to analyze the airplane's turn performance. This has been shown in Figure 6.10 with Mach number as the parameter

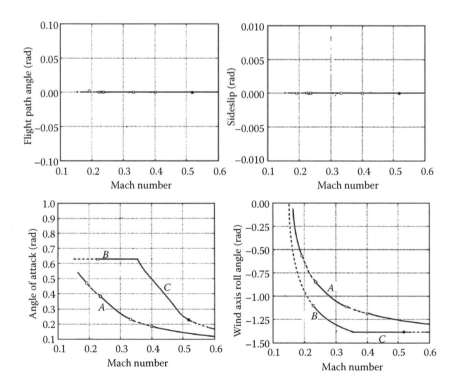

FIGURE 6.10 Constrained states for level turn performance analysis (full line: stable, dashed line: unstable).

on the X axis—this choice will help us correlate these plots better with the following plots of turn rate and radius, which are usually shown with Mach number on the X axis. The results plotted in Figure 6.10 are for the left turn. Qualitatively similar results with sign changes in lateral-direction variables for the right turn can be computed.

Figure 6.10 shows that the sideslip and flight path angles are indeed constrained as desired in Table 6.1. However, note that the $\gamma = 0$ constraint is nonphysical for Cases B and C since the throttle limit has been exceeded. For the AOA and roll angle plots, the three cases in Table 6.1 are marked against the corresponding curves. From the AOA plot, one can notice that all other trim states have a value of AOA below the fixed value of 0.6283 rad for Case B. Likewise, the roll angle for all cases is lower in magnitude than the fixed value of 1.3845 rad for Case C. There are two kinds of instabilities noticed in Figure 6.10— one at an AOA around 0.2 rad, which is common to Cases A and C, and the other around Mach 0.24 observed for Cases A and B. In a turn maneuver, the longitudinal and lateral-directional modes are coupled, hence it is not possible to identify an instability in terms of a pure longitudinal or lateral-directional mode; this can be seen in the following example.

A level turn trim state corresponding to maximum available thrust (third constraint $\eta = 1.0$ in this case) is chosen from the above plots.

Equilibrium state:

$$\underline{x}^* = [M^* \alpha^* \beta^* p^* q^* r^* \phi^* \theta^*]'$$
$$= [0.24 \ 0.372\,\text{rad} \ 0 \ 0.033\,\text{rad/s} \ 0.107\,\text{rad/s} - 0.084\,\text{rad/s}$$
$$- 0.905\,\text{rad} \ 0.236\,\text{rad}]'$$

$$\underline{U}^* = [\eta^* \delta e^* \delta a^* \delta r^*]' = [1.0 \ -0.103\,\text{rad} \ 0.044\,\text{rad} \ 0.013\,\text{rad}]'$$

This equilibrium state corresponds to maximum stable sustained turn rate (STR) of F-18/HARV. From the bifurcation plot, it turns out that this equilibrium state is stable as we conclude from the eigenvalue analysis below. Linearization of equations of aircraft motion evaluated at the above equilibrium state gives the following modal matrix (or system matrix) used for eigenvalue analysis.

System matrix A = :

M	α	β	p	q	r	φ	θ
−0.0987	−0.0341	0.0219	0	0	0	−0.008	−0.0275
−0.9120	−0.2720	0.0000	0	0.9920	0	−0.0845	0.0260
−0.3760	−0.0016	−0.1430	0.3640	0	−0.9320	0.0712	−0.0219
−0.3930	0.4090	−11.40	−0.8810	−0.0552	0.8540	0	0
0.1220	−0.3820	0	0.0804	−0.2260	−0.0314	0	0
0.0130	−0.0299	0.2810	−0.0849	0.0264	−0.1040	0	0
0	0	0	1.0000	0.1900	0.1490	0	0.1440
0	0	0	0	0.6170	−0.7870	−0.1360	0

Eigenvalues of matrix A are

$$
\lambda = \begin{bmatrix}
\underbrace{-0.303 \pm j2.073}_{\text{Coupled Dutch roll}} & \underbrace{-0.056 \pm j0.181}_{\text{Coupled phugoid}} \\
\underbrace{-0.498}_{\text{Coupled roll}} \quad \underbrace{-0.25 \pm j0.585}_{\text{Coupled short period}} \quad \underbrace{-0.0084}_{\text{Coupled spiral}}
\end{bmatrix}
$$

Real parts of all the eigenvalues at this trim condition are negative; hence, we conclude that this level turn trim state is stable. These eigenvalues are distinctly located in the complex plane even though the equilibrium state is coupled in longitudinal and lateral-directional dynamics. One can notionally label the eigenvalues as "short-period-like," "phugoid-like," etc.; however, it is not easy to qualify these modes as representing five typical aircraft dynamics modes from the analysis of the eigenvectors.

Eigenvectors of matrix A are

$E =$

M:	$0.001 \pm j0.0002$	$-0.0263 \pm j0.1446$	0.0136	$-0.0245 \pm j0.0544$	−0.152
α:	$0.0002 \mp j0.0017$	$0.0336 \pm j0.0166$	−0.0135	−0.6603	0.0595
β:	$-0.0464 \mp j0.1583$	$0.0086 \mp j0.0128$	0.0054	$-0.0324 \pm j0.0118$	0.0097
p:	0.8891	$-0.0894 \pm j0.0743$	−0.438	$-0.0894 \mp j0.0743$	−0.0441
q:	$-0.0016 \mp j0.0346$	$-0.0072 \pm j0.118$	0.0922	$-0.0531 \mp j0.4257$	−0.202
r:	$-0.0176 \pm j0.0444$	$-0.0162 \mp j0.0511$	−0.1058	$0.0079 \mp j0.0143$	−0.0274
φ:	$-0.0617 \mp j0.4183$	$0.4074 \mp j0.2788$	0.8869	$-0.3165 \pm j0.2169$	−0.7187
θ:	$-0.0012 \mp j0.0101$	0.8381	−0.0392	$-0.408 \pm j0.1676$	0.6428

The eigenvector corresponding to each eigenvalue is listed in matrix E. The modes and characteristic motions can be figured out by looking at the magnitudes of the entries in the columns of matrix E. Corresponding to the first complex–conjugate pair of eigenvalues, in the first column of matrix E, complex–conjugate pair of eigenvectors are given. Roll rate p has the largest magnitude followed by variables β, r, ϕ, suggesting a Dutch roll mode, but coupled with smaller magnitude of longitudinal variables; that is why, it is referred to as "Dutch-roll like" mode. Similarly, the second column of matrix E corresponds to the second complex–conjugate pair of eigenvalues. Reading the magnitude of variables, pitch angle, θ, has the largest magnitude, followed by bank angle, ϕ, which suggests that this mode is also a coupled mode, but owing to the smaller magnitudes of other lateral-directional variables, this mode may qualify as a "phugoid like" mode.

Homework Exercise: Study other columns of eigenvector matrix E to visualize other modes.

Another important observation that follows by looking at the real part of the eigenvalues is that coupled phugoid and coupled spiral modes are located closest to the imaginary axis, suggesting that these are the eigenvalues more likely to cause instabilities (crossing into the right-half plane) first. One can thus verify by simulation or otherwise that in the bifurcation plots in Figure 6.11, empty squares correspond to coupled spiral mode instabilities (leading to departure, i.e., exponential growth of variables) and solid squares correspond to coupled phugoid mode instabilities (leading to sustained oscillations in variables).

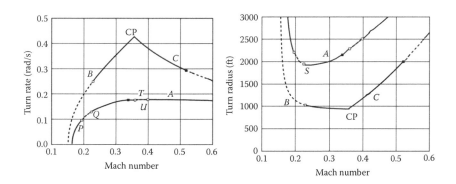

FIGURE 6.11   Plot of turn rate and turn radius against the Mach number (full line: stable, dashed line: unstable).

Figure 6.11 shows the turn performance in terms of the turn rate against the Mach number for the three branches corresponding to Cases A, B, and C in Table 6.1. Curve A in Figure 6.11 marks the sustained, thrust-limited turns. The peak STR is a shade under 0.2 rad/s (point marked "T" in Figure 6.11). Incidentally, there is also a region of instability around this peak value in the Mach number range 0.35–0.4. The lift-limited turns (curve B) are unstable at lower values of Mach number. The maximum instantaneous turn rate (ITR) occurs at the corner point (CP) where the lift-limited and load-factor-limited turn curves meet. The ITR is a little more than twice the value of the STR and occurs around 0.35 Mach, about the same as the Mach number for peak STR. However, the CP is a sharp point with the ITR falling off rapidly on either side unlike the STR, which can be held over a range of Mach numbers with little deterioration. Also, as mentioned previously, at the ITR condition, the thrust available is less than the drag and so the velocity bleeds until the turn can no longer be held.

The tightest sustained turn solution can be read from the plot of turn radius in Figure 6.11 for Case A. The tightest turn radius (TTR) happens to occur at a point marked "S" for a Mach number around 0.25. Tighter turns are possible on branches labeled B and C, but these are not sustained, so they can only be held briefly. The tightest instantaneous turn occurs at the point labeled CP at a Mach number near 0.35, at the intersection of curves B and C. However, there is almost a plateau of turn radius on curve B between Mach number values 0.25–0.35.

In this manner, Figure 6.11 summarizes the airplane turn performance with the key metrics, including stability information. On the basis of results like those in Figure 6.11, it is possible to decide the best operating conditions for an airplane as well as flight conditions where stability augmentation may be required.

Homework Exercise: Figure 6.12 shows a typical turn performance chart for a combat airplane. Based on our study so far, try to identify the various turn performance parameters and constraint curves in this plot.

6.3 NONZERO SIDESLIP TRIM AND STABILITY ANALYSIS

Landing an airplane in a crosswind is a tricky maneuver. One of the methods of crosswind landing is to trim the airplane at a nonzero value of sideslip such that the nose is aligned with the runway. This is shown in the sketch of Figure 6.13.

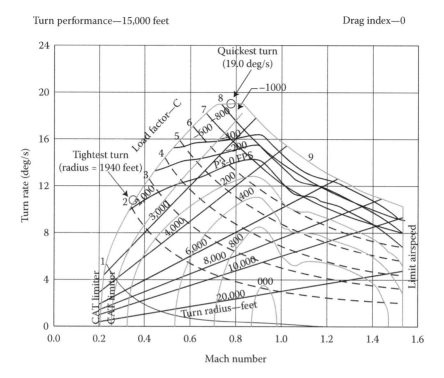

FIGURE 6.12 Typical turn performance diagram for a combat airplane. (From http://web.deu.edu.tr/atiksu/ana45/avionic.html.)

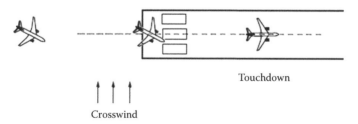

FIGURE 6.13 Sketch showing airplane approaching the runway in a crosswind condition and landing with nonzero sideslip trim. (From https://en.wikipedia.org/wiki/Crosswind_landing#/media/File:Crosswind_landing_sideslip-notext.svg.)

In a sideslip, owing to the airplane's inherent stability, the nose tends to align itself with the relative wind velocity direction. However, as it approaches the runway threshold, the pilot can apply the rudder to counter the stabilizing effect due to sideslip and maintain a nonzero value of sideslip such that the nose is aligned with the runway direction. The nonzero sideslip will also induce a roll, which must be negated by appropriate use of aileron.

The nonzero sideslip trim state can be solved by using the EBA framework. We can impose constraints on the flight path angle and bank angle—usually $\gamma \approx 0$ during landing flare and ideally we would like $\phi = 0$. Additionally, the sideslip angle is constrained to a value depending on the amount of crosswind. We shall demonstrate the procedure with the following constraints imposed:

$$\beta = 1.28\,\text{deg}; \quad \gamma = 0; \quad \phi = 0 \tag{6.2}$$

The elevator deflection is the continuation parameter and the rudder, throttle, and aileron are "freed" in the continuation run so as to satisfy the constraints in Equation 6.2. The foremost is the rudder required to maintain the nonzero sideslip—this is plotted in Figure 6.14. The rudder requirement is almost linear with elevator deflection over the range of interest, that is, −1 to −5 deg elevator. The corresponding schedules of aileron and throttle are also shown in Figure 6.14. Usually, with increasing AOA, owing to the dihedral effect being a function of AOA, larger aileron deflections are required to maintain wings level flight.

Using the control schedules in Figure 6.14, the variation of the airplane trim states as a function of elevator deflection can be computed as shown

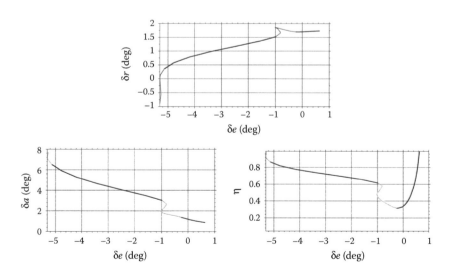

FIGURE 6.14 Schedules of rudder and aileron deflection, and throttle setting, for the nonzero sideslip flight.

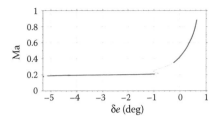
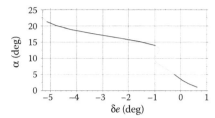

FIGURE 6.15 Bifurcation diagram of the Mach number and AOA with varying elevator deflection as parameter for the nonzero sideslip flight (full line: stable trim; dashed line: unstable trim).

in Figure 6.15 for the Mach number and AOA. Over the range of interest, the Mach number is almost constant around a value of 0.2, whereas the AOA varies between 14 and 21 deg. The trims in this range appear to be stable.

6.4 WING ROCK ONSET AND ITS PREDICTION

As described earlier, *wing rock* is an oscillation, predominantly in the lateral-directional variables, induced when the Dutch roll mode loses stability. The onset of wing rock corresponds to the occurrence of a Hopf bifurcation. Wing rock is a fairly commonly observed phenomenon on military airplanes, hence identifying the occurrence of wing rock is a matter of considerable interest. However, of equal or greater interest is the prediction of the possibility of wing rock during the aircraft design process and ways to avoid the occurrence of wing rock. Therefore, the development of a relatively simple criterion for wing rock onset is a significant issue. In the following, we shall first demonstrate the use of the EBA framework to identify the point of wing rock onset for an example airplane data set. Then we shall present a simplified analytical criterion for predicting onset of wing rock.

Let us consider the F-18 low-AOA model from Chapter 2 for this exercise. We shall carry out an EBA computation with a level flight constraint (zero sideslip, wings level, zero flight path angle). The resulting bifurcation diagram is shown in Figure 6.16. The top plot shows the required variation of the throttle to maintain level flight as a function of the trim Mach number. The various points of instability are marked therein. For instance, the point marked "A" refers to the onset of spiral instability at a low AOA. Points "B" and "C" mark the phugoid instability onset and recovery of stability, respectively, at moderate values of AOA. Finally, the

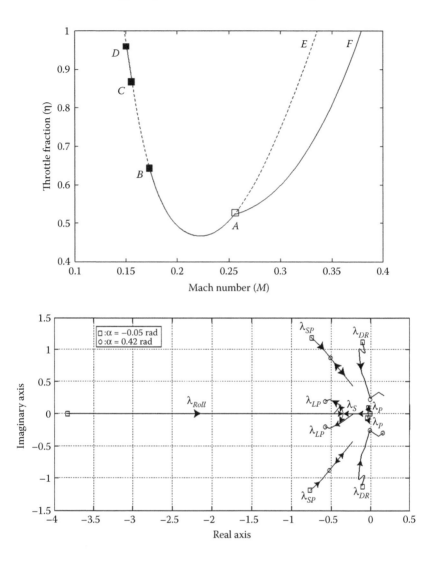

FIGURE 6.16 Bifurcation diagram for the low-angle-of-attack F-18 airplane model and onset of various instabilities.

Hopf bifurcation at the point labeled "D" indicates the onset of Dutch roll instability—that is, wing rock. This can be verified by referring to the bottom plot in Figure 6.16, which shows the movement of the various modal eigenvalues in the complex plane corresponding to the bifurcation diagram.

Moving from lower (−0.05 rad) to higher (0.42 rad) AOA,

- The spiral mode starts off marginally unstable, then moves into the left-half plane where it encounters the roll mode eigenvalue; they collide and separate into a complex conjugate pair signifying a *lateral phugoid* mode.

- The short period mode tends toward the imaginary axis, but remains stable all through. The phugoid mode is initially marginally stable. The eigenvalue pair briefly flit across the imaginary axis but then regain stability.

- The Dutch roll eigenvalue crosses the imaginary axis at an AOA of about 22 deg, following which the Dutch roll mode becomes unstable.

Thus, the onset of wing rock oscillations may be expected around 22 deg AOA for this airplane model.

6.4.1 Analytical Criterion for Wing Rock Onset

Although we have so far avoided writing the decoupled equations for the lateral-directional dynamics or small-perturbation equations about a trim state, for the explicit purpose of obtaining a wing rock onset predictor, we need to consider a small-perturbation lateral-directional model. Under the usual assumptions of small perturbations in the lateral-directional variables alone, about a straight and level flight trim state, the lateral-directional equations of motion may be written as

$$\Delta\dot{\chi} = \left(\frac{g}{V^*}\right)\left\{\left(\frac{\bar{q}S}{W}\right)\Delta C_Y + \Delta\mu\right\} \tag{6.3}$$

$$\Delta\dot{p}_b = \left(\frac{\bar{q}Sb}{I_{xx}}\right)\Delta C_l \tag{6.4}$$

$$\Delta\dot{r}_b = \left(\frac{\bar{q}Sb}{I_{zz}}\right)\Delta C_n \tag{6.5}$$

where p_b, r_b are the body-axis roll and yaw rates, respectively, μ is the wind-axis (velocity vector) roll angle, χ is the wind-axis yaw angle, $\bar{q}$ is the dynamic pressure ($= 1/2 \, \rho \, V^2$), S is a reference area, usually the airplane wing planform area, b is the wing span, g is the acceleration due to gravity, W is the airplane weight, I_{xx}, I_{zz} are the moments of inertia, and V^* is the trim velocity.

Homework Exercise: Derive Equations 6.3 through 6.5 from the equations of motion presented in Chapter 1. (For Equation 6.3, start with Equation 1.28 for $\dot{\chi}$ and take perturbations about a straight and level trim condition assuming zero perturbations in the longitudinal variables. Then use Equation 1.47 for r_w followed by the equation for the side force from Equation 1.59. Ignore the thrust component. For Equations 6.4 and 6.5, use the respective moment equations from Equation 1.78, neglect terms with I_{xz} and take small perturbations. Set the perturbed values of the longitudinal variables to zero.)

Following the aerodynamic model in Equation 2.3 (Chapter 2), the perturbed aerodynamic coefficients are modeled as

$$
\Delta C_Y = C_{Y\beta}\Delta\beta + C_{Yp1}(\Delta p_b - \Delta p_w)\left(\frac{b}{2V^*}\right) + C_{Yp2}\Delta p_w\left(\frac{b}{2V^*}\right)
$$
$$
+ C_{Yr1}(\Delta r_b - \Delta r_w)\left(\frac{b}{2V^*}\right) + C_{Yr2}\Delta r_w\left(\frac{b}{2V^*}\right) \tag{6.6}
$$

$$
\Delta C_l = C_{l\beta}\Delta\beta + C_{lp1}(\Delta p_b - \Delta p_w)\left(\frac{b}{2V^*}\right) + C_{lp2}\Delta p_w\left(\frac{b}{2V^*}\right)
$$
$$
+ C_{lr1}(\Delta r_b - \Delta r_w)\left(\frac{b}{2V^*}\right) + C_{lr2}\Delta r_w\left(\frac{b}{2V^*}\right) \tag{6.7}
$$

$$
\Delta C_n = C_{n\beta}\Delta\beta + C_{np1}(\Delta p_b - \Delta p_w)\left(\frac{b}{2V^*}\right) + C_{np2}\Delta p_w\left(\frac{b}{2V^*}\right)
$$
$$
+ C_{nr1}(\Delta r_b - \Delta r_w)\left(\frac{b}{2V^*}\right) + C_{nr2}\Delta r_w\left(\frac{b}{2V^*}\right) \tag{6.8}
$$

wherein the first term in each right-hand side expression is the *static* term, the second and fourth terms are the *dynamic* terms, and the third and fifth terms are the *flow curvature* terms. Downwash lag model is not used. The subscripts "b" and "w" in Equations 6.6 through 6.8 refer to body-axis and wind-axis quantities, respectively, and the various aerodynamic derivatives are defined as below:

$$C_{Y\beta} = \frac{\partial C_Y}{\partial \beta}\bigg|_* ; \quad C_{l\beta} = \frac{\partial C_l}{\partial \beta}\bigg|_* ; \quad C_{n\beta} = \frac{\partial C_n}{\partial \beta}\bigg|_* ;$$

$$C_{Yp1} = \frac{\partial C_Y}{\partial(p_b - p_w)\left(\dfrac{b}{2V}\right)}\bigg|_* ; \quad C_{lp1} = \frac{\partial C_l}{\partial(p_b - p_w)\left(\dfrac{b}{2V}\right)}\bigg|_* ; \quad C_{np1} = \frac{\partial C_n}{\partial(p_b - p_w)\left(\dfrac{b}{2V}\right)}\bigg|_*$$

$$C_{Yp2} = \frac{\partial C_Y}{\partial p_w\left(\dfrac{b}{2V}\right)}\bigg|_* ; \quad C_{lp2} = \frac{\partial C_l}{\partial p_w\left(\dfrac{b}{2V}\right)}\bigg|_* ; \quad C_{np2} = \frac{\partial C_n}{\partial p_w\left(\dfrac{b}{2V}\right)}\bigg|_*$$

$$C_{Yr1} = \frac{\partial C_Y}{\partial(r_b - r_w)\left(\dfrac{b}{2V}\right)}\bigg|_* ; \quad C_{lr1} = \frac{\partial C_l}{\partial(r_b - r_w)\left(\dfrac{b}{2V}\right)}\bigg|_* ; \quad C_{nr1} = \frac{\partial C_n}{\partial(r_b - r_w)\left(\dfrac{b}{2V}\right)}\bigg|_*$$

$$C_{Yr2} = \frac{\partial C_Y}{\partial r_w\left(\dfrac{b}{2V}\right)}\bigg|_* ; \quad C_{lr2} = \frac{\partial C_l}{\partial r_w\left(\dfrac{b}{2V}\right)}\bigg|_* ; \quad C_{nr2} = \frac{\partial C_n}{\partial r_w\left(\dfrac{b}{2V}\right)}\bigg|_* \qquad (6.9)$$

The "*" indicates that the derivatives are to be evaluated at the trim state.

Note the distinct use of the "1" and "2" derivatives in Equations 6.6 through 6.8. The "1" derivatives refer to the *dynamic* effect due to the relative angular rate between the body and the wind axes, that is, when the airplane is rotating relative to the wind. The "2" derivatives represent the *flow curvature* effect; that is, when the airplane is flying along a curved flight path, and the body- and wind-axis angular rates are identical. For instance, an arbitrary body-axis yaw rate Δr_b can be split into two components: one equal to the wind-axis yaw rate Δr_w, and another equal to the difference between the two, $(\Delta r_b - \Delta r_w)$. The first, Δr_w, multiplies the "2" derivative, and the second, $(\Delta r_b - \Delta r_w)$, multiplies the "1" derivative.

We have assumed small perturbations for the lateral-directional derivatives but the trim AOA α^* is not limited to be small since it is required to be varied to adjudge the point of onset of instability of the modes. However, note that the trim flight path angle is always zero for level flight, that is, $\gamma^* = 0$.

From Equation 1.32, we have the following relations between the perturbed body- and wind-axis roll and yaw rates:

$$\Delta p_b - \Delta p_w = \Delta\dot{\beta}\sin\alpha^*, \quad \Delta r_b - \Delta r_w = -\Delta\dot{\beta}\cos\alpha^* \tag{6.10}$$

$$\Delta p_w = \Delta\dot{\mu} - \Delta\dot{\chi}\sin\gamma^* = \Delta\dot{\mu}, \quad \Delta r_w = \Delta\dot{\chi}\cos\gamma^*\cos\Delta\mu = \Delta\dot{\chi} \tag{6.11}$$

Using these relations, the aerodynamic model in Equations 6.6 through 6.8 can be written as

$$\Delta C_Y = C_{Y\beta}\Delta\beta + C_{Yp1}\Delta\dot{\beta}\sin\alpha^*\left(\frac{b}{2V^*}\right) + C_{Yp2}\Delta\dot{\mu}\left(\frac{b}{2V^*}\right)$$
$$+ C_{Yr1}(-\Delta\dot{\beta})\cos\alpha^*\left(\frac{b}{2V^*}\right) + C_{Yr2}\Delta\dot{\chi}\left(\frac{b}{2V^*}\right) \tag{6.12}$$

$$\Delta C_l = C_{l\beta}\Delta\beta + C_{lp1}\Delta\dot{\beta}\sin\alpha^*\left(\frac{b}{2V^*}\right) + C_{lp2}\Delta\dot{\mu}\left(\frac{b}{2V^*}\right)$$
$$+ C_{lr1}(-\Delta\dot{\beta})\cos\alpha^*\left(\frac{b}{2V^*}\right) + C_{lr2}\Delta\dot{\chi}\left(\frac{b}{2V^*}\right) \tag{6.13}$$

$$\Delta C_n = C_{n\beta}\Delta\beta + C_{np1}\Delta\dot{\beta}\sin\alpha^*\left(\frac{b}{2V^*}\right) + C_{np2}\Delta\dot{\mu}\left(\frac{b}{2V^*}\right)$$
$$+ C_{nr1}(-\Delta\dot{\beta})\cos\alpha^*\left(\frac{b}{2V^*}\right) + C_{nr2}\Delta\dot{\chi}\left(\frac{b}{2V^*}\right) \tag{6.14}$$

Now we can insert the aerodynamic model of Equations 6.12 through 6.14 in the lateral-directional Equations 6.3 through 6.5 to give the complete set of equations as below:

$$\Delta\dot{\chi} = \left(\frac{g}{V^*}\right)\left\{\left(\frac{\bar{q}S}{W}\right)\left[\begin{array}{l}C_{Y\beta}\Delta\beta + C_{Yp1}\Delta\dot{\beta}\sin\alpha^*\left(\frac{b}{2V^*}\right) + C_{Yp2}\Delta\dot{\mu}\left(\frac{b}{2V^*}\right) \\ + C_{Yr1}(-\Delta\dot{\beta})\cos\alpha^*\left(\frac{b}{2V^*}\right) + C_{Yr2}\Delta\dot{\chi}\left(\frac{b}{2V^*}\right)\end{array}\right] + \Delta\mu\right\}$$

$$\tag{6.15}$$

$$\Delta\dot{p}_b = \left(\frac{\bar{q}Sb}{I_{xx}}\right)\left[\begin{array}{l} C_{l\beta}\Delta\beta + C_{lp1}\Delta\dot{\beta}\sin\alpha * \left(\frac{b}{2V^*}\right) + C_{lp2}\Delta\dot{\mu}\left(\frac{b}{2V^*}\right) \\ +C_{lr1}(-\Delta\dot{\beta})\cos\alpha * \left(\frac{b}{2V^*}\right) + C_{lr2}\Delta\dot{\chi}\left(\frac{b}{2V^*}\right) \end{array}\right] \quad (6.16)$$

$$\Delta\dot{r}_b = \left(\frac{\bar{q}Sb}{I_{zz}}\right)\left[\begin{array}{l} C_{n\beta}\Delta\beta + C_{np1}\Delta\dot{\beta}\sin\alpha * \left(\frac{b}{2V^*}\right) + C_{np2}\Delta\dot{\mu}\left(\frac{b}{2V^*}\right) \\ +C_{nr1}(-\Delta\dot{\beta})\cos\alpha * \left(\frac{b}{2V^*}\right) + C_{nr2}\Delta\dot{\chi}\left(\frac{b}{2V^*}\right) \end{array}\right] \quad (6.17)$$

6.4.1.1 Second-Order Form of the Perturbed Lateral-Directional Equations

We next write Equations 6.15 through 6.17 as a set of two second-order differential equations. First of all, the rate derivatives of the side force coefficient, C_{Yp1}, C_{Yp2}, C_{Yr1}, C_{Yr2}, are usually of lesser importance and may be ignored. Then, Equation 6.15 reduces to

$$\Delta\dot{\chi} = \left(\frac{g}{V^*}\right)\left\{\left(\frac{\bar{q}S}{W}\right)C_{Y\beta}\Delta\beta + \Delta\mu\right\} \quad (6.18)$$

Before proceeding further, for ease of algebraic manipulation, let us define some short symbols:

$$\left(\frac{\bar{q}S}{W}\right)C_{Y\beta} \equiv Y_\beta; \quad \left(\frac{\bar{q}Sb}{I_{zz}}\right)C_{n\beta} \equiv N_\beta; \quad \left(\frac{\bar{q}Sb}{I_{xx}}\right)C_{l\beta} \equiv L_\beta;$$

$$\left(\frac{\bar{q}Sb}{I_{zz}}\right)C_{np2}\left(\frac{b}{2V^*}\right) \equiv N_{p2}; \quad \left(\frac{\bar{q}Sb}{I_{zz}}\right)C_{nr1}\left(\frac{b}{2V^*}\right) \equiv N_{r1}; \quad \left(\frac{\bar{q}Sb}{I_{zz}}\right)C_{nr2}\left(\frac{b}{2V^*}\right) \equiv N_{r2};$$

$$\left(\frac{\bar{q}Sb}{I_{xx}}\right)C_{lp2}\left(\frac{b}{2V^*}\right) \equiv L_{p2}; \quad \left(\frac{\bar{q}Sb}{I_{xx}}\right)C_{lr1}\left(\frac{b}{2V^*}\right) \equiv L_{r1}; \quad \left(\frac{\bar{q}Sb}{I_{xx}}\right)C_{lr2}\left(\frac{b}{2V^*}\right) \equiv L_{r2};$$

$$\left(\frac{\bar{q}Sb}{I_{zz}}\right)C_{np1}\left(\frac{b}{2V^*}\right) \equiv N_{p1}; \quad \left(\frac{\bar{q}Sb}{I_{xx}}\right)C_{lp1}\left(\frac{b}{2V^*}\right) \equiv L_{p1} \quad (6.19)$$

So, Equation 6.18 can be compactly written as

$$\Delta\dot{\chi} = \left(\frac{g}{V^*}\right)\{Y_\beta\Delta\beta + \Delta\mu\} \tag{6.20}$$

Combining Equations 1.28 and 1.47, the perturbed wind-axis Euler angle rates can be related to the body-axis angular rates as follows:

$$\Delta\dot{\mu} = \Delta p_b \cos\alpha^* + \Delta r_b \sin\alpha^*, \quad \Delta\dot{\chi} = -\Delta p_b \sin\alpha^* + \Delta r_b \cos\alpha^* + \Delta\dot{\beta} \tag{6.21}$$

Differentiating the first of Equation 6.21 once with respect to time, and using Equations 6.16 and 6.17 on the right-hand side with the short forms of Equation 6.19 gives the following equation for roll:

$$\Delta\ddot{\mu} = L'_\beta\Delta\beta + L'_{p1}\Delta\dot{\beta}\sin\alpha^* + L'_{p2}\Delta\dot{\mu} + L'_{r1}(-\Delta\dot{\beta})\cos\alpha^* + L'_{r2}\Delta\dot{\chi} \tag{6.22}$$

where the "primed" short symbols are defined in the following manner:

$$L'_{(\cdot)} = L_{(\cdot)}\cos\alpha^* + N_{(\cdot)}\sin\alpha^* \tag{6.23}$$

Using $\Delta\dot{\chi}$ from Equation 6.20, we can complete Equation 6.22 as below:

$$\Delta\ddot{\mu} = \left[L'_\beta + \left(\frac{g}{V^*}\right)Y_\beta L'_{r2}\right]\Delta\beta + L'_{p1}\Delta\dot{\beta}\sin\alpha^* + L'_{p2}\Delta\dot{\mu} + L'_{r1}(-\Delta\dot{\beta})\cos\alpha^*$$
$$+ \left(\frac{g}{V^*}\right)L'_{r2}\Delta\mu \tag{6.24}$$

which on collecting like terms, appears as

$$\Delta\ddot{\mu} - \left[L'_\beta + \left(\frac{g}{V^*}\right)Y_\beta L'_{r2}\right]\Delta\beta + \left[L'_{r1}\cos\alpha^* - L'_{p1}\sin\alpha^*\right]\Delta\dot{\beta} - \left(\frac{g}{V^*}\right)L'_{r2}\Delta\mu$$
$$- L'_{p2}\Delta\dot{\mu} = 0 \tag{6.25}$$

Similarly, differentiating the second of Equation 6.21 and using Equations 6.16 and 6.17 yields

$$\Delta\ddot{\chi} = N'_\beta\Delta\beta + N'_{p1}\Delta\dot{\beta}\sin\alpha*$$
$$+N'_{p2}\Delta\dot{\mu} + N'_{r1}(-\Delta\dot{\beta})\cos\alpha* + N'_{r2}\Delta\dot{\chi} + \Delta\ddot{\beta} \tag{6.26}$$

The primed short symbols in Equation 6.26 are defined as below:

$$N'_{(.)} = -L_{(.)}\sin\alpha* + N_{(.)}\cos\alpha* \tag{6.27}$$

Again replacing for $\Delta\dot{\chi}$ and its derivative from Equation 6.20, and collecting like terms in Equation 6.26 gives the following equation for yaw:

$$\Delta\ddot{\beta} + \left[N'_\beta + \left(\frac{g}{V^*}\right)Y_\beta N'_{r2}\right]\Delta\beta - \left[N'_{r1}\cos\alpha* - N'_{p1}\sin\alpha* + \left(\frac{g}{V^*}\right)Y_\beta\right]\Delta\dot{\beta}$$
$$+\left(\frac{g}{V^*}\right)N'_{r2}\Delta\mu + \left[N'_{p2} - \left(\frac{g}{V^*}\right)\right]\Delta\dot{\mu} = 0 \tag{6.28}$$

Equations 6.25 and 6.28 form the set of two second-order lateral-directional small-perturbation equations. For convenience, these are collected together and presented in Table 6.2.

It can be verified that these two equations of second order in the variables $\Delta\mu$ and $\Delta\beta$ are coupled because of the $\Delta\dot{\beta}$, $\Delta\beta$ terms in the $\Delta\mu$ (rolling moment) equation, and the $\Delta\dot{\mu}$, $\Delta\mu$ terms in the $\Delta\beta$ (yawing moment) equation. Note that $\Delta\chi$ (side force) equation has been absorbed into Equations 6.25 and 6.28; hence, there is no separate equation for $\Delta\chi$. In fact, the variable $\Delta\chi$ has itself been eliminated.

TABLE 6.2 Summary of Perturbed Lateral-Directional Dynamics Equations

Yaw
dynamics

$$\Delta\ddot{\beta} + \left[N'_\beta + \left(\frac{g}{V^*}\right)Y_\beta N'_{r2}\right]\Delta\beta - \left[N'_{r1}\cos\alpha* - N'_{p1}\sin\alpha* + \left(\frac{g}{V^*}\right)Y_\beta\right]\Delta\dot{\beta}$$
$$+\left(\frac{g}{V^*}\right)N'_{r2}\Delta\mu + \left[N'_{p2} - \left(\frac{g}{V^*}\right)\right]\Delta\dot{\mu} = 0$$

Roll
dynamics

$$\Delta\ddot{\mu} - \left[L'_\beta + \left(\frac{g}{V^*}\right)Y_\beta L'_{r2}\right]\Delta\beta + \left[L'_{r1}\cos\alpha* - L'_{p1}\sin\alpha*\right]\Delta\dot{\beta}$$
$$-\left(\frac{g}{V^*}\right)L'_{r2}\Delta\mu - L'_{p2}\Delta\dot{\mu} = 0$$

6.4.1.2 Matrix Form of the Perturbed Lateral-Directional Equations

For further analysis, Equations 6.25 and 6.28 need to be cast in matrix form, as below:

$$
\begin{pmatrix} \Delta\ddot{\beta} \\ \Delta\ddot{\mu} \end{pmatrix} + \begin{bmatrix} -\left[N'_{r1}\cos\alpha^* - N'_{p1}\sin\alpha^* + \left(\dfrac{g}{V^*}\right)Y_\beta \right] & \left[N'_{p2} - \left(\dfrac{g}{V^*}\right) \right] \\ \left[L'_{r1}\cos\alpha^* - L'_{p1}\sin\alpha^* \right] & -L'_{p2} \end{bmatrix} \begin{pmatrix} \Delta\dot{\beta} \\ \Delta\dot{\mu} \end{pmatrix}
$$
$$
+ \begin{bmatrix} \left[N'_\beta + \left(\dfrac{g}{V^*}\right)Y_\beta N'_{r2} \right] & \left(\dfrac{g}{V^*}\right)N'_{r2} \\ -\left[L'_\beta + \left(\dfrac{g}{V^*}\right)Y_\beta L'_{r2} \right] & -\left(\dfrac{g}{V^*}\right)L'_{r2} \end{bmatrix} \begin{pmatrix} \Delta\beta \\ \Delta\mu \end{pmatrix} = 0 \tag{6.29}
$$

which can be briefly represented by the following matrix equation:

$$\ddot{y}_{lat} + C_{lat}\,\dot{y}_{lat} + K_{lat}\,y_{lat} = 0 \tag{6.30}$$

where

$$y_{lat} = \begin{pmatrix} \Delta\beta \\ \Delta\mu \end{pmatrix}$$

and C_{lat}, K_{lat} are the damping and stiffness matrix, respectively, from Equation 6.29. Note that the prominent entries in the damping matrix are the so-called damping terms—N'_{r1}, L'_{r1}, N'_{p1}, L'_{p1}, and these do not show up in the stiffness matrix. The main entries in the stiffness matrix are N'_β, L'_β. The flow curvature effect terms with respect to roll—L'_{p2}, N'_{p2}—appear in the damping matrix, whereas the flow curvature terms due to wind-axis yaw rate—N'_{r2}, L'_{r2}—are seen in the stiffness matrix.

6.4.1.3 Static Instability Criterion

The criterion for static instability is given by $det(K_{lat}) = 0$, which is equivalent to the Routh criterion requiring the constant term in the characteristic polynomial to be zero. This condition, labeled S_{lat}, can be easily evaluated from Equation 6.29 to be

$$S_{lat} : \left(\dfrac{g}{V^*}\right)\left[L'_\beta N'_{r2} - N'_\beta L'_{r2} \right] = 0 \tag{6.31}$$

The static instability criterion usually correlates with the onset of spiral mode instability; that is the spiral eigenvalue being at the origin on the complex plane. However, our interest presently is more in the onset of Dutch roll instability. Similar to the case of static instability, there is an exact criterion for the oscillatory or "dynamic" instability given by the Routh discriminant; however, that criterion does not yield a simple expression that can be usefully employed. So, before we can derive a condition for Dutch roll instability onset, we need to further approximate the dynamical model in Equation 6.29.

6.4.1.4 Approximation by Hamiltonian Dynamical System

Typically, at moderate values of AOA α^*, the eigenvalues of the lateral-directional model in Equation 6.27 turn out to be a complex pair of lightly damped poles representing the Dutch roll mode and a second complex pair of low-damped poles, called the *lateral phugoid*, formed by the merger of the roll (rate) and spiral eigenvalues. This is as seen in Figure 6.16. Thus, the lateral-directional dynamics as a whole is poorly damped around the point of onset of the dynamic instability. Therefore, one may justifiably approximate the lateral-directional dynamics as an undamped system. In the traditional way of presenting the lateral-directional dynamic equations, it is quite impossible to identify which terms represent the "damping" and to isolate them in order to arrive at an undamped approximation. However, by transforming the lateral-directional dynamics into second-order equations and presenting them in a matrix form as in Equation 6.29, the damping terms are clearly separated. By setting the supposedly negligible C_{lat} matrix to zero, one can obtain an undamped approximation to the lateral-directional dynamics as follows:

$$\begin{pmatrix} \Delta\ddot{\beta} \\ \Delta\ddot{\mu} \end{pmatrix} + \begin{bmatrix} \left[N'_\beta + \left(\dfrac{g}{V^*}\right)Y_\beta N'_{r2} \right] & \left(\dfrac{g}{V^*}\right)N'_{r2} \\ -\left[L'_\beta + \left(\dfrac{g}{V^*}\right)Y_\beta L'_{r2} \right] & -\left(\dfrac{g}{V^*}\right)L'_{r2} \end{bmatrix} \begin{pmatrix} \Delta\beta \\ \Delta\mu \end{pmatrix} = 0 \qquad (6.32)$$

which is equivalently represented by

$$\ddot{y}_{lat} + K_{lat}\, y_{lat} = 0 \qquad (6.33)$$

Equation 6.33 is in the form of a Hamiltonian system. The onset of dynamic instability in the damped lateral-directional dynamics Equation 6.29 can then be approximated by the dynamic instability of the undamped (Hamiltonian) system in Equation 6.32.

Homework Exercise: Read up some basic introduction to Hamiltonian systems and their dynamics in case you are not already familiar with them.

The eigenvalues of a Hamiltonian system are constrained to be symmetric about the origin in the complex plane. This means that only four possible arrangements of eigenvalues can occur:

1. Two pairs of complex conjugate eigenvalues on the imaginary axis; $\pm i\omega_1, \pm i\omega_2$

2. Two pairs of real eigenvalues; $\pm\sigma_1, \pm\sigma_2$

3. One pair of complex conjugate eigenvalues on the imaginary axis, another pair of real eigenvalues; $\pm i\omega, \pm\sigma$

4. Two pairs of complex conjugate eigenvalues; $\pm(\sigma \pm i\omega), \sigma \neq 0$

Of these, only the first option indicates stable behavior; the other three represent various forms of instability. When the eigenvalues in the first option move along the imaginary axis toward each other with a varying parameter (in our case, the trim AOA α^*) and collide, they can branch out as a pair of complex conjugate eigenvalues (mirror image of each other) as in the fourth option. This is indicated graphically in Figure 6.17 where the arrows indicate the direction of movement of the eigenvalues with varying

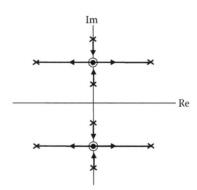

FIGURE 6.17 Dynamic instability mechanism in Hamiltonian systems.

parameter. This is the dynamic instability mechanism in Hamiltonian systems and the onset of dynamic instability occurs at the point where the eigenvalues collide and sit atop each other on the imaginary axis before separating into each half plane. Technically, this is also known as a Hamiltonian Hopf bifurcation mechanism.

6.4.1.5 Dynamic Instability Mechanism

The condition for the onset of dynamic instability in the system of Equation 6.33 can be given by the discriminant of the following quadratic equation in λ^2 being zero:

$$|\lambda^2 I + K_{lat}| = 0 \tag{6.34}$$

Or equivalently, by defining $\lambda^2 = \tau$, by the discriminant of the following quadratic equation in τ being zero:

$$|\tau I + K_{lat}| = 0 \tag{6.35}$$

which may be explicitly written down as follows:

$$\begin{vmatrix} \tau + \left[N'_\beta + \left(\dfrac{g}{V^*} \right) Y_\beta N'_{r2} \right] & \left(\dfrac{g}{V^*} \right) N'_{r2} \\[2ex] - \left[L'_\beta + \left(\dfrac{g}{V^*} \right) Y_\beta L'_{r2} \right] & \tau - \left(\dfrac{g}{V^*} \right) L'_{r2} \end{vmatrix} = 0 \tag{6.36}$$

Expanding Equation 6.36, we get

$$\tau^2 + \left\{ \left[N'_\beta + \left(\frac{g}{V^*} \right) Y_\beta N'_{r2} \right] - \left(\frac{g}{V^*} \right) L'_{r2} \right\} \tau - \underbrace{\left(\frac{g}{V^*} \right) \left[L'_\beta N'_{r2} - N'_\beta L'_{r2} \right]}_{S_{lat}} = 0 \tag{6.37}$$

where the last term with underbraces is the same as S_{lat} in Equation 6.31. The condition that the discriminant be zero gives the dynamic instability criterion:

$$D_{lat} : \left\{ \left[N'_\beta + \left(\frac{g}{V^*} \right) Y_\beta N'_{r2} \right] - \left(\frac{g}{V^*} \right) L'_{r2} \right\}^2 - 4 S_{lat} = 0 \tag{6.38}$$

Equation 6.38 gives the approximate condition for the onset of oscillatory instability in the lateral-directional dynamics, which usually corresponds to a pair of Dutch roll eigenvalues lying on the imaginary axis representing the onset of wing rock. Applying the D_{lat} criterion to the same F-18 airplane data used to compute the results in Figure 6.16, we can get a plot of the variation of the expression in Equation 6.38 with trim AOA. The result is shown in Figure 6.18. The condition in Equation 6.38, $D_{lat} = 0$, is met at only one value of trim AOA, that is at 21.5 deg, which is a very close approximation to the trim AOA of 22 deg for the onset of Dutch roll instability in Figure 6.16. Thus, the D_{lat} criterion may be used as a predictor for wing rock onset.

The advantage of using a wing rock predictor such as the D_{lat} criterion during the aircraft design process is that it is quite simple to evaluate and at the same time fairly accurate. The leading term in the D_{lat} criterion is N_β', which using Equation 6.27 expands as

$$N_\beta' = -L_\beta \sin \alpha^* + N_\beta \cos \alpha^* \tag{6.39}$$

And on using the full form of the symbols from Equation 6.19, we can write Equation 6.39 in terms of the aerodynamic derivatives as

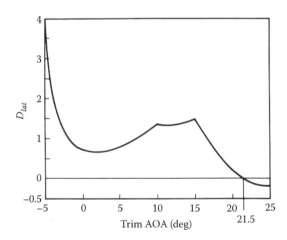

FIGURE 6.18 Plot of the variation of the D_{lat} criterion with trim AOA for the same data used to produce Figure 6.16.

$$\left(\frac{\overline{q}Sb}{I_{zz}}\right)C_{n\beta}\cos\alpha^{\star} - \left(\frac{\overline{q}Sb}{I_{xx}}\right)C_{l\beta}\sin\alpha^{\star} = \left(\frac{\overline{q}Sb}{I_{zz}}\right)\underbrace{\left[C_{n\beta}\cos\alpha^{\star} - \left(\frac{I_{zz}}{I_{xx}}\right)C_{l\beta}\sin\alpha^{\star}\right]}_{C_{n\beta,dyn}}$$

(6.40)

where, as indicated, the term with the underbrace is usually called $C_{n\beta,dyn}$. Thus, $C_{n\beta,dyn}$ is an important parameter related to the possible onset of airplane wing rock. In fact, if the effect of gravity that enters Equation 6.38 via the $(g/V^{\star})$ terms is neglected, and typically the S_{lat} term that represents the spiral mode eigenvalue is quite small, then $C_{n\beta,dyn}$ is by far the predominant term in the expression for D_{lat}. For this reason, the traditional use of $C_{n\beta,dyn}$ as a first-approximation indicator for wing rock onset is justified.

6.5 LATERAL-DIRECTIONAL FEEDBACK CONTROL SYSTEM

In almost every instance, the lateral-directional flight dynamics will be coupled to the airplane longitudinal dynamics. Thus, while the lateral-directional FCS manages the demanded maneuver, the longitudinal FCS will act in consort to maintain the longitudinal flight parameters. Truly coupled lateral-longitudinal flight and its control will be addressed in the next chapter. For the moment, we shall look at the lateral-directional FCS in isolation. For maneuvers that are predominantly lateral or directional, this is a good starting point. For example, an airplane that banks into a turn will lose altitude. So while lateral-directional FCS manages the bank/turn maneuver, the longitudinal FCS will act to maintain altitude/velocity/AOA as desired. For now, we assume that the longitudinal FCS is doing its job and focus solely on the lateral-directional FCS.

The following discussion closely parallels the presentation of the longitudinal FCS in Chapter 5. Indeed, the structures are fundamentally similar, though there are obvious differences in detail. Similar to Figure 5.3 for the longitudinal FCS, the lateral-directional FCS structure may be represented as shown in Figure 6.19 in the "back-stepping" form. For convenience, the block diagram from Figure 5.3 is reproduced at the top of Figure 6.19.

Of the three sequential blocks, the outermost controls the flight path. Together with the longitudinal FCS flight path block, the pilot can command an acceleration, a climb, or a turn. The flight path block computes the desired variation of the navigation variables—velocity V, flight path angle γ, and heading angle χ—required to fly the trajectory commanded by

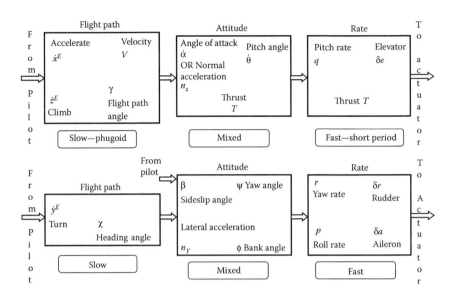

FIGURE 6.19 Generic structure of lateral-directional FCS—sequential, "back-stepping" type.

the pilot. The demand in the navigation variables gives rise to the desired levels of thrust, and the normal and lateral accelerations—these are part of the intermediate attitude block. One more variable input to the attitude block is the sideslip angle β. This is mostly required to be zero except in the odd case, such as the crosswind landing example discussed earlier.

The attitude block calculates the change in the attitude angles—the pitch, yaw, and roll Euler angles—in order to fly the maneuver commanded by the pilot. The rates of change of the attitude angles can be used to derive the demanded body-axis rates, p, q, and r. These are met by appropriate deflections of the aerodynamic controls—elevator, rudder, and aileron—and the thrust (throttle). This is part of the innermost rate block.

Please note that Figure 6.19 presents some kind of an ideal, sensible FCS structure of the "back-stepping" kind. In practice, many variations of this theme are possible with regard to the blocks, their interconnections, the choice of commanded and computed variables, etc.

The back-stepping FCS structure is predicated on the assumption that the rate loop dynamics will be faster than that of the attitude loop, which in turn will be faster than the flight path loop dynamics. These timescales are to be set during the control design process.

The lateral-directional FCS structure in Figure 6.19 can be used to examine the two maneuvers studied in this chapter—the turn and the nonzero sideslip trims. For a zero-β turn, the turn maneuver decides the rate of change of heading angle, which calls for a certain lateral acceleration profile. This requires a certain variation of bank angle while sideslip angle is held at zero. The aileron is used to control the bank angle and the rudder holds the sideslip to zero. Now, take the case of wings level, nonzero sideslip flight. The sideslip angle can be prescribed as input to the attitude block and held by use of the rudder. The ailerons may be used to hold wings level despite the aerodynamic influence of the sideslip angle.

However, it must be noted that the lateral and directional variables are coupled in many ways and it is often difficult to distinguish one effect from the other. There are also many variants of the lateral-directional control effectors—spoilers, differential horizontal tail, and even roll control using rudder. So, the above discussion should be used only as an elementary theoretical presentation—there is more to this business than meets the eye.

Homework Exercise: Prepare a chart for the lateral-directional FCS structure in Figure 6.19 similar to that presented in Figure 5.4 for the longitudinal FCS clearly marking the various levels and the surrogate commands and controls at each level.

Figure 6.20 shows a typical three-loop structure for a lateral-directional flight control law. The outer (flight path) loop commands the heading. Change in heading is obtained by commanding a lateral acceleration (Latax) or bank angle. The middle (attitude) loop commands the Latax/bank angle and the sideslip. The roll and yaw rates required to meet these demands are calculated. The inner (rate) loop enforces the demanded roll and yaw rates by suitably commanding the aileron and rudder deflections.

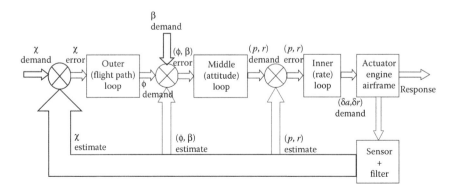

FIGURE 6.20   Typical three-loop lateral-directional feedback control law.

The control demands are input to the actuator/airframe block and the variables being fed back are sensed/estimated by the filter block.

Each of the control loops in Figure 6.20 may be closed using one among a variety of control design techniques, as for example the PID law in Equation 5.1 or the dynamic inversion law discussed in Section 5.6.

Homework Exercise: Sketch out a dynamic inversion control law algorithm for the feedback control law structure in Figure 6.20 along similar lines to that presented in Section 5.6 and Figure 5.12 for the longitudinal case.

EXERCISE PROBLEMS

1. Determine a zero sideslip level turn equilibrium state for low-angle-of-attack F-18 model in Chapter 2.

2. Linearize the equations of aircraft motion. Carry out modal analysis of the equilibrium state computed in exercise 1.

3. Simulate linear equations of motion developed in exercise 2 using step inputs in aileron and rudder and observe the response around a stable and an unstable level trim state.

4. Use the system matrix in Section 6.2 and an appropriate set of initial conditions to simulate the linear modes as described based on analysis of eigenvectors.

5. For the low alpha F-18 model, determine the angle of attack at which the onset of wing rock motion takes place in level flight. Also, determine the angles of attack at which spiral and phugoid instabilities occur in level flight.

6. Simulate various unstable modes in exercise 5 using the linear model of aircraft at post-instability trim angles of attack. Use step or impulse control inputs to appropriate controls or appropriate set of initial conditions.

7. In Section 6.2, we found that coupled phugoid and coupled spiral mode eigenvalues being close to imaginary axis were the most likely to cause instabilities for the F-18/HARV model. Is it the same case for the low AOA F-18 model? Verify!

8. Derive expressions for damping derivatives in Equation 6.9 in terms of wing parameters [3].

REFERENCES

1. Nolan, R. C., Wing rock prediction method for a high performance fighter aircraft, Thesis, Air Force Institute of Technology, AFIT/GAE/ENY/92J-02, 1992.
2. https://www.slideshare.net/Project_KRIT/goman-modeling-and-analysis-of-aircraft-spin
3. Sinha, N. K., and Ananthkrishnan, N., *Elementary Flight Dynamics with an Introduction to Bifurcation and Continuation Methods*, CRC Press, Boca Raton, Florida, USA, 2013.

CHAPTER 7

Coupled Lateral–Longitudinal Flight Dynamics

IN CHAPTER 4, WE HAVE DISCUSSED airplane trim, stability, and maneuvers limited to the longitudinal plane. Then, in Chapter 6, we looked at trim and stability of motions out of the longitudinal plane, which involved changes in the lateral and directional variables. However, more often than not, the lateral-directional motion is coupled with the longitudinal dynamics leading to a quite general six degree of freedom motion involving changes in all the variables. In some instances, such a coupled lateral–longitudinal motion may be flown intentionally, such as a descending turn where the airplane spirals down with every horizontal loop. At other times, coupled motions may be inadvertent—a classic example is *spin*. Even motions that are predominantly lateral-directional in nature may take on a slightly different character due to coupling with the longitudinal variables. One example is that of limit cycle oscillations called *wing rock*, which we analyzed in Chapter 6 using a lateral-directional approximation. However, there are instances where the nature of the *wing rock* limit cycles may be significantly altered due to coupling with the longitudinal dynamics. For instance, whereas the traditional *wing rock* limit cycles usually emerge at a supercritical Hopf bifurcation (HB) and grow in the standard manner as sketched in Figure 7.1a, the large-amplitude *wing rock* limit cycles may be the result of a subcritical HB followed by a

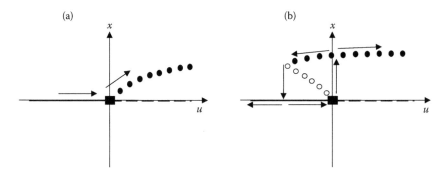

FIGURE 7.1 (a) Standard wing rock limit cycles at a supercritical Hopf bifurcation and (b) large-amplitude wing rock limit cycles at a subcritical Hopf bifurcation followed by a fold bifurcation (solid lines: stable equilibrium solution branch; dashed lines: unstable equilibrium solution branch; empty circles: unstable limit cycles; solid circles: stable limit cycles).

fold bifurcation (FB) of the branch of limit cycles, as depicted in Figure 7.1b. The onset of *wing rock* in the two cases in Figure 7.1 is distinct. In the traditional wing rock, the oscillations slowly build up as a parameter (usually the angle of attack) is increased. On the other hand, in case of large-amplitude *wing rock*, there is a sudden, finite-amplitude oscillation at onset. Also, the bifurcation diagram for large-amplitude *wing rock* in Figure 7.1 shows possible hysteresis, which is absent in the traditional *wing rock*.

There are three major sources of coupling between the lateral and longitudinal dynamics ("lateral" as used here is taken to mean "lateral-directional"). The first is the kinematic coupling in the airplane translational dynamics. With reference to Equation 1.59, the terms of interest are:

$(-p \sin \alpha + r \cos \alpha)$ in the equation for $\dot{\beta}$, in which the first term approximates to $-p\alpha$

$(p \cos \alpha + r \sin \alpha) \sin \beta$ in the equation for $\dot{\alpha}$, which approximates to $p\beta$

Notice that these terms, $-p\alpha$ and $p\beta$, are nonlinear. What they suggest is, if an airplane flying with a nonzero angle of attack has a roll rate, then it creates sideslip. Likewise, in the presence of a nonzero sideslip angle, a roll rate can create an angle of attack. Physically, this phenomenon is very relevant for rapidly rolling airplanes. In these cases, an initial angle of attack may be converted to a sideslip angle and then to a negative angle of attack, and so on. If the timescale of the rolling maneuver is short, then the airplane's inherent stability in the longitudinal and directional sense will not be able to suppress the angle of attack or sideslip angle created within that short time frame. On the other hand, if the rolling motion is relatively slower, then, while the above kinematic coupling effect still exists, it is not so dominant. The airplane's inherent stability in the short period and Dutch roll modes will, at least partly, kill the angle of attack or sideslip angle created by the kinematic coupling mechanism.

Homework Exercise: Usually, a rapidly rolling airplane will roll approximately about its body X axis, whereas an airplane with a relatively slower roll rate will tend to roll about its velocity vector. Based on the above discussion, try to explain why this may be so.

The second source of coupling is due to inertia coupling terms in the equations of rotational motion, Equation 1.78. These are the terms below:

$$
\begin{bmatrix}
-pqI_{xz} - qr(I_{yy} - I_{zz}) \\
(p^2 - r^2)I_{xz} - pr(I_{zz} - I_{xx}) \\
qrI_{xz} - pq(I_{xx} - I_{yy})
\end{bmatrix}
$$

And they contribute to the rolling moment, pitching moment, and yawing moment equations, respectively. These terms are nonlinear as well. Essentially, they imply that angular velocity components about two of the body axes will create a moment about the third axis. For instance, consider the term $-pr(I_{zz} - I_{xx})$ in the pitching moment equation. This suggests that an airplane having a roll rate p and yaw rate r (both lateral-directional variables) will experience a pitching moment (longitudinal dynamics), which will induce a pitch rate q.

The kinematic and inertial coupling effects are common to all airplanes and are independent of the airplane configuration but for the moments of inertia. However, since all the terms in these cases are nonlinear, as long as the variables involved—p, q, r, α, β—are relatively small, their products may be considered negligible. Hence, the impact of these coupling effects is not heavily felt in mild maneuvers. However, for example, in flight at high angles of attack, or in rapid rolling maneuvers, these terms may become quite significant and can have a major impact on the airplane dynamics.

We shall examine a couple of such instances below where kinematic and inertial coupling effects can dominate the dynamics.

The third source of lateral–longitudinal coupling is the aerodynamics. This effect is very airplane dependent and varies widely from one airplane to another, so it is difficult to make generalizations. There may also be aerodynamic coupling effects due to deflection of control surfaces. For example, the effect due to deflection of right and left aileron, or right and left elevator, may not be symmetric. The asymmetry in aerodynamic force/moment can create a coupling effect. In addition, there may be coupling effects because of the engines, for example, due to rotating components. There may also be lateral–longitudinal coupling effects due to the control system, often deliberately introduced.

Since the coupling effects are typically nonlinear, their influence is largely felt in flight regimes where either the rotation rate is large or the angle of attack is high or both. Figure 2.13 shows the different flight regimes on a map of rotation rate versus angle of attack. Nonlinear dynamics at stall and in the post-stall regimes are seen at high angles of attack. Large roll rates at lower angles of attack are responsible for the roll coupling problem. Spin is a phenomenon that involves both high angles of attack and large rotation rates.

In this chapter, we shall apply much of what we have learned throughout this book to problems of coupled lateral–longitudinal airplane maneuvers. We shall investigate coupled roll maneuvers using the pseudo-steady-state (PSS) model presented in Chapter 1 with the aircraft data in Chapter 2. We shall use the F-18 HARV model to study airplane dynamics at high angles of attack, especially spin. Control techniques based on bifurcation theory and dynamic inversion presented earlier in this book will be demonstrated.

7.1 INERTIA COUPLED ROLL MANEUVERS

Military airplanes are usually required to have a certain minimum roll rate capacity, given in terms of nondimensional roll rate, $pb/2V$. At these roll rates, the kinematic coupling effect seen earlier is obviously present. In addition, modern combat aircraft are usually slender, fuselage-heavy configurations. As a result, the axial moment of inertia I_{xx} is significantly smaller than the transverse moments of inertia, I_{yy} and I_{zz}. Consequently, the inertia coupling terms in the pitch and yaw equations, as marked below, are significant, implying significant coupling between the yawing (directional) motion and pitching (longitudinal) motion, especially in the presence of large roll rate, p.

$$
\begin{bmatrix}
-pqI_{xz} - \underbrace{qr(I_{yy} - I_{zz})}_{\text{smaller}} \\[2ex]
(p^2 - r^2)I_{xz} - \underbrace{pr(I_{zz} - I_{xx})}_{\text{large}} \\[2ex]
qrI_{xz} - \underbrace{pq(I_{xx} - I_{yy})}_{\text{large}}
\end{bmatrix}
$$

As explained in Chapter 1, the set of equations used to study airplane dynamics in constant-velocity rolling maneuvers is as listed in Table 1.9 and reproduced below.

$$
[0] + \begin{bmatrix} -D\cos\beta + Y\sin\beta \\ D\sin\beta + Y\cos\beta \\ -L \end{bmatrix} + \begin{bmatrix} T\cos\beta\cos\alpha \\ -T\sin\beta\cos\alpha \\ -T\sin\alpha \end{bmatrix}
$$

$$
= m \begin{bmatrix} 0 \\ ((-p\sin\alpha + r\cos\alpha) + \dot{\beta})V \\ ((p\cos\alpha + r\sin\alpha)\sin\beta - (q - \dot{\alpha})\cos\beta)V \end{bmatrix} \tag{1.59}
$$

$$
\begin{bmatrix} \mathcal{L} \\ M \\ N \end{bmatrix} = \begin{bmatrix} I_{xx}\dot{p} - I_{xz}\dot{r} \\ I_{yy}\dot{q} \\ -I_{zx}\dot{p} + I_{zz}\dot{r} \end{bmatrix} + \begin{bmatrix} -pqI_{zx} - qr(I_{yy} - I_{zz}) \\ (p^2 - r^2)I_{zx} - pr(I_{zz} - I_{xx}) \\ qrI_{xz} - pq(I_{xx} - I_{yy}) \end{bmatrix} \tag{1.78}
$$

The first of Equation 1.59 is what constrains the velocity to remain constant; hence, $\dot{V} = 0$. That leaves five differential equations in the five state variables—α, β, p, q, r. This is called the PSS model (see Reference 1). The aerodynamic model and the aero data used for simulations have been presented in Chapter 2 (Tables 2.5 and 2.6).

In vector form, the PSS model can be represented as follows:

$$
\underline{\dot{x}} = \underline{f}(\underline{x}, \underline{U}) \tag{7.1}
$$

where $\underline{x} = [\alpha, \beta, p, q, r]$ is the vector of state variables of aircraft and $\underline{U} = [\delta a, \delta r, \delta e]$ is the vector of control inputs.

Homework Exercise: Carry out a bifurcation analysis of the PSS model with data as above to identify trim and stability in longitudinal flight. Since thrust data are not available, it is not possible to judge whether the trims so obtained correspond to level flight or not. It may simply be assumed that the thrust can be suitably set so as to assure level flight trims. In any case, that information is not crucial to the rest of the analysis.

We shall select a trim state at $\alpha^* = -2$ deg as the initial state from which to initiate the roll maneuver. At low angles of attack, the linear aerodynamic model in Table 2.6 is adequate. Then, the nonlinearity enters the equations of motion only through the kinematic and inertia coupling terms. The initial trim state can be identified from the bifurcation analysis in the Homework Exercise above or otherwise as follows:

$$x^* = [\alpha, \beta, p, q, r] = \left[-2 \deg, 0, 0, -0.1837 \deg/s, 0 \right]$$
$$u^* = [\delta a^*] = [0]$$
$$p^* = [\delta r^*, \delta e^*] = [0, 2 \deg]$$

Aileron deflection δa is the primary parameter for roll maneuvers. With the other parameters held fixed, a standard bifurcation analysis may be carried out with aileron deflection as the continuation parameter. Results from such an SBA computation are shown in Figure 7.2.

Examining the plot of roll rate p versus aileron deflection δa, for small δa, the relation between PSS roll rate p and δa is approximately linear and the states are stable. Negative aileron deflection gives positive roll rates and vice versa, which is as expected, following the sign convention established in Chapter 2. For both positive and negative δa, with increasing aileron deflection, the primary PSS branch encounters a saddle-node bifurcation (SNB) (e.g., point marked "A" in Figure 7.2) where it folds over. Between points "A" and "B," the PSS solutions are unstable; likewise for negative values of roll rate. Point "B" is another SNB point where the branch of PSS solutions folds over once again giving rise to a set of stable, large-amplitude roll rate states ("B–E"). Similarly, for negative PSS roll rate solutions, the saddle-node point at "D" gives the large-amplitude branch "D–C."

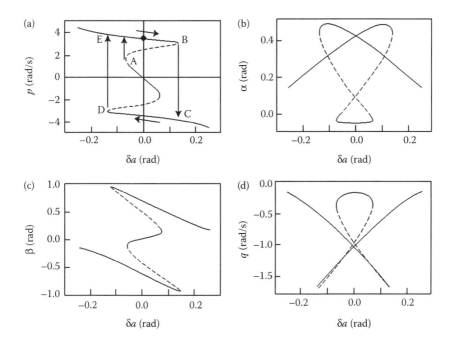

FIGURE 7.2   Bifurcation diagram of PSS variables ((a) roll rate p, (b) angle-of-attack α, (c) sideslip angle β, and (d) pitch rate q) as functions of aileron deflection at fixed $\delta e = 2$ deg, $\delta r = 0$ (solid lines: stable solutions; dashed lines: unstable solutions).

Several interesting observations can be made from the bifurcation diagram in Figure 7.2. First of all, notice that stable PSS roll rate solutions are possible only between $\delta a = 0$ and the critical aileron deflection $(\delta a_{cr} = -0.07$ rad) corresponding to the saddle-node point "A." Further, stable roll rate states are possible on the large-amplitude branch "B–E." Likewise for negative PSS roll rate states. Between the points "A" and "B," no stable positive roll rate solutions exist; likewise, the negative roll rate branch terminating at point "D" is unstable. Second, for $\delta a = 0$, there are five possible PSS roll rate solutions—one is the default solution $p = 0$, two others are unstable, and the other two are on the large-amplitude branches "B–E" and "D–C." The last two solutions suggest that a large (pseudo-) steady-state roll rate is possible even with ailerons centered, $\delta a = 0$. These solutions are called *autorotation* states.

In practice, as the aileron deflection is gradually increased starting at $\delta a = 0$, the PSS roll rate increases until the saddle-node point "A" is encountered. On further increasing δa, the roll rate shows a "jump" to the large-amplitude branch "B–E," as marked by the up-arrow in

Figure 7.2. A sudden increase in roll rate of this magnitude, well beyond the nearly linear increase that may be expected, can be fairly disconcerting to the pilot. From the post-jump point, if δa is decreased in an attempt to cut down the roll rate, it can be seen that the fall in roll rate is marginal. Eventually, for $\delta a = 0$, the PSS roll rate is still on the large-amplitude branch at the point marked by the filled circle in Figure 7.2. The airplane is then in a state of *autorotation*—rolling despite the ailerons being centered.

Further applying aileron in the same sense, gradually brings the PSS roll rate to the point "B." Over this stretch, the sense of the roll rate is actually opposite to what is expected; a phenomenon called *roll reversal*. Moving beyond point "B," there is a jump marked "B–C" taking the roll rate to the other large-amplitude branch with the sign of the roll rate being flipped. If the aileron inputs are applied gradually, in a quasi-steady manner, then the variation of the PSS roll rate may resemble a hysteresis loop "B–C–D–E–B" as indicated in Figure 7.2. Of course, the aileron inputs may be applied in a more vigorous, transient manner to break out of the large-amplitude state or a hysteresis cycle. Nevertheless, phenomena such as *autorotation* and *roll reversal* due to the bifurcation structure of Figure 7.2 are not unfamiliar. Of course, these phenomena may also be caused by other physical effects, including aeroelastic effects that are not considered in this book. However, the picture painted by the bifurcation diagram in Figure 7.2 is quite general and applicable to many airplanes since it is only a function of the kinematic and inertial coupling terms, which are common to all airplane models. The aerodynamic model used to generate the diagram of Figure 7.2 is linear and does not largely influence the qualitative nature of the bifurcations.

Examining the other PSS variables in Figure 7.2, the sideslip angle is well limited on the primary stable branch. However, the sideslip values on the stable, large-amplitude branches are huge—almost unrealistic. Not surprisingly, owing to the kinematic coupling effects, the large-amplitude PSS angle of attack solutions are also large. The primary PSS branch shows a slightly negative value of angle of attack suggesting that the airplane is undergoing a rolling pull-down maneuver. Correspondingly, the PSS pitch rate solutions are also negative. In practice, at the large values of angle of attack and sideslip for the large-amplitude PSS branch, nonlinear aerodynamic modeling is necessary, which may change the results in Figure 7.2 numerically. However, bifurcations such as those at point "A," which occur at low values of α and β are unlikely to be affected by the modeling of aerodynamic nonlinearities. Hence, the critical condition identified by the bifurcation at point "A" is quite realistic. In practice, the allowable aileron deflection may be limited by the critical value ($\delta a_{cr} = -0.07$ rad) and correspondingly the PSS roll rate will also be limited by the value at point "A," which is approximately 2 rad/s.

We can check the predictions made by the bifurcation analysis in Figure 7.2 by carrying out a time simulation of the airplane dynamics for a specific sequence of control inputs. To carry out this simulation, the following set of equations from Chapter 1 are used:

Velocity (wind-axis):

$$
\begin{bmatrix} -mg\sin\gamma \\ mg\sin\mu\cos\gamma \\ mg\cos\mu\cos\gamma \end{bmatrix} + \begin{bmatrix} -D\cos\beta + Y\sin\beta \\ D\sin\beta + Y\cos\beta \\ -L \end{bmatrix} + \begin{bmatrix} T\cos\beta\cos\alpha \\ -T\sin\beta\cos\alpha \\ -T\sin\alpha \end{bmatrix}
$$

$$
= m \begin{bmatrix} \dot{V} \\ ((-p\sin\alpha + r\cos\alpha) + \dot{\beta})V \\ ((p\cos\alpha + r\sin\alpha)\sin\beta - (q - \dot{\alpha})\cos\beta)V \end{bmatrix} \tag{1.59}
$$

Angular velocity (body-axis):

$$
\begin{bmatrix} \mathcal{L} \\ M \\ N \end{bmatrix} = \begin{bmatrix} I_{xx}\dot{p} - I_{xz}\dot{r} \\ I_{yy}\dot{q} \\ -I_{zx}\dot{p} + I_{zz}\dot{r} \end{bmatrix} + \begin{bmatrix} -pqI_{zx} - qr(I_{yy} - I_{zz}) \\ (p^2 - r^2)I_{zx} - pr(I_{zz} - I_{xx}) \\ qrI_{xz} - pq(I_{xx} - I_{yy}) \end{bmatrix} \tag{1.78}
$$

Orientation (body-axis)

$$
\begin{bmatrix} \dot{\phi} \\ \dot{\theta} \end{bmatrix} = \begin{bmatrix} 1 & \tan\theta\sin\phi & \tan\theta\cos\phi \\ 0 & \cos\phi & -\sin\phi \end{bmatrix} \begin{bmatrix} p \\ q \\ r \end{bmatrix} \tag{1.25}
$$

Of these, velocity is set to a fixed value and the $\dot{V}$ equation in Equation 1.59 is not used, leaving a set of seven differential equations in seven state

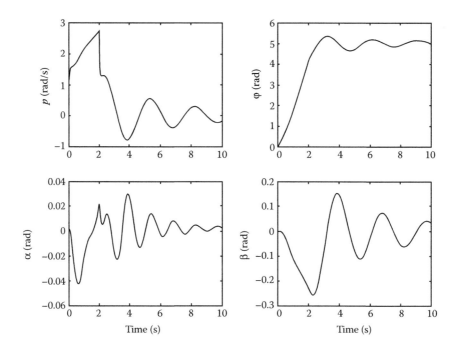

FIGURE 7.3 Time simulation results showing jump in roll rate for 2 s pulse in aileron (amplitude −6 deg) and elevator (amplitude 2 deg) at zero rudder deflection.

variables. A sample simulation with rudder deflection fixed at $\delta r = 0$ and a two-second aileron pulse of magnitude $\delta a = -6$ deg and two-second elevator pulse of magnitude 2 deg together during $t = 0$–2 s is shown in Figure 7.3. The sharp buildup in roll rate between $t = 0$ and $t = 2$ s is clearly visible. Withdrawal of the aileron pulse at $t = 2$ s leads to a sudden drop in roll rate followed by a gradual settling down to the steady-state value of $p = 0$. For this sharp (not gradual) aileron input, hysteresis and other phenomena described above are not seen.

Homework Exercise: Provide other types of control inputs for this airplane model and try to capture the different nonlinear phenomena discussed previously.

7.1.1 Zero-Sideslip Roll Maneuvers

As we have seen repeatedly, airplane roll and yaw dynamics are inevitably coupled. An aileron input, as in the previous example, in addition to creating a rolling moment that gives rise to a roll rate, also causes a yawing moment. This is because of the differential drag created due to up-going and down-going aileron surfaces. On the other hand, sideslip influences the roll dynamics through the dihedral effect. Because of this coupling between the roll and yaw dynamics, it is normal to apply the aileron and rudder in conjunction. While the aileron commands the roll maneuver, the rudder is used to control the sideslip and hence the yawing motion. For instance, one can study roll maneuvers with an added constraint that requires the sideslip to be maintained at a fixed value (usually zero) at all the pseudo-steady states.

Let us consider an additional constraint equation for zero sideslip,

$$\beta = 0 \tag{7.2}$$

with the PSS equations used earlier for bifurcation analysis of the roll maneuver. With the constraint condition included, we need to release one of the parameters to be used to satisfy the constraint Equation 7.2, and employ the extended bifurcation analysis (EBA) technique. The variables in the PSS model for the EBA computation are as follows:

$$x^* = [\alpha, \beta, p, q, r]; \quad p_1^* = \delta r^*$$
$$u^* = [\delta a^*]$$
$$p_2^* = [\delta e^*] = [2 \text{ deg}]$$

In this formulation, $p_1^* = \delta r^*$, the rudder deflection is the parameter freed to allow the constraint on sideslip in Equation 7.2 to be met while $u^* = \delta a^*$ is used as the continuation parameter. Results of the EBA computation are presented in Figure 7.4.

The variation of the rudder as a function of the aileron deflection, plotted in Figure 7.4, is called the aileron–rudder interconnect (ARI) law. It is seen that with increasing aileron deflection designed to give a positive roll rate, increasing left rudder (positive) is needed to maintain zero sideslip. In fact, stable zero-sideslip PSS solutions are seen up to the transcritical bifurcation in Figure 7.4 at a roll rate of approximately 2 rad/s. The p–δa relationship up to this point is approximately linear, which is desirable. Since the sideslip is constrained, there is no notable buildup in angle of attack and the pitch rate is also nearly constant. Thus, effectively, the $\beta = 0$ constraint counteracts the kinematic coupling effect.

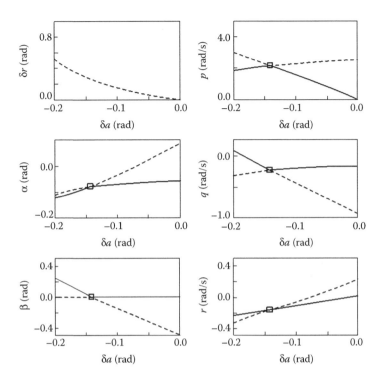

FIGURE 7.4 Zero sideslip ARI law (top left) and corresponding PSS solutions (solid lines: stable solutions; dashed lines: unstable solutions; empty square: transcritical bifurcation).

Beyond the transcritical bifurcation point, subject to the rudder position limit that is not modeled in this simulation, continued zero-sideslip PSS solutions are possible. However, these solutions are no longer stable, hence cannot be sustained in flight. Instead, an alternative branch of stable PSS solutions emerges from the transcritical bifurcation, which shows increasing sideslip angle with further aileron deflection. Hence, practically, it is no longer feasible to maintain zero-sideslip roll maneuvers beyond the limit set by the appearance of the transcritical bifurcation. The post-transcritical stable states show a nearly constant value of roll rate or a slight decrease with increasing aileron deflection. Hence, no purpose is served by flying the airplane at values of aileron deflection beyond the transcritical bifurcation.

The maximum value of roll rate in Figure 7.4 is very similar to the maximum roll rate seen in the bifurcation diagram of Figure 7.2 without any constraint. The difference is that in case of Figure 7.4, the effect of going beyond the critical value of aileron deflection is benign (sideslip begins to

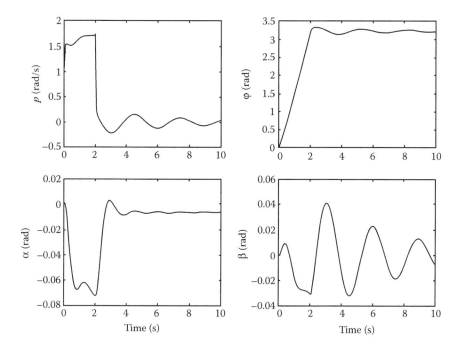

FIGURE 7.5   Time simulation results for 2 s pulse in aileron (amplitude −6 deg) and elevator (amplitude 2 deg) and rudder as per ARI ($\delta a - \delta r$) law for zero-sideslip roll maneuver as shown in Figure 7.4.

build up while roll rate marginally declines) as against the catastrophic jump in case of Figure 7.2. Also the p–δa slope in Figure 7.4 between the origin and the bifurcation point is lesser than in case of Figure 7.2.

A similar simulation as was done in case of Figure 7.3 is carried out but with the ARI law from Figure 7.4 used to schedule the rudder deflection as a function of the aileron input. The results are shown in Figure 7.5. The roll rate in Figure 7.5 for the same aileron input of $\delta a = -6$ deg is lower since the p–δa slope in Figure 7.4 is smaller, but the jump phenomenon seen in Figure 7.3 has vanished. In general, the response in Figure 7.5 is smoother.

7.1.2 Velocity-Vector Roll Maneuvers

Another way of negating the effects of kinematic coupling during a roll maneuver is to constrain the airplane to roll about its velocity vector. As discussed earlier, a rapidly rolling airplane tends to roll about its body X axis, whereas a slower roll maneuver is closer to a roll about the velocity vector. The difference between a body-axis and a velocity-vector roll (VVR) is shown in the sketch in Figure 7.6. In a body-axis roll, the trim angle of

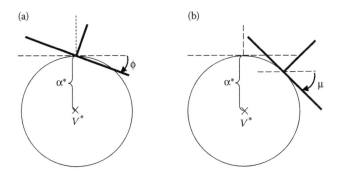

FIGURE 7.6   Aircraft in (a) body-axis roll and in (b) velocity-vector roll.

attack is converted into sideslip due to kinematic coupling, whereas in a VVR, the trim angle of attack is ideally maintained unchanged.

Analysis of an airplane constrained to roll about its velocity vector can be performed by adding a constraint condition:

$$\frac{\underline{V} \cdot \underline{\omega}}{||\underline{V} \cdot \underline{\omega}||} = 1 \tag{7.3}$$

where $\underline{V}$ and $\underline{\omega}$ are the airplane's translational and rotational velocity vectors, respectively, with components along the body axes given by $\underline{V} = u\hat{i} + v\hat{j} + w\hat{k}$ and $\underline{\omega} = p\hat{i} + q\hat{j} + r\hat{k}$. The condition in Equation 7.3 requires the rotational velocity vector to be aligned with the airplane's translational velocity. The condition in Equation 7.3 can be approximately written as

$$\frac{p + q\beta + r\alpha}{\sqrt{p^2 + q^2 + r^2}} - 1 = 0 \tag{7.4}$$

The formulation of the EBA problem for the VVR maneuver is identical to that for the zero-sideslip roll case except for using Equation 7.4 instead of the zero-sideslip constraint in Equation 7.2. The states and parameters are selected as follows:

$$x^* = [\alpha, \beta, p, q, r]; \quad p_1^* = \delta r^*$$
$$u^* = [\delta a^*]$$
$$p_2^* = [\delta e^*] = [2 \text{ deg}]$$

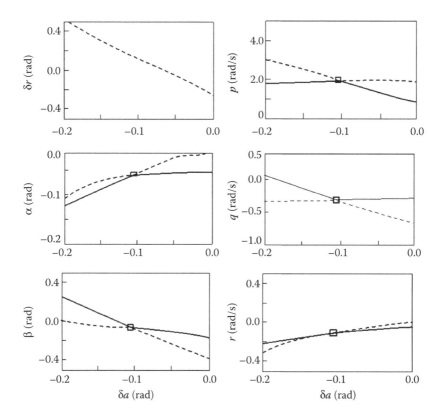

FIGURE 7.7   Velocity-vector roll ARI law (top left) and corresponding PSS solutions (solid lines: stable solutions; dashed lines: unstable solutions; empty square: transcritical bifurcation).

Results of this EBA computation are plotted in Figure 7.7. By and large, the plots are similar to those in Figure 7.5 for the zero-sideslip case. The occurrence of the transcritical bifurcation and the peak stable PSS roll rate value of nearly 2 rad/s are common. In Figure 7.7, the PSS sideslip states are close to zero, but not precisely. In fact, given the close correspondence between the results for the zero-sideslip formulation and the VVR case, it can be adjudged that the zero-sideslip constraint gives a roll maneuver that closely approximates a VVR.

Homework Exercise: Figure out why the PSS roll rate at zero aileron deflection is not zero in Figure 7.7. Can you think of a way to modify the VVR formulation to avoid this feature?

Time simulation results for the VVR maneuver are plotted in Figure 7.8 for the same control inputs as used previously but with the rudder varied

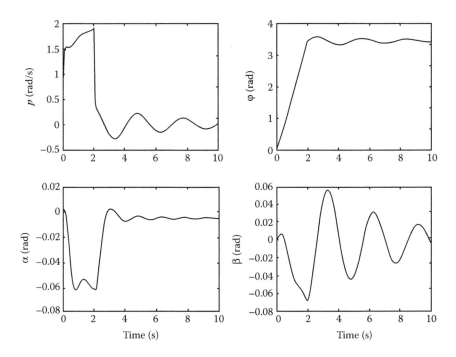

FIGURE 7.8   Time simulation results for 2 s pulse in aileron (amplitude −6 deg) and elevator (amplitude 2 deg) and rudder as per ARI ($\delta a - \delta r$) interconnect law for VVR maneuver as shown in Figure 7.7.

as per the ARI law shown in Figure 7.7. The results are quite similar to the simulation plots in Figure 7.5 for the zero-sideslip case, which is understandable given the discussion above about the commonality between the two constraint conditions.

7.2 HIGH AOA FLIGHT DYNAMICS AND SPIN

So far, we have examined rapid roll maneuvers; these are usually initiated at low values of angle of attack. On the other hand, we can examine airplane trim and stability at high angles of attack without necessarily having large rate maneuvers. Usually, this leads to stall and post-stall maneuvers that depend largely on the nature of the nonlinear and often asymmetric aerodynamic phenomena that occur in the post-stall regime. Finally, we have the phenomenon of spin, which usually involves high angles of attack and large yaw/roll rates. Among other classifications of spin, a useful one is the spin type—steep or flat. As pictured in Figure 7.9, flat spin usually occurs at higher angles of attack

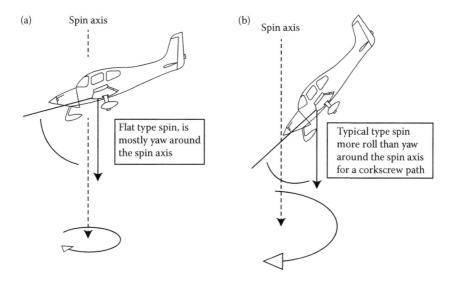

FIGURE 7.9 Sketch of airplane in spin—flat spin (a) and steep spin (b). (From www.myflyingstuff.com.)

and has a predominantly yawing motion, whereas steep spin is found at moderate-to-high angles of attack and features both rolling and yawing motion.

In the following, we shall carry out a bifurcation analysis of the airplane dynamics meant to uncover phenomena at high angles of attack. The F-18 HARV model presented in Chapter 2 will be used for the computation. The aerodynamic data are as shown in Figures 2.15 through 2.17. The full airplane dynamic equations with eight states as discussed in Chapter 1 will be employed. An SBA computation is carried out with elevator deflection as the continuation parameter while other controls (aileron, rudder and throttle) are held fixed. An initial steady state is selected as below:

$$x^* = [Ma,\ \alpha,\ \beta,\ p,\ q,\ r,\ \varphi,\ \theta]^* = [0.2,\ 0.22\ \text{rad},\ 0,\ 0,\ 0,\ 0,\ 0,\ 0.22\ \text{rad}]$$
$$u^* = [\eta,\ \delta e,\ \delta a,\ \delta r]^* = [0.38, -0.0167\ \text{rad},\ 0,\ 0]$$

It may be noted that the initial state corresponds to a level trim, though it is not necessary to choose one as the starting state. In general, the other states in the bifurcation diagram will not be level trims. The bifurcation diagram is plotted in Figure 7.10 showing four of the key states—angle of attack, roll and yaw rates, and pitch angle.

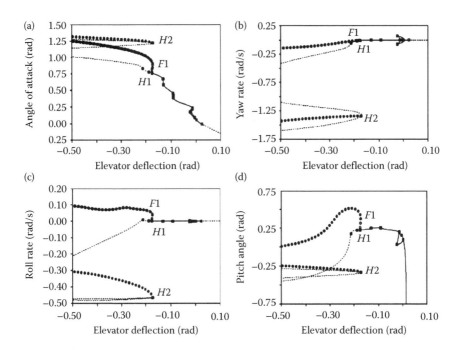

FIGURE 7.10 Bifurcation diagram of aircraft state variables ((a) angle-of-attack, (b) yaw rate, (c) roll rate, and (d) pitch angle) as function of elevator deflection at fixed $\eta = 0.38$, $\delta a = \delta r = 0$ (solid lines: stable equilibrium states; dashed lines: unstable equilibrium states; filled circles: stable limit states; empty circles: unstable limit states; solid squares: Hopf bifurcation point; empty squares: pitchfork bifurcation point).

The dynamics at lower angles of attack is quite similar to what has already been discussed in Chapter 6 (Figures 6.2 and 6.3). As the trim angle of attack is increased, stall occurs near 35 deg (~0.6 rad) angle of attack. However, no dramatic phenomenon is seen at stall in this case— stable steady states exist beyond stall as well. The first notable instability is the subcritical HB labeled $H1$ near 43 deg (~0.75 rad) angle of attack. The unstable limit cycles originating at $H1$ turn over at an FB labeled $F1$ giving rise to a branch of stable limit cycles. The oscillations in the periodic states on this branch turn out to be predominantly in pitch and angle of attack, hence the phenomenon is called *pitch bucking*.

At further higher angles of attack, there is a turning point near AOA of 70 deg (~1.2 rad) and located very close to it is another HB labeled $H2$. This one is a supercritical HB and the branch of stable limit cycles generated at $H2$ represents oscillatory spin states. At these high angles of attack, the spin is usually of the flat type with a large yaw rate, as seen in Figure 7.10, of the order of nearly 90 deg/s. The roll rate is relatively smaller. The nose is pitched down about 20 deg below the horizon and the angle of attack is around 70 deg, which makes the flight path angle near about 90 deg. Thus, the airplane is rapidly spinning in yaw as it descends nearly vertically. In addition, there is an oscillation about the mean value of each variable. To get a better picture of the dynamics, a time simulation is carried out for one particular spin state on the oscillatory branch in Figure 7.10. The time history plots are shown in Figure 7.11.

Homework Exercise: Analyze the time history in Figure 7.11 and correlate the information with the bifurcation diagram in Figure 7.10. Also estimate the rate of descent from the plot of distance along Z axis versus time (approximately 200 ft/s).

Previously, we have noted that the steady-state spin solutions in Figure 7.10 are unstable. Instead, stable periodic solutions (limit cycles) have emerged at the Hopf bifurcation point $H2$. Now, we can investigate the nature of the instability of the spin steady states. The SBA analysis in Figure 7.10 also computes the eigenvalues and eigenvectors at each steady state. Choosing one such unstable spin steady state from Figure 7.10, as below:

$$x^* = [Ma, \alpha, \beta, p, q, r, \varphi, \theta]^*$$
$$= \big[0.175, \ 1.263 \, \text{rad}, \ 0.03 \, \text{rad}, \ -0.47 \, \text{rad/s}, \ 0.0487 \, \text{rad/s},$$
$$-1.5 \, \text{rad/s}, \ 0.03 \, \text{rad}, \ -0.3 \, \text{rad}\big]$$
$$u^* = [\eta, \ \delta e, \ \delta a, \ \delta r]^* = [0.38, \ -0.39 \, \text{rad}, \ 0, \ 0]$$

The 8×8 Jacobian matrix at this steady state can be picked up from the SBA data. The eigenvalues of the Jacobian matrix turn out to be

$$\lambda_s = [-0.4873 \mp j2.4826, \ 0.044 \mp j2.351, \ -0.2349,$$
$$-0.1779, \ -0.1553 \mp j1.6374]$$

As can be seen, there are three complex conjugate pairs and two real eigenvalues. Only one of the complex conjugate pairs has a positive real part indicating instability; the other eigenvalues all indicate stable modes. However, owing to the strong coupling between the lateral and longitudinal modes, it is difficult to clearly relate these eigenvalues to any of the specific airplane dynamics modes that we have described earlier under more

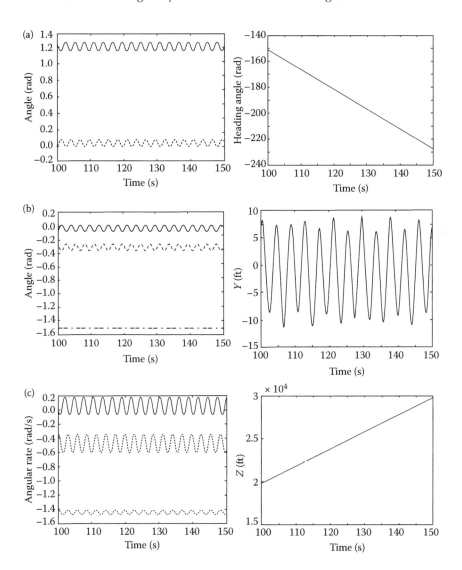

FIGURE 7.11 Time simulation of spin dynamics from initial conditions obtained from bifurcation analysis results (Figure 7.10). (a) Angle of attack α (——) and sideslip angle β (- - - -). (b) Roll angle φ (——), pitch angle θ (----), and flight path angle γ (---). (c) Roll p (- - - -), pitch q (——), and yaw r (---) rates.

benign trim conditions. However, a few educated guesses are possible. At these high angles of attack and low velocities, the phugoid-like mode is likely to be well damped and have a relatively high frequency. Also, the roll-like mode is usually stable under most conditions and the spiral-like mode is expected to be stable at high angles of attack. These are likely to be the two negative, real eigenvalues above. The modes that are susceptible to loss of stability are the short period-like mode and the Dutch roll-like mode.

To investigate further, we can examine the eigenvectors corresponding to these eigenvalues. The eigenvectors is this case are listed below:

Eigenvectors of A:

$E=$

M:	$-0.0001 \pm j0.0003$	$-0.0030 \pm j0.0007$	0.0984	-0.0487	$0.0006 \pm j0.0015$
α:	$-0.0261 \pm j0.0061$	$0.0906 \pm j0.2306$	-0.3091	0.2479	$-0.1210 \pm j0.0289$
β:	$0.0363 \pm j0.2594$	$-0.1552 \pm j0.1810$	0.0503	-0.0425	$0.0136 \pm j0.0228$
p:	-0.7669	0.6339	-0.2719	0.1391	$-0.0782 \pm j0.0153$
q:	$-0.0673 \pm j0.5028$	$-0.2884 \pm j0.4898$	0.0856	-0.0501	$0.0592 \pm j0.2592$
r:	$-0.1042 \pm j0.0231$	$0.1651 \pm j0.0207$	0.8458	-0.9234	$-0.0625 \pm j0.0385$
φ:	$0.0429 \pm j0.2665$	$-0.1698 \pm j0.1992$	-0.0233	0.0206	-0.6924
θ:	$-0.0313 \pm j0.0071$	$0.0703 \pm j0.2361$	-0.3084	0.2438	$-0.0968 \pm j0.6423$

Homework Exercise: Use the eigenvector data presented above in conjunction with the eigenvalues listed earlier to arrive at some conclusions about the modal stabilities. For example, the first column shows relatively large magnitudes of β, p, r, and φ and corresponds to a complex conjugate eigenvalue pair; hence it may be considered to be the Dutch roll-like mode. Likewise, the second column shows the largest magnitude of q among all the columns; hence this is likely to be the short period-like mode. From the list of eigenvalues, this corresponds to the unstable mode, so it is quite likely that the instability leading to the $H2$ point is either a short period-like mode or some form of a coupled short period–Dutch roll kind of dynamics.

7.2.1 Analytical Criterion for Spin Susceptibility

As we have seen, spin steady states (either equilibrium or periodic, stable or unstable) are high-angle-of-attack, large yaw/roll rate solutions, yet at nominal values of elevator deflection. The most natural manner of arranging a bifurcation diagram that matches these features is to have a pair of SNB points such that the branch of steady states folds back on itself, something in the shape of the letter "Z." Looking more closely at the bifurcation diagram in Figure 7.12, the pair of SNBs, $S1$ and $S2$, can be identified. Beyond $S2$, depending on the nature of the steady states, their stability and further bifurcations, the spin solution may take one of different forms—either equilibrium or periodic or even chaotic, either stable or unstable, perhaps even multiple spin types (modes). However, at the root of all this is

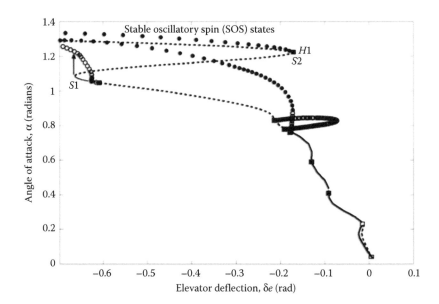

FIGURE 7.12 Bifurcation diagram for F-18 HARV airplane model showing the pair of saddle-node bifurcations, S1 and S2, which act as an indicator of potential spin states.

the requirement for a second SNB, S2. Thus, the occurrence of a "second" saddle-node point in the high AOA, large-rate regime may be considered as an indicator of potential spin states. Note that this approach does not focus on the point of departure from which the airplane may enter into the spin state. For example, Figure 7.13 shows two possible ways in which the airplane may depart from its low AOA state to enter into the spin state at high AOA. In one case, the airplane jumps into spin from the first SNB point, S1. In the other case, an HB on the low-AOA branch gives rise to a branch of

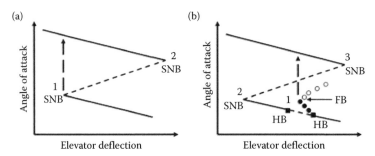

FIGURE 7.13 Sketch of two possible ways in which an airplane may enter into the spin state, (a) jump at SNB point, and (b) jump at FB point.

stable limit cycles that folds over at an FB. The airplane can jump from the FB point to the high AOA spin states. Other scenarios are possible as well.

7.2.1.1 Derivation of the Criterion

The derivation of the spin susceptibility criterion follows the work in Reference 2. A reduced-order model for equilibrium spin states is considered, as follows:

$$\dot{p} = L_{p2}(\alpha)p_w + L_{r1}(\alpha)r_b + L_0(\delta)$$

$$\dot{q} = \left(\frac{I_{zz} - I_{xx}}{I_{yy}}\right)p_b r + M(\alpha) + M_0(\delta) \qquad (7.5)$$

$$\dot{r} = N_{p2}(\alpha)p_w + N_{r1}(\alpha)r_b + N_0(\delta)$$

where the symbols on the right-hand side are as defined in Equation 6.19 and the aerodynamic terms modeling the flow curvature effects due to yaw rate r_w have been ignored. The dependence on α has been explicitly noted at several places but this will be dropped in future for convenience. The terms L_0, M_0, N_0 represent the effect due to the controls—aileron, rudder, and elevator. Additionally, two constraints are assumed:

$$\beta = 0 \quad \text{and} \quad q = 0 \qquad (7.6)$$

and p_w has been approximated to p_b. Thus, after dropping the subscript "b," α, p, r are the three independent variables under consideration.

Homework Exercise: A better substitution may be to approximate p_w by using Equation 1.47 as follows:

$$p_w = \underbrace{(p_b \cos\alpha + r_b \sin\alpha)\cos\beta}_{1} + \underbrace{(q_b - \dot{\alpha})\sin\beta}_{0}$$

Redo the following derivation with this approximation instead.

The aim of the exercise is to investigate the possibility of a trim state corresponding to the saddle-node point S2 in Figure 7.12. The first step therefore is to solve for the trim states of Equation 7.5. Setting the left-hand sides of Equation 7.5 to zero and solving for p, r and using the result in the equation for pitching moment, we get

$$A(L_{r1}N_0 - N_{r1}L_0)(N_{p2}L_0 - L_{p2}N_0) + \Delta^2(M(\alpha) + M_{\delta e}\delta e) = 0 \qquad (7.7)$$

where

$$A = \left(\frac{I_{zz} - I_{xx}}{I_{yy}}\right); \quad \Delta = L_{p2}N_{r1} - N_{p2}L_{r1} \tag{7.8}$$

And $M_0(\delta)$ has been modeled as $M_{\delta e}\delta e$. Equation 7.7 is a single, unified trim equation that relates the trim angle of attack to the elevator setting, given a certain choice of the aileron and rudder deflections.

The second step is to find the condition under which a trim state corresponds to an SNB point. Mathematically, the condition for a steady state given by a solution of Equation 7.7 to be SNB point is given by

$$\frac{\partial \delta e}{\partial \alpha} = 0 \tag{7.9}$$

This requires differentiating Equation 7.7 with respect to α and then imposing the condition in Equation 7.9. Doing so, we obtain the following condition:

$$A[(L_{r1,\alpha}N_0 - N_{r1,\alpha}L_0)(N_{p2}L_0 - L_{p2}N_0) + (L_{r1}N_0 - N_{r1}L_0)(N_{p2,\alpha}L_0 - L_{p2,\alpha}N_0)]$$

$$+ \Delta^2 M_\alpha(\alpha) - 2A\frac{\Delta_\alpha}{\Delta}(L_{r1}N_0 - N_{r1}L_0)(N_{p2}L_0 - L_{p2}N_0) = 0 \tag{7.10}$$

where the subscript "α" denotes the partial derivative with respect to α, and δe has been eliminated by using Equation 7.7. Given N_0, L_0 for a fixed setting of aileron and rudder, Equation 7.10 is a function of α alone. Representing the condition in Equation 7.10 symbolically by $G = 0$, a plot of G versus α will reveal points where the condition $G = 0$ is satisfied. These points are the steady states of Equation 7.5, which are also SNB points, satisfying the criterion in Equation 7.9. Thus, Equation 7.10 provides a condition for the possible existence of high AOA spin states. By itself, the $G = 0$ test does not indicate whether the spin states are equilibria/periodic, or stable/unstable, etc. Nor does it reveal how the airplane may depart from a regular low-AOA trim state in order to enter a possible spin state. Hence, it only indicates the susceptibility of the airplane to spin. As long as a

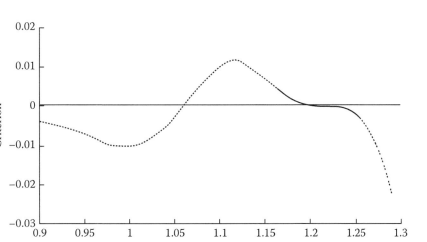

FIGURE 7.14 Plot of "*G*" as function of angle of attack for F-18/HARV data for the same L_0, N_0 as used for the plot in Figure 7.12.

spin state exists, it is possible that the airplane may enter it in one way or another. In this way, the criterion in Equation 7.10 is different from traditional spin onset criteria, which focus more on the nature of instability at the point of entry from which departure takes place.

When the criterion in Equation 7.10 is plotted as in Figure 7.14 for the same airplane data as in Figure 7.12, we see that there are two distinct points of α where the function G crosses zero. These are observed at $\alpha \approx 1.05$ rad and $\alpha \approx 1.21$ rad, which correspond closely to the SNB points labeled *S1* and *S2* in Figure 7.12, respectively. Of these, it is the occurrence of the second SNB point that signifies potential onset of spin. Note that the existence of a single zero crossing of the G function is not adequate to suggest the existence of possible spin states. As described with reference to Figures 7.12 and 7.13, it is the second SNB point that signifies susceptibility to spin.

An alternative form of the criterion in Equation 7.10 is possible where the zero condition is written in terms of angle of attack and yaw rate, typically the two most key variables used to describe a spin state. A sample plot of this form of the criterion is shown in Figure 7.15.

Homework Exercise: Look up Reference 2 for the derivation and use of the spin susceptibility criterion in the form as plotted in Figure 7.15.

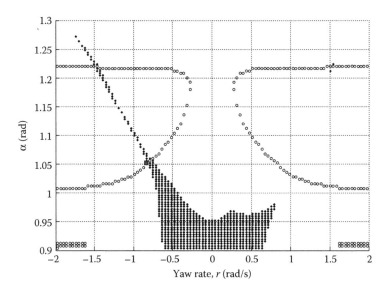

FIGURE 7.15 Spin susceptibility criterion in terms of a plot of α and r—points where the "+" and "o" signs intersect are possible SNB points. The second "connected" SNB point near $\alpha = 1.22$ rad and $r = -1.5$ rad/s is the predicted SNB point at spin onset.

7.3 BIFURCATION TAILORING/TRIM SHAPING AS CONTROL STRATEGY

Bifurcations are usually responsible for onset of nonlinear behavior as a result of loss or exchange of stability. One of the ways of avoiding a certain bifurcation and hence preventing a subsequent loss of stability or a departure from controlled flight phenomenon is to reconstruct the bifurcation curves so as to avoid or postpone certain bifurcation points. This was initially proposed as a way to avoid jump phenomena in inertia-coupled roll maneuvers [3]. This procedure has since gone by the names of "bifurcation tailoring" or "trim shaping" [4,5]. In fact, the basic idea of adjusting the rudder deflection in sync with the aileron input so as to avoid undesirable dynamics as, for example, in rapid roll maneuvers has been used for a long time—it is called the ARI law. However, the use of bifurcation methods provided a firm footing and a systematic method for devising the ARI law. In general, beyond roll maneuvers, the idea of modifying equilibrium solutions so as to either avoid onset of instability or push undesirable bifurcation points outside the limits of control deflections can be used as a control strategy to prevent the airplane from entering into certain nonlinear motions. In the following, both for the PSS model and F-18 low-angle-of-attack model, some examples are presented to demonstrate this aspect.

7.3.1 Linear ARI Law for Jump Prevention

Earlier, we have seen scheduling of rudder with aileron based on certain constraints on the dynamics of aircraft such as zero sideslip or enforcing a VVR. In some sense, the ARIs so obtained as the schedule of rudder deflection as a function of the primary continuation parameter (in these cases, the aileron) do prevent an undesirable jump phenomenon by arresting the growth in sideslip angle, a potential cause for the jump. From the bifurcation point of view, what is physically seen as arresting the growth of sideslip can be viewed as a problem of avoiding (via some control interconnect laws) the SNB point that leads to jump. To understand this better, let us consider the equilibrium (trim) surfaces in case of a two-parameter dynamical system. For example, think of the roll rate as the state variable with the aileron and rudder as the two parameters. Figure 7.16 shows a sketch of an equilibrium surface plotted for state variable x versus two parameters u and p. One may consider u to be the primary continuation parameter and p as the secondary varied parameter. Usually, one would consider the equilibrium solutions of x with varying u for different fixed values of p. These amount to cuts of the surface in Figure 7.16 by planes parallel to the x–u plane. One can imagine that certain cuts would show

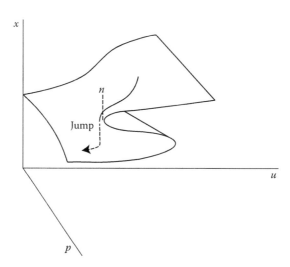

FIGURE 7.16 Equilibrium surface for a dynamical system in two parameters showing jump.

a curve of equilibrium states with the folded structure having two SNB points and hence the potential for jump. Yet other cuts may yield smooth equilibrium curves with no folds. One can also imagine scheduling p as a function of u such that we forge a path on the equilibrium surface, which avoids the folded regions, thus preventing jumps.

This bifurcation theoretic strategy is applied to construct a linear ARI law, $(\delta r = k\delta a)$ now to avoid jump in roll rate. Using the same data as was used for Figure 7.2, the SNB point "A" in Figure 7.2 is first tracked using continuation in two-parameter $\delta a - \delta r$ control space. A critical value of the linear ARI constant k is thus found to be approximately 2.04. PSS roll-rate solutions for five different values of the ARI constant k are shown in Figure 7.17. The curves numbered "4" and "5" show an SNB and therefore indicate a possible jump phenomenon. Curve "3" is the critical one showing a transcritical bifurcation. For curves labeled "1" and "2," the fold has vanished and the roll rate increases smoothly with changing aileron without a jump. Thus, any value of k less than the critical value is likely to result in a jump, whereas selecting a higher value of k than the critical one yields a jump-free roll steady-state solution.

Let us choose the constant $k = 2.05$ for the linear ARI law and carry out a bifurcation analysis of the PSS model for rolling maneuvers. The results are plotted in Figure 7.18. From the plots, it is observed that with the use of the linear ARI, the SNB vanishes, which avoids jump and the associated nonlinear behavior in the roll maneuver as seen above. Maximum roll rate is also marginally better as compared to no-ARI case

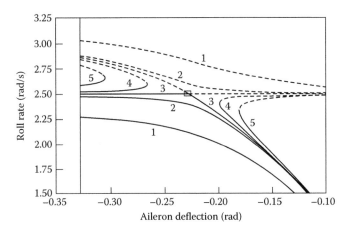

FIGURE 7.17   Pseudo-steady state roll rate versus aileron deflection for various ARI constant k values.

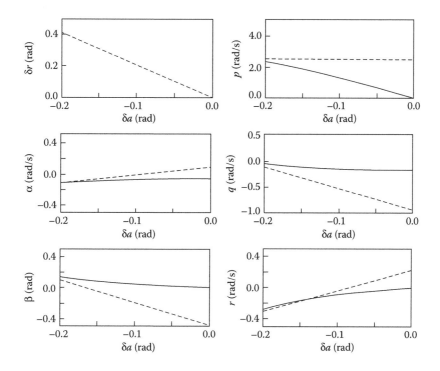

FIGURE 7.18 Jump avoidance via a linear ARI based on bifurcation surfaces in two-parameter space.

(Figure 7.2). The only disadvantage with this solution being that roll rate response to aileron deflection is sluggish now and the maximum roll rate needs nearly twice the aileron deflection input as compared to previous cases. Note that the sideslip is maintained close to zero.

Homework Exercise: Compare the linear ARI results in Figure 7.18 with the zero-sideslip roll maneuver results in Figure 7.4. Is it possible to make a judgment about which is the better approach—computationally and flight dynamically?

7.3.2 Trim Shaping for Level Flight Trims

Next, another example of bifurcation tailoring is presented for shaping of the bifurcation curve for the low-angle-of-attack F-18 model (see data in Chapter 2) in level flight trim. Bifurcation diagrams for this model with the level trim constraint are shown in Figure 7.19.

Notice the pair of HBs labeled "1" and "2" in Figure 7.19. One may expect to experience limit cycle oscillations past the HB with label "1."

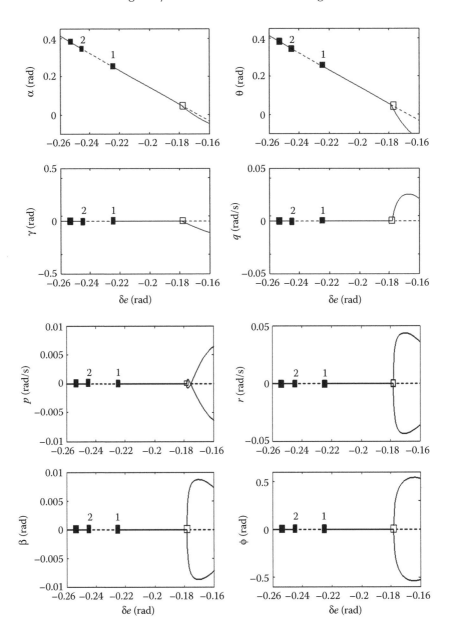

FIGURE 7.19 Bifurcation plots for low-alpha F-18 model in level flight trim condition (solid lines: stable trims; dashed lines: unstable trims; solid square: Hopf bifurcation; empty square: pitchfork bifurcation).

This most probably signifies a longitudinal instability, which may be verified by examining the Phugoid eigenvalues. Thus, effectively, point "1" may be taken to be the limit of allowable angle of attack for this model.

One may wish to stretch the limit of accessible angle of attack by reshaping the trim curve while still maintaining level flight trim so that the unstable patch between the HB points labeled "1" and "2" could be avoided. $\alpha - \delta e$-curve shaping in the continuation framework can be done by using a constraint of the following form:

$$\alpha + 2.28\delta e_0 + 0.28 = 0 \quad \text{for } \alpha \geq 0.2 \text{ rad} \tag{7.11}$$

and no $\alpha - \delta e$ constraint for $\alpha < 0.2$ rad. Along with this, the level flight constraints are imposed: $\gamma = 0$, $\beta = 0$, $\phi = 0$. To account for the four constraints, the EBA procedure requires four free parameters. Three of them are obviously the throttle, aileron, and rudder—needed to maintain the level flight trim condition. An additional parameter is introduced as the feedback gain k_α via the feedback law, $\delta e = \delta e_0 - k_\alpha \alpha$.

Figure 7.20 shows the results of the EBA computation for the formulation above. The variation of the gain parameter k_α is also shown there. It can also be confirmed that the level flight constraint is met at all trim states. The trim curve for angle of attack shows a kink at 0.2 rad. The HBs at points "1" and "2" no longer appear on the bifurcation diagram of Figure 7.2 but in the process, the shape of the $\alpha - \delta e$ trim curve has been altered.

Homework Exercise: Compare the approach in Figure 7.20 with that presented in Section 5.5 using gain scheduling.

7.3.3 Rolling Pull-Down Maneuver with Zero Sideslip

One can imagine that there is some common ground between trim shaping and the EBA approach. In either case, some additional constraints or relationships are imposed, but the solution approach may be somewhat different. For some problems, a combination of the two—EBA and trim shaping—may be required. For instance, consider once again the problem of rapid rolling maneuvers. Let the requirement now be to have a rolling pull-down maneuver with a given normal acceleration or pitch rate and a zero-sideslip constraint. For example, let us specify the constraints as

$$\beta = 0; \quad q + 0.18 = 0 \tag{7.12}$$

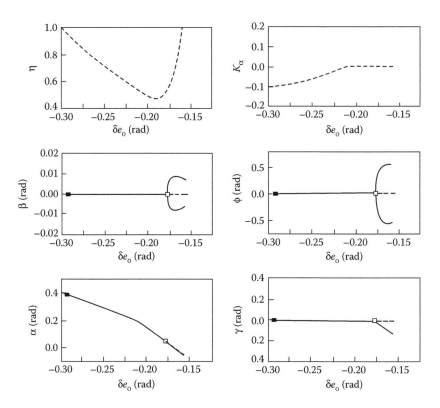

FIGURE 7.20 Bifurcation plots for low alpha F-18 model in level flight trim condition with trim shaping (solid lines: stable trims; dashed lines: unstable trims; solid square: Hopf bifurcation; empty square: pitchfork bifurcation).

where q is in rad/s. The constraints in Equation 7.12 can be met by letting both the rudder and elevator to be "free parameters" in the EBA procedure with aileron deflection as the primary continuation parameter. The PSS model is used once again. The resulting bifurcation diagram for roll rate and the schedules of elevator and rudder with the aileron parameter are as shown in Figure 7.21. The EBA solution for the constraints in Equation 7.12 is the one with the pair of transcritical bifurcations. In this case, the stable PSS roll-rate solution increases almost linearly with aileron deflection up to the first transcritical bifurcation and then marginally decreases with further aileron input.

In practice, one may prefer a somewhat smoother trim curve for the roll rate with aileron, such as the one labeled "2" in Figure 7.21. On the other hand, the one labeled "4" with a SNB may be unacceptable. Thus, in this instance, the EBA procedure gives the critical solution with two

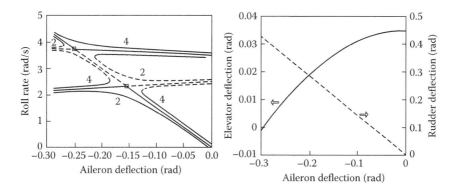

FIGURE 7.21   Unfolding of PSS solution branches for $\beta = 0$; $q + 0.18 = 0$ constraints by detuning of rudder and elevator schedules with aileron.

transcritical bifurcation points. This solution may need further "tailoring" to modify it to the shape of solution "2" in Figure 7.21. A similar approach can be used for the case of the zero-sideslip pseudo-steady states computed in Figure 7.4.

7.4 CONTROL PROTOTYPING FOR RECOVERY FROM SPIN

Recent advances in nonlinear control theory have helped devise control algorithms for nonlinear systems for various objectives. Two popular nonlinear techniques, one based on dynamic inversion (discussed in Chapters 5 and 6) known as NDI and the other one known as sliding mode control (SMC), have recently been found to be useful for control prototyping for different objectives in aircraft dynamics. Both these techniques, based on continuous feedback linearization [6], are used for the control of spin in the following.

A schematic of NDI control algorithm is shown in Figure 7.22. In principle, it is very similar to the NDI framework presented separately for the longitudinal and lateral dynamics in Chapters 5 and 6. The algorithm is based on separating the natural timescales of aircraft dynamics as an inner loop representing fast dynamics via fast aircraft variables (p, q, r) and an outer loop for the slower dynamics comprising of slower variables (α, β, μ). Unique combination of values prescribed to commanded outer-loop angular variables (α_c, β_c, μ_c) corresponding to different aircraft states are obtained from bifurcation analysis results. As there are only three variables to be commanded in both inner and outer loop each, the control inputs directly influencing these variables (δe, δr, δa) are chosen as the ones to be computed as a result of inversion; the fourth control, throttle

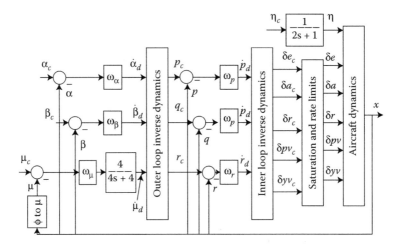

FIGURE 7.22 NDI control law. (Adapted from Snell, S. A., Enns, D. F., and Garrard, W. L., Jr., *Journal of Guidance, Control, and Dynamics*, 15(4), 976–984, 1992.)

deflection, is passed through a filter that models lag in throttle response and used as a direct input. The control deflections computed by inversion are passed through saturation and rate limiter blocks before being input to the aircraft dynamics block.

In the NDI control framework, control prototyping for aircraft switching between two states can be carried out. For example, the spin recovery problem can be thought of as a problem of switching states of aircraft from a spin state to another safe state, for example, a stable low-angle-of-attack level flight trim state [8]. One of the spin states computed earlier in Figure 7.10 is used in this example and the target states are three level trim states as given in Table 7.1 also selected from the bifurcation results; variables not listed in Table 7.1 are all zero.

Spin state:

$$\underline{x}^* = [M^* \quad \alpha^* \quad \beta^* \quad p^* \quad q^* \quad r^* \quad \phi^* \quad \theta^*]'$$
$$= [0.175 \quad 1.263 \, \text{rad} \quad 0.03 \, \text{rad} \quad -0.47 \, \text{rad/s} \quad 0.0487 \, \text{rad/s}$$
$$-1.5 \, \text{rad/s} \quad 0.03 \, \text{rad} \quad -0.3 \, \text{rad}]'$$
$$\underline{U}^* = [\eta^* \quad \delta e^* \quad \delta a^* \quad \delta r^*] = [0.38 \quad -0.39 \, \text{rad} \quad 0 \quad 0]'$$

For the spin recovery problem, it is assumed that all the states are available for feedback, the bandwidths ω_p, ω_q, ω_r in the inner loop are set as

TABLE 7.1   Level Flight Trim States Used for Spin Recovery

Variable	Trim A	Trim B	Trim C
M	0.14	0.16	0.2
α	41.83 deg (0.73 rad)	28.65 deg (0.5 rad)	17.12 deg (0.3 rad)
θ	41.83 deg (0.73 rad)	28.65 deg (0.5 rad)	17.12 deg (0.3 rad)
η	1.39	0.91	0.54
δe	−8.59 deg (−0.15 rad)	−5.73 deg (−0.1 rad)	−2.86 deg (−0.05 rad)

10 rad/s, and in the outer loop bandwidths are selected as $\omega_\alpha = \omega_\beta = 2$ rad/s, $\omega_\mu = 1.5$ rad/s. Thrust vectoring control angles δpv, δyv are zero in this case. The NDI-computed control inputs are plotted in Figure 7.23, and recovery from oscillatory spin state (seen in the simulation up to $t = 50$ s) to each of the level trim states in Table 7.1 is shown in Figure 7.24. Level trims states A and B in Table 7.1 are achievable as seen from the simulation results in Figure 7.24; however, the trim state C could not be achieved. An oscillatory response in the lateral variables around state C is observed and this may not be acceptable and may need enhanced stabilization from the inner loop of the NDI control. Figure 7.23 shows saturation of the rudder

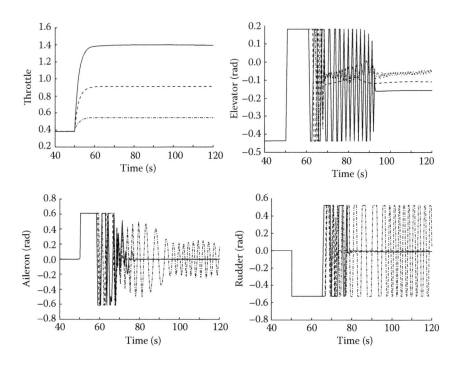

FIGURE 7.23   Computed NDI control inputs for spin recovery.

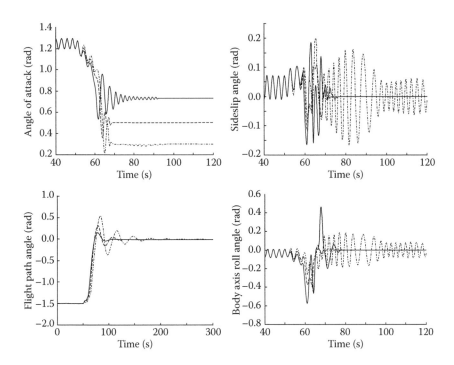

FIGURE 7.24 Spin recovery using the NDI control law in Figure 7.22.

which may be leading to these oscillations. In practice, recommended recovery from spin is initiated first by applying rudder to arrest yaw rate and simultaneously applying elevator to decrease the angle of attack. In addition, reaching level trim state demands the use of aileron to control roll rate and throttle to adjust speed and/or pitch angle. These are well reflected in the computed control commands via dynamic inversion in Figure 7.23. Inclusion of thrust vectored stabilizing moments in the inner loop of NDI control algorithm is one solution to stabilizing the aircraft at Trim C (see Reference 3).

The discontinuous high-frequency control inputs (known as "chattering" in control terminology) computed using the NDI control algorithm may not be desirable; also, a daisy chaining algorithm [7] is required to compute the additional thrust vectoring angles. Besides, NDI control laws are known to have issues with robustness. Sliding mode–based controllers are supposedly robust (can handle uncertainties in aerodynamic derivatives particularly at higher angles of attack) and are also devoid of chattering. The technique is based on designing an attracting sliding surface "S" in the domain of attraction of the

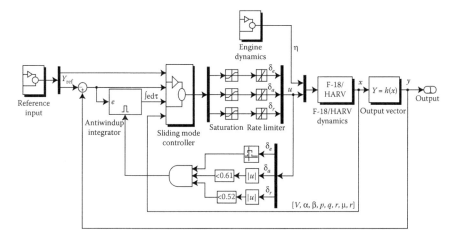

FIGURE 7.25 Simulink block diagram of sliding mode controller.

desired final state and on a reachability condition that requires that switching from an initial state to the sliding surface is possible. For detailed derivation of a sliding mode–based controller, the reader may see Reference 9. The sliding mode controller developed therein is shown in Figure 7.25. All states are assumed to be available for feedback here as well; however, robustness (with respect to model uncertainties) of the SMC is guaranteed.

7.4.1 Spin Recovery Using Sliding Mode Control

The spin recovery example of the F-18/HARV model is attempted using the sliding mode controller in Figure 7.25. Once the transients are settled and the aircraft is in an oscillatory spin state, the reachability condition is activated in the form of a step input to the commanded variables (α_c, β_c, μ_c) so as to bring the aircraft on the sliding surface in the domain of attraction of the desired state, that is, level flight trim state at $\alpha = 0.3$ rad. The aircraft now slides on the surface to eventually reach the level trim state. Control commands computed and recovery simulation results are shown in Figure 7.26. Control commands plotted look more or less the same as the ones computed using NDI laws and plotted in Figure 7.23; the difference being that there is no rudder saturation and chattering in computed control commands is much reduced in this case. SMC technique can thus be looked upon as a good substitute for NDI control algorithms, a technique popular for control prototyping in aircraft flight dynamics.

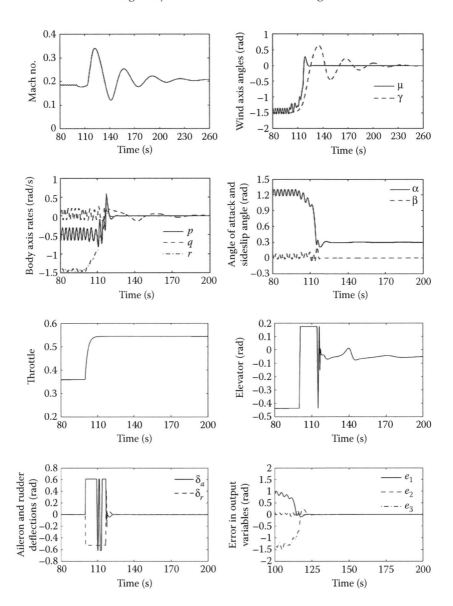

FIGURE 7.26   Spin recovery maneuver and sliding mode control commands.

7.5 CAREFREE MANEUVERING USING SLIDING MODE CONTROLLER

In general, the philosophy of recovering airplanes from undesired motions to safer states can be extended to designing maneuvers. For a recovery maneuver, as in the case of spin recovery, the initial state is the one from where to recover, and the final state is usually a low-angle-of-attack level flight trim state. Airplane maneuvers are usually initiated from a level flight trim state and the desired state corresponds to the maneuver, both the states being available from a bifurcation analysis of the aircraft model. One can also choose different states for maneuvering based on flight envelopes of aircraft which consists of attainable equilibrium region (AER) [10,11]. AER allows carefree maneuvering of aircraft within the available control authority.

Computation of AER and an example maneuver using sliding mode controller and AER is detailed later but before that a minimum radius turn maneuver is designed using the sliding mode controller. Minimum radius turn state available from bifurcation analysis results presented in Chapter 6 is used as the desired state.

7.5.1 Minimum Radius Turn Maneuver Using Sliding Mode Controller

In this maneuver, aircraft is required to perform a sustained horizontal loop of minimum radius at a constant speed and zero sideslip angle, starting from a straight and level flight condition. The two states required are thus a straight and level trim state and a level turn trim state. The *final* minimum-radius level turn state and corresponding trimmed control inputs (taken from Chapter 6) are

$$\underline{x} = [Ma^*, \ \alpha^*, \ \beta^*, \ p^*, \ q^*, \ r^*, \ \phi^*, \ \theta^*]$$
$$= [0.24, \ 0.37 \ \text{rad}, \ 0.0, \ 0.033 \ \text{rad/s} \ 0.107 \ \text{rad/s},$$
$$-0.084 \ \text{rad/s}, \ -0.91 \ \text{rad}, \ 0.237 \ \text{rad}]$$
$$\underline{U} = [\eta^*, \ \delta e^*, \ \delta a^*, \ \delta r^*] = [1.0, \ -0.103 \ \text{rad}, \ 0.044 \ \text{rad}, \ 0.0134 \ \text{rad}]$$

The *initial* level flight trim state and trimmed control inputs are

$$\underline{x} = [Ma^*, \ \alpha^*, \ \beta^*, \ p^*, \ q^*, \ r^*, \ \phi^*, \ \theta^*]$$
$$= [0.24, \ 0.189 \ \text{rad}, \ 0.0, \ 0.0, \ 0.0, \ 0.0, \ 0.0, \ 0.189 \ \text{rad}]$$
$$\underline{U} = [\eta^*, \ \delta e^*, \ \delta a^*, \ \delta r^*] = [0.386, \ -0.0168 \ \text{rad}, \ 0.0 \ \text{rad}, \ 0.0 \ \text{rad}]$$

Step input to the commanded state variables (α_c, β_c, μ_c) from initial straight and level trim state to final minimum radius level turn state is given as input to the SMC.

The simulation results for this maneuver are presented in Figure 7.27 along with the computed control command histories. Fast convergence to the desired state can be seen in the aircraft variables (Figure 7.27) within

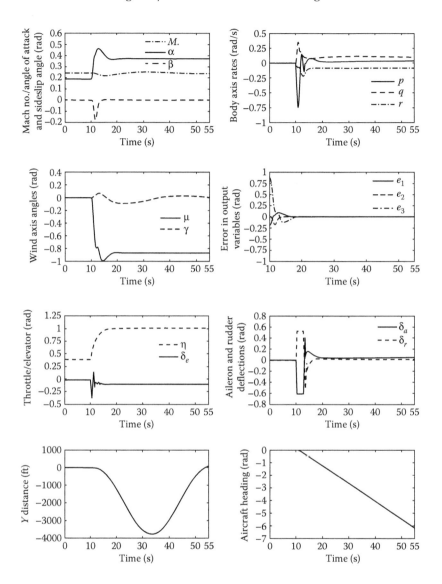

FIGURE 7.27   Minimum radius turn maneuver with ±25% uncertainty in aerodynamic derivatives.

the control position and rate limits given in Table 2.4. Negative step aileron input simultaneously with a positive rudder step input effectively results in a negative roll rate. A doublet-type response follows for both aileron and rudder to settle the aircraft to a constant desired bank angle and zero sideslip angle. Commanded throttle input follows a first-order response and elevator doublet results in desired final angle of attack. Aircraft left heading completes a full circle of 2π rad (360 deg), which is also represented by sideways drift from initial straight line path of approximately 4000 ft, the diameter of the horizontal loop. Turn radius plotted in Figure 6.11 shows the minimum radius of turn to be approximately 2000 ft for maximum throttle ($\eta = 1.0$), very nearly half the diameter of about 4000 ft computed here by the SMC algorithm. Thus, it turns out that the maneuver designed based on the sliding mode controller in conjunction with bifurcation analysis (steady state) results is very close to optimal.

Homework Exercise: Examine the data in Figure 2.17 carefully to appreciate the moments generated in totality by aileron and rudder in Figure 7.27.

7.5.2 Maneuver Design Based on AER

In Khatri et al. [11], a method based on EBA is described to compute flight envelope based on available control authority. The flight envelop thus computed is not only useful for carefree access to equilibrium states inside the envelope but also for devising recovery strategies from loss of control scenarios. In this example, the desired maneuver is to execute a maximum amplitude cycle in sideslip angle in the horizontal (level, zero flight path angle) plane holding a constant angle of attack and restricting the bank angle to zero. Following this, the aircraft is required to perform a high-angle-of-attack pitch-reversal maneuver in the vertical plane at full throttle.

To design the maneuver, first the maximum attainable sideslip angle and angle of attack are to be computed under the required condition described above within the available control limits as given in Table 2.4. A continuation of the aircraft dynamic equations along with the following constraints:

$$\dot{\underline{x}} = f(\underline{x}, \delta e, \eta, \delta a, \delta r) = 0; \quad \phi = 0; \quad \gamma = 0; \quad \alpha = 0.05236 \text{ rad} \quad (7.13)$$

is first carried out with elevator deflection as the continuation parameter and aileron, rudder deflections, and throttle as the free parameters to take care of the additional constraint equations. The continuation is carried out until one of the control saturates, in this case, rudder, which also results in maximum sideslip angle, thus marking the first point of the boundary of AER, "1," shown in Figure 7.28a. The second continuation consists of solving

$$\dot{\underline{x}} = f(\underline{x}, \delta e, \eta, \delta a, \delta r_{max} = 0.523) = 0; \quad \phi = 0; \quad \gamma = 0; \quad (7.14)$$

with elevator as the continuation parameter again and throttle and aileron deflection as the free parameters. This continuation is carried out until the aileron saturates at the second point on the AER, marked "2." The third continuation is carried out from point "2," which solves the following equations:

$$\dot{x} = f(\underline{x}, \delta e, \eta, \delta a_{max} = 0.61 \text{ rad}, \delta r) = 0; \quad \phi = 0; \quad \gamma = 0; \quad (7.15)$$

with throttle and rudder varying as the free parameters and elevator varying as the continuation parameter. The limit is reached when throttle reaches the maximum value of 1.0, point marked "3" in Figure 7.28. From the starting point "3," continuation of the following equations:

$$\dot{x} = f(\underline{x}, \delta e, \eta = 1.0, \delta a, \delta r) = 0; \quad \phi = 0; \quad \alpha = 0.534 \text{ rad} \quad (7.16)$$

with aileron and elevator as the free parameters and rudder as the continuation parameter completes the bounded region marked by the left AER boundary. The right AER boundary is a mirror image of the left side as shown in Figure 7.28. For the second part of the maneuver, to obtain the attainable set of angles of attack in symmetric flight, the following equations:

$$\dot{x} = f(\underline{x}, \delta e, \eta = 1.0, \delta a, \delta r) = 0; \quad \phi = 0; \quad \beta = 0 \quad (7.17)$$

are solved with elevator varying as the continuation parameter and aileron and rudder as the free parameters. The results of this continuation are plotted in Figure 7.28b showing maximum achievable angle of attack to be 0.96 rad. It is seen from that the angle of attack response to elevator input is sluggish beyond the angle of attack 0.8 rad; therefore, a maximum angle of attack of 0.75 rad is selected for showing the pitch-reversal maneuver.

The desired maneuver is initiated from a level trim state at $\alpha = 0.07$ rad, which corresponds to maximum achievable sideslip angle of 0.38 rad from Figure 7.28. However, maximum sideslip angle is chosen to be 0.35 rad to be on the safer side and to remain inside the AER when constructing the maneuver. Commanded sideslip angle in the horizontal plane is thus given as

$$\alpha_c = 0.07 \text{ rad}, \quad \beta_c = 0.35\sin(0.2\pi t) \text{ rad}, \quad \phi_c = 0 \quad (7.18)$$

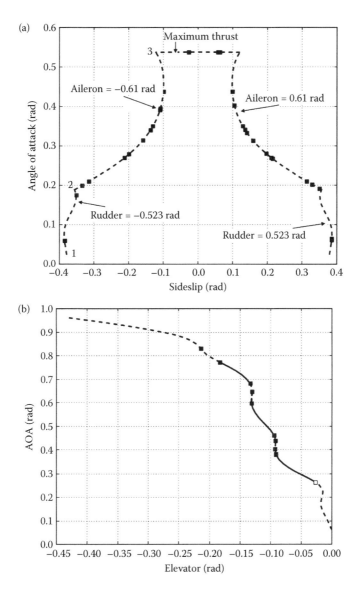

FIGURE 7.28 Flight envelope of F-18/HARV two-dimensional AER. (a) $\alpha - \beta$ in level flight trim condition ($\phi = 0$; $\gamma = 0$). (b) $\alpha - \delta e$ at maximum thrust in symmetric flight ($\phi = 0$; $\beta = 0$; $\eta = 1.0$) based on available control authority. (Adapted from Khatri, A. K., Singh, J., and Sinha, N. K., *Journal of Guidance, Control, and Dynamics*, 36(6), 1829–1834, 2013.)

for the nose pointing maneuver followed by

$$\alpha_c = 0.07 + 0.34(1 - \cos(0.25\pi t)), \quad \beta_c = 0, \quad \phi_c = 0 \tag{7.19}$$

for pitch-reversal maneuver after the cycle of nose pointing maneuver is over.

A variant of the sliding mode controller described in Reference 8 is used here for constructing the maneuver. Simulation results for this maneuver are plotted in Figure 7.29 and computed control commands are plotted in Figure 7.30.

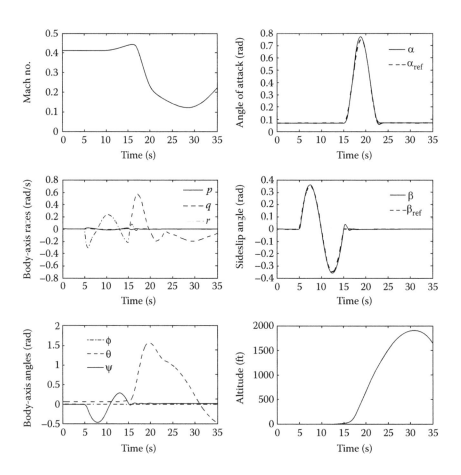

FIGURE 7.29 Simulation results of the maneuver based on AER.

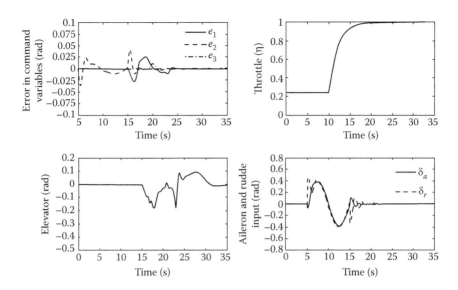

FIGURE 7.30 Control commands for the nose pointing and pitch-reversal maneuver.

Homework Exercise: Study the sequence of control commands and resulting airplane motions in Figures 7.29 and 7.30.

EXERCISE PROBLEMS

1. The F-18/HARV data show an order of magnitude difference between the axial and transversal moments of inertia and is therefore prone to inertia-coupling effects in flight. Construct a model for F-18/HARV at a suitable angle of attack in level flight trim to study its roll dynamics using bifurcation analysis technique.

2. Design feedback law for Dutch roll and spiral mode stability augmentation systems.

3. Construct a linear (state-space) model of F-18/HARV trimmed at $\alpha = 5$ deg in level flight. Design a feedback law for augmenting Dutch roll stability of the aircraft at this trim condition.

4. Using AOA and pitch rate feedback and setting short period mode specifications as that of one of the eigenvalues in Figure 5.8, carry

out gain scheduling over the complete range of level flight trim conditions for F-18/HARV.

5. Read about various techniques of gain scheduling.

6. Study the control commands computed in Section 7.5 for each maneuver and discuss the sequence of control inputs as required by a pilot to execute the maneuvers in the absence of a controller. Are they the same?

REFERENCES

1. Schy, A. A., and Hannah, M. E., Prediction of jump phenomena in roll-coupled maneuvers of airplanes, *Journal of Aircraft*, 14(4), 375–382, 1977.
2. Paranjape, A. A., and Ananthkrishnan, N., Analytical criterion for aircraft spin susceptibility, *Journal of Aircraft*, 47(5), 1804–1807, 2010.
3. Ananthkrishnan, N., and Sudhakar, K., Prevention of jump in inertia-coupled roll maneuvers of aircraft, *Journal of Aircraft*, 31(4), 981–983, 1994.
4. Charles, G. A., Bernardo, M. D., Lowenberg, M. H., Stoten, D. P., and Wang, X. F., Bifurcation tailoring of equilibria: A feedback control approach, *Latin American Journal of Applied Research*, 31(3), 199–210, 2001.
5. Sinha, N. K., Bifurcation Analysis and Control of Aircraft Flight Dynamics in Constrained Trims, PhD Thesis, IIT Bombay, 2002.
6. Slotine, J. J. E., and Li, W., *Applied Nonlinear Control*, Prentice-Hall, Englewood Cliffs, New Jersey, 1991.
7. Snell, S. A., Enns, D. F., and Garrad, W. L., Jr., Nonlinear inversion flight control for a supermaneuverable aircraft, *Journal of Guidance, Control, and Dynamics*, 15(4), 976–984, 1992.
8. Raghvendra, P. K., Sahai, T., Kumar, P. A., Chauhan, M., and Ananthkrishnan, N., Aircraft spin recovery, with and without thrust vectoring, using nonlinear dynamic inversion, *Journal of Aircraft*, 42(6), 1492–1503, 2005.
9. Khatri, A. K., Singh, J., and Sinha, N. K., Aircraft maneuver design using bifurcation analysis and sliding mode control techniques, *Journal of Guidance, Control, and Dynamics*, 35(5), 1435–1449, 2012.
10. Goman, M. G., Khramtsovsky, A. V., and Kolesnikov, E. N., Evaluation of aircraft performance and maneuverability by computation of attainable equilibrium sets, *Journal of Guidance, Control, and Dynamics*, 31(2), 329–339, 2008.
11. Khatri, A. K., Singh, J., and Sinha, N. K., Accessible regions for controlled aircraft maneuvering, *Journal of Guidance, Control, and Dynamics*, 36(6), 1829–1834, 2013.

Dynamics and Control of a 10-Thruster Flight Vehicle

IN THIS CHAPTER, WE shall apply many of the lessons learned in the previous chapters of this book to a flight vehicle that is somewhat different from the ones usually encountered in textbooks or at airports. This is a 10-thruster kill vehicle (KV), usually called a divert attitude control system (DACS). As pictured in Figure 8.1, it consists of 10 thrusters—four divert thrusters mounted near the center of gravity (CG) of the vehicle and six attitude thrusters placed at the rear of the vehicle. The divert thrusters (two of them are pictured in the firing mode in Figure 8.1) provide the lateral acceleration to maneuver in the yaw and pitch planes, respectively. The attitude thrusters are used to nullify the moment created by firing the divert thrusters and more generally for attitude control. Together, the divert and attitude thrusters may also be used to regulate the net thruster area, which controls the gas pressure in the combustion chamber, and hence regulate the thrust. The layout and components of a typical DACS system are shown in Figure 8.1.

8.1 FLIGHT DYNAMICS OF THE 10-THRUSTER DACS

By and large, the 6 DOF model for the flight dynamics of the DACS vehicle is the same as that presented in Chapter 1. There are chiefly two points of difference—First, the DACS vehicle is usually meant to fly in the outer

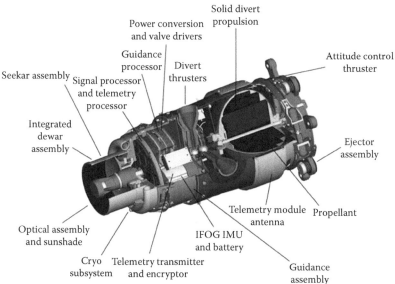

FIGURE 8.1 Top image shows a DACS in flight firing two of its divert thrusters. (From www.raytheon.com.) Bottom image shows a cut-away view of a typical DACS system. (From http://news.ifeng.com/mil/3/detail_2008_02/27/477418_4. shtml.)

atmosphere, so aerodynamic forces and moments may not be so prominent. Second, unlike the thrust in airplanes that is predominantly along the airplane body X axis, the DACS thrusters are all aligned to the body Y and Z axis. In fact, no thrust component is usually available along the body X axis for such a 10-thruster configuration. Other configurations that use thrust vectoring nozzles in place of the attitude thrusters may also provide an X direction thrust, but these are not discussed here.

Let us first assemble the relevant equations from Chapter 1. Then we shall modify them to account for the thrust from the 10 DACS thrusters.

Navigation equations:
From Equation 1.59,

$$\dot{V} = \frac{1}{m}(T\cos\alpha\cos\beta - D\cos\beta + Y\sin\beta) - g\sin\gamma \tag{8.1}$$

From Equations 1.28, 1.47, and 1.59,

$$\dot{\gamma} = \frac{1}{mV}[\cos\mu(T\sin\alpha + L)$$
$$ - \sin\mu(-T\cos\alpha\sin\beta + D\sin\beta + Y\cos\beta)] - \frac{g}{V}\cos\gamma \tag{8.2}$$

$$\dot{\chi} = \frac{1}{mV\cos\gamma}[\sin\mu(T\sin\alpha + L)$$
$$ + \cos\mu(-T\cos\alpha\sin\beta + D\sin\beta + Y\cos\beta)] \tag{8.3}$$

Attitude equations:
Likewise, from Equations 1.28, 1.47, and 1.59,

$$\dot{\mu} = \frac{1}{\cos\beta}(p\cos\alpha + r\sin\alpha) + \frac{\tan\beta}{mV}(T\sin\alpha + L) - \frac{g}{V}\cos\mu\cos\gamma\tan\beta$$
$$ + \frac{\tan\gamma}{mV}[\sin\mu(T\sin\alpha + L) + \cos\mu(-T\cos\alpha\sin\beta + D\sin\beta + Y\cos\beta)]$$
$$\tag{8.4}$$

From Equation 1.59,

$$\dot{\alpha} = q - (p\cos\alpha + r\sin\alpha)\tan\beta - \frac{1}{mV\cos\beta}(T\sin\alpha + L)$$
$$ + \frac{g}{V\cos\beta}\cos\mu\cos\gamma \tag{8.5}$$

$$\dot{\beta} = (p\sin\alpha - r\cos\alpha) + \frac{1}{mV}(-T\cos\alpha\sin\beta + D\sin\beta + Y\cos\beta)$$

$$+ \frac{g}{V}\sin\mu\cos\gamma \tag{8.6}$$

Rate equations:

For the DACS configuration as pictured in Figure 8.1, the cross products of inertia are all zero and, due to symmetry, the moment of inertia about the yaw and pitch axes is equal. Then, from Equation 1.78,

$$\dot{p} = \frac{\mathcal{L}}{I_{xx}} \tag{8.7}$$

$$\dot{q} = \frac{(I_{zz} - I_{xx})}{I_{yy}}pr + \frac{M}{I_{yy}} \tag{8.8}$$

$$\dot{r} = \frac{(I_{xx} - I_{yy})}{I_{zz}}pq + \frac{N}{I_{zz}} \tag{8.9}$$

In Equations 8.1 through 8.6, T refers to only the X-component of the thrust vector. For the 10-thruster DACS, we shall allow all three components of the thrust vector to be included in the equations of motion. In general, the thrust components are available along the vehicle body axis. Since the wind-axis orientation angles and velocity components are used in Equations 8.1 through 8.6, we shall first transform the body-axis thrust vector components into their wind-axis components. This can be easily done using the transformation matrices in Equation 1.12 as follows:

$$\begin{pmatrix} T_x^W \\ T_y^W \\ T_z^W \end{pmatrix} = \begin{bmatrix} \cos\beta\cos\alpha & \sin\beta & \cos\beta\sin\alpha \\ -\sin\beta\cos\alpha & \cos\beta & -\sin\beta\sin\alpha \\ -\sin\alpha & 0 & \cos\alpha \end{bmatrix} \begin{pmatrix} T_x^B \\ T_y^B \\ T_z^B \end{pmatrix} \tag{8.10}$$

The T in Equations 8.1 through 8.6 is actually the component T_x^B. It must be replaced by all three body-axis components of thrust according to Equation 8.10. By doing this, Equations 8.1 through 8.6 may be updated as below with thrust components along all three body axes.

$$\dot{V} = \frac{1}{m}\left(\left[T_x^B \cos\alpha\cos\beta + T_y^B \sin\beta + T_z^B \sin\alpha\cos\beta\right] - D\cos\beta + Y\sin\beta\right)$$
$$- g\sin\gamma \tag{8.11}$$

$$\dot{\gamma} = \frac{1}{mV}\left[\begin{array}{l} \cos\mu\left(\left[T_x^B \sin\alpha - T_z^B \cos\alpha\right] + L\right) \\ -\sin\mu\left(-T_x^B \cos\alpha\sin\beta + T_y^B \cos\beta - T_z^B \sin\beta\sin\alpha \right) \\ +D\sin\beta + Y\cos\beta) \end{array}\right] - \frac{g}{V}\cos\gamma$$

$$\tag{8.12}$$

$$\dot{\chi} = \frac{1}{mV\cos\gamma}\left[\begin{array}{l} \sin\mu\left(\left[T_x^B \sin\alpha - T_z^B \cos\alpha\right] + L\right) \\ +\cos\mu\left(\begin{array}{l}\left[-T_x^B \cos\alpha\sin\beta + T_y^B \cos\beta - T_z^B \sin\beta\sin\alpha \right] \\ +D\sin\beta + Y\cos\beta \end{array}\right) \end{array}\right]$$

$$\tag{8.13}$$

$$\dot{\mu} = \frac{1}{\cos\beta}(p\cos\alpha + r\sin\alpha) + \frac{\tan\beta}{mV}\left(\left[T_x^B \sin\alpha - T_z^B \cos\alpha\right] + L\right)$$
$$- \frac{g}{V}\cos\mu\cos\gamma\tan\beta$$
$$+ \frac{\tan\gamma}{mV}\left[\begin{array}{l} \sin\mu\left(\left[T_x^B \sin\alpha - T_z^B \cos\alpha\right] + L\right) \\ +\cos\mu\left(\left[-T_x^B \cos\alpha\sin\beta + T_y^B \cos\beta - T_z^B \sin\beta\sin\alpha\right]\right) \\ +D\sin\beta + Y\cos\beta) \end{array}\right]$$

$$\tag{8.14}$$

$$\dot{\alpha} = q - (p\cos\alpha + r\sin\alpha)\tan\beta - \frac{1}{mV\cos\beta}\left(\left[T_x^B \sin\alpha - T_z^B \cos\alpha\right] + L\right)$$
$$+ \frac{g}{V\cos\beta}\cos\mu\cos\gamma \tag{8.15}$$

$$\dot{\beta} = (p\sin\alpha - r\cos\alpha) + \frac{1}{mV}\left(\begin{array}{l}\left[-T_x^B \cos\alpha\sin\beta + T_y^B \cos\beta - T_z^B \sin\beta\sin\alpha\right] \\ +D\sin\beta + Y\cos\beta \end{array}\right)$$
$$+ \frac{g}{V}\sin\mu\cos\gamma \tag{8.16}$$

Equations 8.11 through 8.16 constitute the general form of the *navigation* and *attitude* equations for the 10-thruster DACS vehicle. The *rate* equations are unchanged from Equations 8.7 through 8.9 since there is no influence of the thrust components on the equations per se. The moment contribution due to the thrusters will be accounted for within the rolling, pitching, and yawing moment terms in Equations 8.7 through 8.9.

8.2 MODELING THE THRUSTER FORCES AND MOMENTS

It is required to sum up the forces from the various thrusters to set the correct values of the thrust components in Equations 8.11 through 8.16 and to model the moments due to the thruster firings in Equations 8.7 through 8.9. For this purpose, an accurate picture of the DACS configuration and the layout of the thrusters is necessary. Figure 8.2 shows the typical location and orientation of the thrusters for a 10-thruster DACS vehicle. Of the four divert thrusters near the vehicle CG, two are oriented along the Y axis and two along the Z axis. Usually, one of the Y-thrusters and/or one of the Z-thrusters is fired at a time to alter the vehicle flight path in the lateral/normal sense. In Figure 8.2, the six attitude thrusters at the rear of the vehicle are arranged such that two lie along the Y axis (yaw) and four along the Z axis (pitch). A pair of pitch thrusters in the same sense

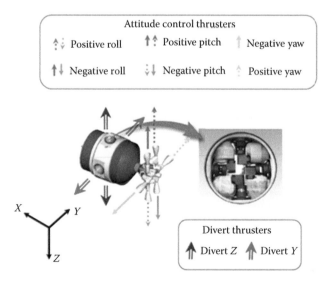

FIGURE 8.2 Force and moment components on a typical DACS thruster configuration.

are used to provide a pitching moment, whereas an opposite pair of pitch thrusters can give a rolling moment. Thus, at a time, up to four attitude thrusters may be fired simultaneously to provide the desired moments— one for yaw, a pair for roll, and an additional one for pitch. However, the details of which thrusters are to be fired, in which sequence, and to what extent depend on the thrust management system, which is beyond the scope of this discussion.

To evaluate the forces and moments due to the various thrusters, a numbering convention is desirable and the one followed here is shown in Figure 8.3. Each set of thrusters—the divert (DCS) and the attitude (ACS)—is numbered clockwise as shown in Figure 8.3. Thus, DCS1 and DCS3 are used for normal acceleration, whereas DCS2 and DCS4 give a lateral acceleration. Either the pair of ACS1/ACS2 or the pair ACS4/ACS5 can be used to pitch in the appropriate sense. ACS3 or ACS6 provides a yawing moment in either sense. Either of the pairs ACS1/ACS4 or ACS2/ACS5 may be used to achieve the desired rolling moment.

The net thrust along each of the body axes can be written by inspection of Figure 8.3:

$$T_x^B = 0$$
$$T_y^B = T_{DCS4} + T_{ACS6} - T_{DCS2} - T_{ACS3} \tag{8.17}$$
$$T_z^B = T_{DCS1} + T_{ACS1} + T_{ACS2} - T_{DCS3} - T_{ACS4} - T_{ACS5}$$

Note that the thrust component in the X direction is zero for this configuration.

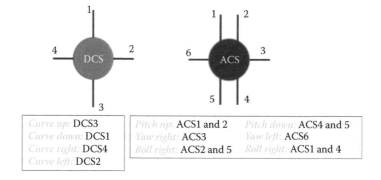

FIGURE 8.3 DACS thruster numbering convention and corresponding force/moment.

Likewise, the net moments due to the thrusters along the three axes may be modeled as

$$M_x^T = (T_{ACS5} + T_{ACS2} - T_{ACS1} - T_{ACS4}) \times RollArm_{ACS}$$
$$M_y^T = (T_{DCS3} - T_{DCS1}) \times PitchArm_{DCS} + (T_{ACS1} + T_{ACS2} - T_{ACS4} - T_{ACS5})$$
$$\times PitchArm_{ACS}$$
$$M_z^T = (T_{DCS4} - T_{DCS2}) \times YawArm_{DCS} + (T_{ACS3} - T_{ACS6}) \times YawArm_{ACS}$$

(8.18)

where the superscript "T" indicates that the moment is due to the thrusters as against the aerodynamic moment. Also, the various "arms" in Equation 8.18 are the distances from the point of application of thrust to the instantaneous CG location. The thrust due to any individual thruster is found by using the formula

$$T = C_f P_c A_t$$ (8.19)

where A_t is the thruster throat area, P_c is the combustor chamber pressure, and C_f is the thrust coefficient. Thus, the thrust of an individual thruster depends on the setting of its throat area, and the combustor chamber pressure, which is common to all the thrusters. A higher chamber pressure will give a larger thrust but will also burn out the propellant at a faster rate, reducing the time of powered flight. (The propellant mass burn rate is usually modeled as $\dot{r} = aP_c^n$.) One solution is to fire the thrusters in pulses, raising the pressure to a higher "fire level" during those intervals, and settling to a lower "hold level" pressure at other times. This is part of the internal pressure control (IPC) algorithm, which is beyond the present scope.

8.3 MODELING THE CHANGE IN CG AND MOMENTS OF INERTIA

Since the propellant constitutes a significant proportion of the mass of the vehicle, as the propellant is consumed, there are large changes in m that appear in Equations 8.11 through 8.16. Likewise, there are changes in the moments of inertia that appear in Equations 8.7 through 8.9. Additionally, the change in the propellant mass creates a shift of the vehicle CG, so the moment arms in Equations 8.17 and 8.18 vary in flight. Hence, it is important to model the effect of propellant mass consumption on the overall vehicle mass, CG, and moments of inertia.

The mass of the DACS vehicle is divided into the propellant mass and other mass that includes the structure, systems, and everything else (collectively called "hardware"). While the propellant mass is variable in flight, the hardware mass is fixed. The rate at which the propellant mass is consumed is

$$\dot{m}_g = \rho_P A_b \dot{r} \tag{8.20}$$

where ρ_p is the solid propellant density, A_b is the instantaneous burn area, and $\dot{r} = a P_c^n$ as mentioned earlier. Then the instantaneous CG of the DACS vehicle can be calculated as

$$CG_{DACS} = \frac{(m_{hardware} \times CG_{hardware}) + (m_{propellant} \times CG_{propellant})}{m_{hardware} + m_{propellant}} \tag{8.21}$$

where $CG_{hardware}$ is fixed and $CG_{propellant}$ changes depending on the manner in which the propellant burns (the burn surface and burn rate).

Likewise, for the moment of inertia calculation, the individual moment of inertia contributions from hardware and the propellant, respectively, are estimated. These are first found about their respective CG, and then shifted to the instantaneous CG by parallel axis theorem in case of the pitch and yaw axis moments of inertia. Thus, for the hardware components:

$$I_{xx}(hardware) = I_{xx(initial)}(hardware)$$
$$I_{yy}(hardware) = I_{yy(initial)}(hardware) + m_{hardware}(CG_{DACS} - CG_{hardware})^2$$
$$I_{zz}(hardware) = I_{zz(initial)}(hardware) + m_{hardware}(CG_{DACS} - CG_{hardware})^2$$

$$\tag{8.22}$$

where the subscript "*initial*" refers to the point just before the propellant starts burning and CG_{DACS} is the instantaneous DACS CG from Equation 8.21.

In case of the propellant, the moment of inertia calculation depends very much on the grain geometry, so it will vary from case to case. If we assume a fairly standard annular tube geometry with inner and outer radius, R_1 and R_2, respectively, and length h, then the moments of inertia about the three DACS body axes may be written as below, where the parallel axis theorem is again employed in case of the pitch and yaw moments of inertia.

$$I_{xx}(propellant) = \frac{m_{propellant}\left(R_1^2 + R_2^2\right)}{2}$$

$$I_{yy}(propellant) = \frac{1}{12}m_{propellant}\left\{3\left(R_1^2 + R_2^2\right)+h^2\right\}$$
$$+ m_{propellant}(CG_{DACS} - CG_{propellant})^2 \qquad (8.23)$$

$$I_{yy}(propellant) = \frac{1}{12}m_{propellant}\left\{3\left(R_1^2 + R_2^2\right)+h^2\right\}$$
$$+ m_{propellant}(CG_{DACS} - CG_{propellant})^2$$

The net moment of inertia about each axis is the sum of the moments of inertia due to hardware and the propellant.

8.4 MODELING THE AERODYNAMIC FORCES AND MOMENTS

For flight in the 30–50 km altitude range, aerodynamic drag alone is modeled as described below. For flight at higher altitudes, the aerodynamic forces and moments may be considered negligible.

The aerodynamic drag is modeled as a sum of three terms representing pressure drag, base drag, and friction drag, all at zero angle of attack, as follows:

$$C_D = C_{DP} + C_{DB} + (C_{D_f})_{Blunt} \qquad (8.24)$$

Each of the terms in Equation 8.24 is modeled as below.
Pressure drag:

$$C_{DP} = C\sin^N\theta + \frac{K}{\theta^A}\left(\frac{R_N}{R_B}\right)^B \qquad (8.25)$$

where θ is the nose cone half-angle, and R_N and R_B are the nose and base radius, respectively, of the DACS vehicle. The coefficients in Equation 8.25 are given by the following expressions, which are a function of the free stream Mach number M_∞.

$$C = 1.944 + 1.872M_\infty^{-1} - 17.0M_\infty^{-2} + 38.194M_\infty^{-3}$$
$$N = 1.931 + 0.8635M_\infty^{-1} - 8.063M_\infty^{-2} + 12.205M_\infty^{-3}$$
$$K = 11.433 + 34.96M_\infty^{-1} - 921.54M_\infty^{-2} + 2607.3M_\infty^{-3}$$
$$A = 0.5359 + 0.09964M_\infty^{-1} + 10.769M_\infty^{-2} - 104.21M_\infty^{-3} + 209.43M_\infty^{-4}$$
$$B = 3.296 + 2.997M_\infty^{-1} - 74.378M_\infty^{-2} + 154.67M_\infty^{-3}$$

$$(8.26)$$

Base drag:

$$C_{DB} = (1 + \sin\theta)(C_{DB})_{ref} \tag{8.27}$$

where θ is again the nose cone half-angle, and $(C_{DB})_{ref}$ is given by

$$(C_{DB})_{ref} = e^{\left(-1.706 - 0.33M_\infty + 0.00566M_\infty^2\right)} \tag{8.28}$$

Friction drag:

$$\frac{(C_{D_f})_{Blunt}}{(C_{D_f})_{Sharp}} = 1.0 - \left[(0.8 + 0.052M_\infty)\frac{R_N}{R_B}\right] \tag{8.29}$$

where R_N and R_B are again the nose and base radius, respectively, and $(C_{D_f})_{Sharp}$ is given by

$$(C_{D_f})_{Sharp} = \frac{0.0776}{(R_{\infty L})^{0.2}}\left(\frac{u_e}{u_\infty}\right)^{1.8}\left(\frac{P_e}{P_\infty}\right)^{0.8}\left(\frac{T_e}{T_\infty}\right)^{-0.58}\left(\frac{h^*}{h_e}\right)^{-0.58}\cot\theta \tag{8.30}$$

The various terms in Equation 8.30 are modeled as follows:

$$\left(\frac{u_e}{u_\infty}\right) = \left[1 - \frac{1.4}{M_\infty^2}(M_\infty\sin\theta)^{2.9}\right]^{0.5} \tag{8.31}$$

$$\left(\frac{P_e}{P_\infty}\right) = 1 + 2.8 M_\infty^2 \sin^2\theta\left[\frac{2.5 + 8M_\infty\sin\theta}{1 + 16M_\infty\sin\theta}\right] \tag{8.32}$$

$$\left(\frac{T_e}{T_\infty}\right) = 1 + 0.0966 M_\infty\sin\theta + 0.2276(M_\infty\sin\theta)^2 \tag{8.33}$$

$$\left(\frac{h^*}{h_e}\right) = 0.5 + 0.5\left(\frac{T_W}{T_\infty}\right)\left(\frac{T_e}{T_\infty}\right)^{-1} + 0.0374 M_e^2 \tag{8.34}$$

and $R_{\infty L}$ is the free stream Reynolds number based on vehicle length. Among the terms in Equation 8.34,

$$M_e = M_\infty\left(\frac{u_e}{u_\infty}\right)\left(\frac{T_e}{T_\infty}\right)^{-0.5} \tag{8.35}$$

$$\left(\frac{T_W}{T_\infty}\right) = 1 + (rm) \tag{8.36}$$

where $r = \mathrm{Pr}^{0.33}$ and $m = (\gamma_{air} - 1)M_\infty^2/2$, Pr being the Prandtl number, and γ_{air} the ratio of specific heats.

8.5 CONTROL AND GUIDANCE FRAMEWORK FOR 10-THRUSTER DACS

We shall follow a similar approach to that presented in Chapters 5 and 6 to obtain a flight control law for the 10-thruster DACS vehicle. However, there are two points of difference that we must note between the control of conventional aircraft presented earlier and the DACS flight control to be discussed below. The first is the lack of thrust along the forward (body X axis) direction; as a result, it is not possible to regulate the DACS velocity. Thus, among the navigation variables (V, γ, χ), there cannot be a control loop for the velocity, V. Second, the DACS has separate control actuators for maneuvering and for attitude control. This is in contrast to conventional airplanes where the same control actuator is used for both purposes. For example, airplanes use the elevator to regulate the characteristics of the short-period mode, which primarily involves the pitch dynamics. At the same time, to initiate a pull-up or push-down maneuver, the elevator must again be employed to pitch the airplane to a new angle of attack that creates the lift required to carry out the maneuver. Similarly, perturbations in the roll rate may be damped out by using the aileron in order to keep wings level. However, for a turn maneuver, the same aileron must be deployed to create a bank angle that initiates a turn. In contrast, for the DACS, maneuvers involving normal or lateral acceleration are obtained by use of the divert thrusters, and attitude control is provided separately by the attitude thrusters. Hence, the structure of the control law will be slightly different for the DACS as compared to the traditional one used for airplanes.

A guidance and control framework for the 10-thruster DACS vehicle is shown in Figure 8.4. The "DACS plant" has two kinds of control inputs— thrust commands, T_c, to be provided for by the divert thrusters, and moment commands, M_c, to be obtained by the use of the attitude thrusters. The input to the controller comes from a guidance law. The guidance law estimates the target's position and trajectory and calculates the maneuvers required to put the DACS on a collision course with the target. These maneuvers are usually in the form of normal and lateral acceleration commands. In response to these commands, the "flight path controller" evaluates the thrust to be commanded from the divert thrusters. The moment commanded from the attitude thrusters comes from a two-loop structure—an inner "rate controller" that evaluates the moments required in response to rate commands, and an outer "attitude controller" that evaluates the rate commands required to maintain the desired attitude of the DACS vehicle. The desired attitude is usually one where the vehicle nose is aligned with the velocity vector. In addition, the attitude thrusters

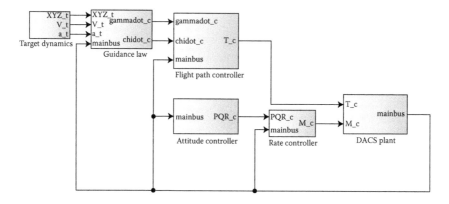

FIGURE 8.4   Framework for DACS vehicle guidance and control.

must also be fired every time the divert thrusters are fired in order to null any moment that may be created due to the divert thruster firing. This "null moment" command may be given as part of a feedforward loop not indicated in Figure 8.4.

The key point to note in Figure 8.4 is that the flight path controller can directly command the divert thrust to obtain the desired acceleration. This is in contrast to conventional airplanes where the lateral/normal acceleration command will first require a change in the attitude angles, which calls for angular rates, and hence a moment command. The other point, as mentioned earlier, is that the velocity cannot be commanded in the absence of a forward thrust force.

8.6 DACS CONTROL LAW

Each of the controller blocks in Figure 8.4 may be computed by any of the popular control design methods. In this text, following the lead in Chapter 5, we shall present the outline of an inversion-based control law. In the interests of simplicity, we shall assume the aerodynamic forces and moments to be negligible in deriving the control law, but not in simulating the "DACS plant." Also, the thrust component along the body X axis T_x is absent.

8.6.1 Navigation Equations and Flight Path Controller

Under the above assumptions, Equations 8.11 through 8.13 for the navigation variables appear as follows:

$$\dot{V} = \frac{1}{m}\left(\left[T_y^B \sin\beta + T_z^B \sin\alpha\cos\beta\right]\right) - g\sin\gamma \tag{8.37}$$

$$\dot{\gamma} = \frac{1}{mV}\left[\cos\mu\left(\left[-T_z^B \cos\alpha\right]\right) - \sin\mu\left(\left[T_y^B \cos\beta - T_z^B \sin\beta\sin\alpha\right]\right)\right] - \frac{g}{V}\cos\gamma \tag{8.38}$$

$$\dot{\chi} = \frac{1}{mV\cos\gamma}\left[\sin\mu\left(-T_z^B \cos\alpha\right) + \cos\mu\left(T_y^B \cos\beta - T_z^B \sin\beta\sin\alpha\right)\right] \tag{8.39}$$

Of these, the velocity V cannot be controlled, hence Equation 8.37 need not be considered for the flight path controller.

The input to the flight path controller block from the guidance law is actually the lateral and longitudinal velocity vector slew rates, $\dot{\chi}$ and $\dot{\gamma}$, respectively. So, define the vectors $z_1 = [\dot{\chi}, \dot{\gamma}]^T$ and $\mu_1 = [T_y^B, T_z^B]^T$. Then, Equations 8.38 and 8.39 can be cast in matrix form as below:

$$z_1 = F_1 + B_1 \mu_1 \tag{8.40}$$

with the matrices F_1 and B_1 defined as

$$F_1 = \begin{bmatrix} 0 \\ -\dfrac{g}{V} \cos\gamma \end{bmatrix} \tag{8.41}$$

$$B_1 = \frac{1}{mV} \begin{bmatrix} \cos\mu\cos\beta/\cos\gamma & -(\sin\mu\cos\alpha + \cos\mu\sin\beta\sin\alpha)/\cos\gamma \\ -\sin\mu\cos\beta & -\cos\mu\cos\alpha + \sin\mu\sin\beta\sin\alpha \end{bmatrix} \tag{8.42}$$

Because the vector z_1 is already defined in terms of the rates, Equation 8.40 has effectively become a static equation, not a "dynamic" one. That is, it is written directly in terms of the state vector z_1, not the time derivative of the state vector. Hence, the inversion of Equation 8.40 for the flight path controller is also a static one.

Note that z_1 in Equation 8.40 is the output that is obtained when the input μ_1 is applied. For the controller design, we must first replace z_1 with the commanded value z_{1c} that is received from the guidance law. Then, Equation 8.40 is written as

$$z_{1c} = F_1 + B_1 \mu_1 \tag{8.43}$$

On inverting, Equation 8.43 yields the flight path controller law:

$$\mu_1 = B_1^{-1}(z_{1c} - F_1) \tag{8.44}$$

where z_{1c} is now the input to the flight path controller block in Figure 8.4 and μ_1 is the output of the block, representing the thrust demanded along the two axes in order to meet the commanded z_{1c}.

8.6.2 Attitude Equations and Attitude Controller

On neglecting the aerodynamic forces and setting $T_x = 0$, Equations 8.14 through 8.16 for the DACS attitude dynamics appear as under:

$$\dot{\mu} = \frac{1}{\cos\beta}(p\cos\alpha + r\sin\alpha) + \frac{\tan\beta}{mV}\left(\left[-T_z^B\cos\alpha\right]\right) - \frac{g}{V}\cos\mu\cos\gamma\tan\beta$$

$$+ \frac{\tan\gamma}{mV}\left[\sin\mu\left(\left[-T_z^B\cos\alpha\right]\right) + \cos\mu\left(\left[T_y^B\cos\beta - T_z^B\sin\beta\sin\alpha\right]\right)\right]$$

(8.45)

$$\dot{\alpha} = q - (p\cos\alpha + r\sin\alpha)\tan\beta - \frac{1}{mV\cos\beta}\left(\left[-T_z^B\cos\alpha\right]\right)$$

$$+ \frac{g}{V\cos\beta}\cos\mu\cos\gamma$$

(8.46)

$$\dot{\beta} = (p\sin\alpha - r\cos\alpha) + \frac{1}{mV}\left(\left[T_y^B\cos\beta - T_z^B\sin\beta\sin\alpha\right]\right) + \frac{g}{V}\sin\mu\cos\gamma$$

(8.47)

The attitude dynamics regulates the attitude angles $z_2 = [\mu,\alpha,\beta]^T$ and it does so by manipulating the angular rates $\mu_2 = [p,q,r]^T$ in Equations 8.45 through 8.47. As before, we first cast Equations 8.45 through 8.47 in matrix form as below:

$$\dot{z}_2 = F_2 + A_2\mu_1 + B_2\mu_2$$

(8.48)

with the matrices F_2, A_2, and B_2 defined as

$$F_2 = \frac{g}{V}\cos\gamma\begin{bmatrix} -\cos\gamma\tan\beta \\ \cos\mu/\cos\beta \\ \sin\mu \end{bmatrix}$$

(8.49)

$$A_2 = \frac{1}{mV}\begin{bmatrix} \tan\gamma\cos\mu\cos\beta & -(\tan\beta\cos\alpha + \tan\gamma\sin\mu\cos\alpha \\ & +\tan\gamma\cos\mu\sin\alpha\sin\beta) \\ 0 & \cos\alpha/\cos\beta \\ \cos\beta & -\sin\beta\sin\alpha \end{bmatrix}$$

(8.50)

$$B_2 = \begin{bmatrix} \cos\alpha/\cos\beta & 0 & \sin\alpha/\cos\beta \\ -\cos\alpha\tan\beta & 1 & -\sin\alpha\tan\beta \\ \sin\alpha & 0 & -\cos\alpha \end{bmatrix} \qquad (8.51)$$

Note that μ_1 from the flight path controller also enters the right-hand side of Equation 8.48. However, here, it is a known parameter having already been solved in Equation 8.44.

As written, μ_2 in Equation 8.48 is the input and z_2 are the states. For the controller design, we need to command the values of z_2 (called z_{2c}) and then calculate the μ_2 required to meet that command. Typically, for the DACS vehicle, the commanded states are $z_{2c} = [0,0,0]^T$. The desired closed-loop dynamics $\dot{z}_{2d}$ is modeled as

$$\dot{z}_{2d} = -K_{z2}(z_2 - z_{2c}) \qquad (8.52)$$

where K_{z2} are the gains (also called "bandwidth" in the dynamic inversion literature), which are to be selected or tuned by the control designer. Using Equation 8.52 on the left-hand side of Equation 8.48, we can derive the inversion control law for the attitude loop as follows:

$$\mu_2 = B_2^{-1}(\dot{z}_{2d} - F_2 - A_2\mu_1) \qquad (8.53)$$

where μ_2 are the commanded values of the roll, pitch, and yaw rates required to achieve the desired values of the attitude angles.

8.6.3 Rate Equations and Rate Controller

The rate equations for the DACS vehicle are reproduced from Equations 8.7 through 8.9 below, where the rolling, pitching, and yawing moment terms on the right-hand sides are due to the thruster moments in Equation 8.18, the aerodynamic moments having been neglected.

$$\dot{p} = \frac{1}{I_{xx}} M_x^T \qquad (8.54)$$

$$\dot{q} = \frac{(I_{zz} - I_{xx})}{I_{yy}} pr + \frac{1}{I_{yy}} M_y^T \qquad (8.55)$$

$$\dot{r} = \frac{(I_{xx} - I_{yy})}{I_{zz}} pq + \frac{1}{I_{zz}} M_z^T \qquad (8.56)$$

For the rate dynamics, the states are $z_3 = [p,q,r]^T$ and the controls are $\mu_3 = \left[M_x^T, M_y^T, M_z^T \right]^T$. Casting Equations 8.54 through 8.56 in matrix form,

$$\dot{z}_3 = F_3 + B_3 \mu_3 \qquad (8.57)$$

where the matrices are as defined below:

$$F_3 = \begin{bmatrix} 0 \\ \dfrac{(I_{zz} - I_{xx})}{I_{yy}} pr \\ \dfrac{(I_{xx} - I_{yy})}{I_{zz}} pq \end{bmatrix} \qquad (8.58)$$

$$B_3 = \begin{bmatrix} \dfrac{1}{I_{xx}} & 0 & 0 \\ 0 & \dfrac{1}{I_{yy}} & 0 \\ 0 & 0 & \dfrac{1}{I_{zz}} \end{bmatrix} \qquad (8.59)$$

If the commanded states are $z_{3c} = [p_c, q_c, r_c]^T$, which are obtained from the output μ_2 of the attitude controller, the desired closed-loop dynamics $\dot{z}_{3d}$ may be modeled as

$$\dot{z}_{3d} = -K_{z3}(z_3 - z_{3c}) \qquad (8.60)$$

where K_{z3} is the gain matrix for the rate loop. Inserting Equation 8.60 as the left-hand side of Equation 8.57 yields the rate-loop inversion control law as

$$\mu_3 = B_3^{-1}(\dot{z}_{3d} - F_3) \qquad (8.61)$$

In this manner, the control laws for the three control blocks—flight path controller, attitude controller, and rate controller—in Figure 8.4 may be derived. Of course, the gain matrices K_{z2} and K_{z3} must be suitably selected. For the sample simulations presented later, these values are set at

$$K_{z2} = diag(7,9,9) \quad \text{and} \quad K_{z3} = diag(48,45,45) \tag{8.62}$$

8.6.4 Issue of Invertibility

Before we close this discussion on the inversion-based control law for the 10-thruster DACS vehicle, we need to briefly address a crucial question. It may be noticed that the computation of the inversion-based control laws requires inversion of the matrices B_1, B_2, and B_3. Invertibility can be assured under all conditions provided the determinant of each of these three matrices does not become zero under any possible combination of the state variables.

Homework Exercise: Derive expressions for the determinant of each of the three matrices B_1, B_2, and B_3.

It can be verified that the determinants of matrices B_2 and B_3 never become zero under any physical circumstance. However, in case of B_1, the condition that the determinant be zero may be written as

$$\frac{\cos\alpha\cos\beta}{\cos\gamma} = 0 \tag{8.62}$$

which is possible when either $\alpha = \pm 90$ deg or $\beta = \pm 90$ deg. While either condition is fairly unlikely to be encountered during a normal DACS flight, the possibility must be kept in mind.

8.6.5 The Question of Stability

Additionally, the question can arise about whether the closed-loop system obtained with the inversion-based control laws as derived here with the structure shown in Figure 8.4 is indeed stable. One way in which this can be checked is to carry out a numerical closed-loop stability analysis as demonstrated in Section 5.7. Another way out is to theoretically establish the stability of each of the three loops in Figure 8.4. This is usually done by using *Lyapunov's second method*, which was briefly mentioned toward the end of Section 3.2.1. Since the *second method* has not been described in this text, we shall not present the theoretical stability analysis of the inversion control loops here. However, the interested and informed reader is encouraged to carry out such an exercise.

Homework Exercise: Try to establish the stability of each of the three loops in the inversion-based control law derived above. One possible choice of the Lyapunov function is $V = (1/2)(z - z_c)^T(z - z_c)$, where z is the state vector in each of the three loops derived above and z_c is the commanded state vector. Thus, $(z - z_c)$ is the tracking error.

Sometimes, the theoretical stability analysis does not give a definite result—it does not clearly establish the system to be stable, yet it does not imply that it is unstable. In that case, one can either try an alternative Lyapunov function or modify the control law in such a way that the loop becomes provably stable.

Homework Exercise: In case of the rate-loop controller, try a modified version of the inversion-based control law, as below, where $\dot{z}_{3c}$ is the derivative of the commanded state vector:

$$\mu_3 = B_3^{-1}(\dot{z}_{3d} - \dot{z}_{3c} - F_3) \tag{8.63}$$

8.7 DACS GUIDANCE LAW

The input to the flight path controller in Figure 8.4 comes from a guidance law. Based on the relative kinematics between the DACS and the target, the guidance law estimates the rates $\dot{\chi}$ and $\dot{\gamma}$ needed in order to achieve a collision. Therefore, the DACS needs a seeker that can sense the target position and motion. During simulations, a target model is required and a seeker model with a filter is needed to provide realistic estimates of what might be sensed in flight. A simple target model may assume it to be traveling on a purely ballistic trajectory, the only acceleration being due to gravity. The guidance law is usually derived based only on the relative vehicle kinematics, ignoring any external effect other than gravity.

A sketch of a typical DACS–target engagement scenario is shown in Figure 8.5. The position and velocity of the DACS (KV) and target are marked and the line connecting their instantaneous positions is the line of sight (LOS). The rate at which the LOS reorients itself is called the LOS rate. Typically, the LOS rate and the closing velocity (relative velocity between the KV and target) are the key parameters in a guidance problem.

The scenario pictured in Figure 8.5 is a true three-dimensional engagement. Oftentimes, the guidance problem is broken down into two two-dimensional engagements, each commanding one of the two rates, $\dot{\chi}$

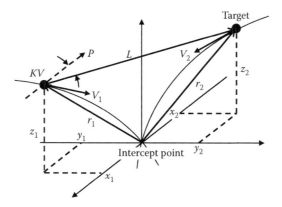

FIGURE 8.5 Typical intercept trajectories of DACS (KV) and target in three-dimensional space. (Adapted from L. LaCroix and S. Kurzius, *Peeling the Onion: A Heuristic Overview of Hit-to-Kill Missile Defense in the Twenty-First Century*, edited by Manijeh Razeghi, Gail J. Brown, *Proceedings of SPIE* 5732, Quantum Sensing and Nanophotonic Devices II, pp. 225–249, 2005.)

and $\dot{\gamma}$. We would prefer to treat the DACS guidance problem as a single three-dimensional engagement instead. Also, we shall formulate it in the nature of a regulation problem in control, which will permit us to use the same dynamic inversion technique that we have employed in this text to illustrate various control laws. The use of the dynamic inversion formulation to derive a generalized three-dimensional guidance law appears to be novel.

8.7.1 Derivation of Dynamic Inversion-Based Guidance Law

The key to obtaining a satisfactory guidance law is the LOS rate. Usually, a zero LOS rate indicates that the KV and target are on a collision course. Thus, regulating the LOS rate to zero is a good starting point for deriving a guidance law. With reference to Figure 8.5, the normalized LOS vector from the KV to the target can be represented in index notation as

$$\lambda_i = \frac{R_i}{r} \tag{8.64}$$

where $i = 1, 2, 3$ are the Cartesian components and r is the magnitude of the LOS vector defined by

$$r^2 = R_j R_j \tag{8.65}$$

Differentiating λ_i in Equation 8.64 with respect to time gives the LOS rate:

$$\dot{\lambda}_i = \frac{r\dot{R}_i - R_i\dot{r}}{r^2} \tag{8.66}$$

In order to regulate the LOS rate, we would like a differential equation in the LOS rate. This requires Equation 8.66 to be differentiated with respect to time once more, which gives the rate of change of LOS rate as

$$\ddot{\lambda}_i = \frac{r^2\ddot{R}_i - r\ddot{r}R_i - 2r\dot{r}\dot{R}_i + 2\dot{r}^2R_i}{r^3} \tag{8.67}$$

which after a good deal of algebraic manipulation yields

$$\ddot{\lambda}_i = F_{\lambda i} - B_{\lambda ij}a_{Tj} + B_{\lambda ij}a_{Dj} \tag{8.68}$$

where a_{Tj}, a_{Dj} are the components of the target and KV acceleration, respectively. The other terms on the right-hand side of Equation 8.69 are

$$F_{\lambda i} = \frac{-\dot{R}_j\dot{R}_jR_i - 2r\dot{r}\dot{R}_i + 3\dot{r}^2R_i}{r^3} = -\dot{\lambda}_j\dot{\lambda}_j\lambda_i - 2\frac{\dot{r}}{r}\dot{\lambda}_i \tag{8.69}$$

$$B_{\lambda ij} = \frac{R_iR_j - r^2\delta_{ij}}{r^3} = \frac{1}{r}(\lambda_i\lambda_j - \delta_{ij}) \tag{8.70}$$

Homework Exercise: Derive the LOS acceleration in the form Equation 8.68 with the terms in Equations 8.69 and 8.70.

Thus, the LOS acceleration has been written entirely in terms of the LOS range and LOS rate, and the accelerations of the target and the KV (DACS). Of the various terms in Equation 8.68, the LOS range and LOS rate may presumably be sensed by the KV seeker and the target acceleration may be either estimated or ignored, assuming it to be only due to gravity. The KV acceleration a_{Dj} is the unknown we would like to solve for. At first glance, this looks like a straightforward calculation, except that there is a little hitch. Solving for a_{Dj} in Equation 8.68 would require the coefficient matrix $B_{\lambda ij}$ to be inverted, but as it turns out, $B_{\lambda ij}$ is singular. This can be verified formally or understood from the fact that the three components λ_i of the LOS vector are not all independent as they must satisfy the relation $\lambda_j\lambda_j = 1$. So a third, independent condition is required before we can solve for a_{Dj}. This can be obtained by constraining the direction of the KV acceleration, either normal to the LOS vector or normal to the KV velocity vector. Such conditions are common even in traditional guidance schemes and adjectives such as "true" or "pure" are used to distinguish one constraint from another.

Homework Exercise: Review some of the traditional guidance laws and their implementation.

In the following, we shall choose the constraint that the KV acceleration be normal to its velocity vector. Thus, a third, independent condition in addition to Equation 8.68 is obtained as follows:

$$V_{Dj} \cdot a_{Dj} = 0 \tag{8.71}$$

Then, Equations 8.68 and 8.71 form the basis for deriving the guidance law.

8.7.1.1 In Terms of Azimuth and Elevation Angles

Since only two components of λ_i are independent, it is better to define two new independent variables in place of the three components of λ_i. Besides, the new variables can be physically more meaningful or easier to appreciate. So, two new independent variables are selected as the azimuth angle ξ and the elevation angle η, which are defined in terms of λ_i as follows:

$$\begin{aligned} \xi &= \tan^{-1} \frac{\lambda_2}{\lambda_1} \\ \eta &= \sin^{-1} \lambda_3 \end{aligned} \tag{8.72}$$

We shall now write Equation 8.68 in terms of variables ξ and η. This can be easily done by differentiating Equation 8.72 with respect to time and using Equations 8.66 through 8.70 as appropriate. First, differentiating Equation 8.72 gives

$$\begin{aligned} \dot{\xi} &= \frac{1}{1+\lambda_2^2/\lambda_1^2} \cdot \frac{\lambda_1\dot{\lambda}_2 - \lambda_2\dot{\lambda}_1}{\lambda_1^2} = \frac{\lambda_1\dot{\lambda}_2 - \lambda_2\dot{\lambda}_1}{\cos^2 \eta} \\ \dot{\eta} &= \frac{1}{\sqrt{1-\lambda_3^2}} \dot{\lambda}_3 = \frac{\dot{\lambda}_3}{\cos \eta} \end{aligned} \tag{8.73}$$

where $\lambda_1^2 + \lambda_2^2 + \lambda_3^2 = 1$ has been used.

Differentiating Equation 8.73 once again,

$$\ddot{\xi} = \frac{\lambda_1 \ddot{\lambda}_2 - \lambda_2 \ddot{\lambda}_1}{\cos^2 \eta} + 2 \tan \eta \dot{\eta} \dot{\xi}$$

$$\ddot{\eta} = \frac{\ddot{\lambda}_3}{\cos \eta} + \tan \eta \dot{\eta}^2 \qquad (8.74)$$

In matrix form, Equation 8.74 appears as

$$\begin{pmatrix} \ddot{\xi} \\ \ddot{\eta} \end{pmatrix} = F' + M' \ddot{\lambda}_i \qquad (8.75)$$

where

$$F' = \tan \eta \dot{\eta} \begin{bmatrix} 2\dot{\xi} \\ \dot{\eta} \end{bmatrix} \qquad (8.76)$$

$$M' = \frac{1}{\cos^2 \eta} \begin{bmatrix} -\lambda_2 & \lambda_1 & 0 \\ 0 & 0 & \cos \eta \end{bmatrix} \qquad (8.77)$$

and $\ddot{\lambda}_i$ is available from Equation 8.68. Substituting for $\ddot{\lambda}_i$, we have

$$\begin{pmatrix} \ddot{\xi} \\ \ddot{\eta} \end{pmatrix} = F' + M'(F_{\lambda i} - B_{\lambda ij} a_{Tj} + B_{\lambda ij} a_{Dj}) \qquad (8.78)$$

where a_{Dj} now enters the picture. To this, add Equation 8.71 as an additional condition:

$$V_{Dj} \cdot a_{Dj} = 0 \qquad (8.71)$$

and we have three independent equations that relate the KV acceleration to the LOS acceleration components. Collecting Equations 8.71 and 8.78 into a single matrix equation and dropping the "i, j" subscripts, we get

$$
\begin{pmatrix} \ddot{\xi} \\ \ddot{\eta} \\ 0 \end{pmatrix} = \begin{bmatrix} F' + M'(F_\lambda - B_\lambda a_T) \\ 0 \end{bmatrix} + \begin{bmatrix} M'B_\lambda \\ V_D \end{bmatrix} a_D \tag{8.79}
$$

Equation 8.79 is the "forward" equation—given the KV acceleration a_D satisfying the constraint $V_D \cdot a_D = 0$, it may be used to calculate the LOS accelerations $\ddot{\xi}$ and $\ddot{\eta}$. To derive the guidance law, Equation 8.79 can be considered as a dynamical system for a control regulation problem. The aim is to regulate the LOS rates, $\dot{\xi}$ and $\dot{\eta}$, to zero. That is, the commanded LOS rates, $\dot{\xi}_c$ and $\dot{\eta}_c$, are set to zero and the desired LOS accelerations are modeled as

$$
\begin{pmatrix} \ddot{\xi}_d \\ \ddot{\eta}_d \end{pmatrix} = \begin{bmatrix} -K_{\dot{\xi}} & 0 \\ 0 & -K_{\dot{\eta}} \end{bmatrix} \begin{pmatrix} \dot{\xi} - \dot{\xi}_c \\ \dot{\eta} - \dot{\eta}_c \end{pmatrix} \tag{8.80}
$$

where $K_{\dot{\xi}}$ and $K_{\dot{\eta}}$ are the positive gain elements (also called the band-widths). In the usual manner of dynamic inversion, Equation 8.80 is sub-stituted on the left-hand side of Equation 8.79, giving

$$
\begin{pmatrix} -K_{\dot{\xi}}(\dot{\xi} - \dot{\xi}_c) \\ -K_{\dot{\eta}}(\dot{\eta} - \dot{\eta}_c) \\ 0 \end{pmatrix} = \begin{bmatrix} F' + M'(F_\lambda - B_\lambda a_T) \\ 0 \end{bmatrix} + \begin{bmatrix} M'B_\lambda \\ V_D \end{bmatrix} a_D \tag{8.81}
$$

Inverting Equation 8.81 yields the acceleration that must be com-manded from the KV:

$$
a_D = \begin{bmatrix} M'B_\lambda \\ V_D \end{bmatrix}^{-1} \left[\begin{pmatrix} -K_{\dot{\xi}}(\dot{\xi} - \dot{\xi}_c) \\ -K_{\dot{\eta}}(\dot{\eta} - \dot{\eta}_c) \\ 0 \end{pmatrix} - \begin{bmatrix} F' + M'(F_\lambda - B_\lambda a_T) \\ 0 \end{bmatrix} \right] \tag{8.82}
$$

which is the guidance law! If the accelerations commanded by Equation 8.82 are indeed achieved, then the DACS (KV) will be able to push the LOS rate to zero, which should put it on a collision course with the target.

8.7.1.2 A Question of Invertibility

As with all inversion-based laws, the question of whether the coefficient matrix in Equation 8.82 is invertible must be addressed. A little calculation reveals that the condition for the determinant of the 3×3 matrix

$$\begin{bmatrix} M'B_\lambda \\ V_D \end{bmatrix}^{-1}$$

in Equation 8.82 to be singular works out to

$$V_{D1}\lambda_1 + V_{D2}\lambda_2 + V_{D3}\lambda_3 = 0 \tag{8.83}$$

which is the same as $V_D \cdot \lambda = 0$, where V_D is the DACS velocity vector and λ is the LOS range vector. In other words, the matrix in question is not invertible and hence the guidance law in Equation 8.82 is not admissible when the DACS-to-target line of sight is perpendicular to the DACS velocity vector. Since this is a highly unlikely situation, and if this were to indeed happen, then the target would most probably be outside the field of view of the DACS seeker, making the entire exercise a nonstarter; we can safely assume that the guidance law is acceptable under the usual operating conditions.

Homework Exercise: Derive the condition in Equation 8.83.

8.7.1.3 Decoding the Inversion-Based Guidance Law

At first glance, the guidance law in Equation 8.82 seems quite complex, so it will be worthwhile to break it down to a simpler form and also place it in the context of traditional guidance laws, which are usually of the proportional-navigation (P-N) form. To do this, let us assume a two-dimensional (planar) engagement scenario. This is easily done by setting the elevation angle to zero, that is, $\eta = 0$. Then, with regard to Figure 8.5, the engagement takes place entirely in the X–Y plane. When the guidance law in Equation 8.82 is restricted to the X–Y plane, it takes the following form:

$$a_D = (K_{\dot\xi} r + 2V_c)\dot\xi \begin{bmatrix} -\sin\xi \\ \cos\xi \end{bmatrix} \tag{8.84}$$

where it is assumed that the commanded LOS azimuth angle rate is zero, that is, $\dot\xi_c = 0$, and the target acceleration is ignored, that is, $a_T = 0$ is assumed in Equation 8.82. In Equation 8.84, a_D is the vector of components of the KV acceleration in the X–Y plane, r is the LOS range, and V_c is the closing velocity defined as the negative of the LOS rate.

Homework Exercise: Derive the planar form of the guidance law in Equation 8.84.

The magnitude of the acceleration in Equation 8.84 can be cast in the following form:

$$\left|a_{Dj}\right| = \underbrace{\left(K_{\dot{\xi}} \frac{r}{V_c} + 2 \right) V_c}_{\dot{N}_c} \dot{\xi}$$

(8.85)

where the factor r/V_c is also referred to "time-to-go" or t_{go}. As written in Equation 8.85, the KV acceleration is proportional to the closing velocity V_c and the LOS rate $\dot{\xi}$. This is precisely of the form of a P-N guidance law with the coefficient (the bracketed term in Equation 8.85) called the "navigation ratio" or gain. Thus, the inversion-based guidance law is inherently of the form of a P-N law with variable navigation ratio, N_c. In fact, as seen from Equation 8.85, the navigation ratio is a function of the LOS range or time-to-go. As the LOS range or time-to-go tends to zero, N_c tends to a fixed value of 2. At larger distances from the target, the guidance law in Equation 8.85 commands a larger acceleration. That is, the acceleration demand is higher in the initial stages of the engagement and the guidance law tends to a P-N law with a gain of 2 in the end stages. In other words, the guidance law in Equation 8.85 can also be referred to as a "variable-gain" P-N law.

8.8 SIMULATION OF 10-THRUSTER DACS FLIGHT WITH GUIDANCE AND CONTROL

In this section, we shall apply the guidance and control law derived above to simulate the closed-loop dynamics of an example DACS vehicle. Sample data for a DACS vehicle and the engagement geometry assumed for the simulation is as listed in Table 8.1. The target is assumed to be flying a ballistic trajectory with the only acceleration being that due to gravity. With the target velocity being fixed at 3 km/s, the relative velocity between the target and DACS is very close to 5 km/s at the closing stages. The initial separation between the target and DACS is taken to be nearly 50 km.

TABLE 8.1   Parameter Values for DACS–Target Engagement Simulation

Parameter	Value
DACS mass	$m = 110$ kg
DACS inertia	$I_{xx} = 1.0346$ kg-m², $I_{yy} = 2.3605$ kg-m², $I_{zz} = 2.3605$ kg-m²
Initial velocity and orientation	$V_D = 2$ km/s, $\chi = 0, \gamma = 0$
Initial attitude angles	$\mu = 0, \alpha = 0, \beta = 0$
Initial rates	$p = 0, q = 0, r = 0$
Target velocity	$V_T = 3$ km/s
Initial DACS–target separation	$\Delta x = 50$ km, $\Delta y = 2$ km, $\Delta z = 0$
Controller gains	$K_{z2} = diag(7,9,9)$ and $K_{z3} = diag(48,45,45)$
Guidance law gains	$K_\xi = 1$ and $K_\eta = 1$
DACS acceleration limit	$5g$ (each axis)

Thus, the DACS flight time is approximately 10 s. The controller and guidance law gains are taken as specified in Table 8.1. Additionally, a limit on the DACS acceleration permissible is set as $5g$ along each axis.

Since the engagement scenario is a three-dimensional one, the guidance law in the form as in Equation 8.82 is used. However, even though the target is assumed to be accelerating under the influence of gravity, the guidance law is run assuming zero target acceleration.

Figure 8.6 shows the trajectories followed by the target and the KV DACS during this engagement. Along the X axis, the target and DACS approach each other. For the Y axis with initial separation of 2 km, the DACS needs to maneuver to put itself on collision course with the target. In the vertical direction, since DACS does not budget for the acceleration of the target due to gravity, there is a lag initially. However, the guidance law is able to compensate for it and DACS hits the target at the end around $t = 10$ s despite the target acceleration a_T being set to zero in Equation 8.82.

Figure 8.7 shows the LOS rates in terms of the rate of change of the azimuth and elevation angles. Since the DACS and target are at the same altitude, there is almost no error in the elevation; hence, the elevation rate required is nearly zero at all times. The slightly nonzero LOS rate in elevation in Figure 8.7 is due to the lag in H in Figure 8.6, which we have seen is because of the acceleration due to gravity. On the other hand, there is a large error in the azimuth as seen by the maneuver in Y in Figure 8.6. Therefore, the guidance law commands an azimuth rate and $\dot{\xi}$ is driven to zero by the closing stages. The build-up in LOS rate after $t = 10$ s must be

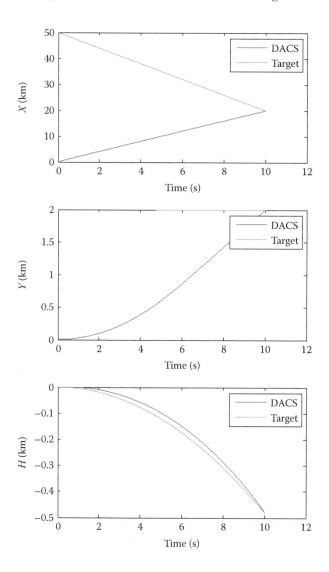

FIGURE 8.6   Simulation of closed-loop DACS vehicle engaging target trajectories.

ignored since the simulation has continued after the target and DACS have crossed each other.

Notice that $\dot{\xi}$ reduces slowly in the initial stages and falls more rapidly in the intermediate stage. The reason for this will become apparent when we examine the DACS body-axis accelerations in Figure 8.8. Following Equation 8.82 (or more clearly from Equation 8.85), the demanded DACS body-axis accelerations are proportional to the LOS rate. However, because

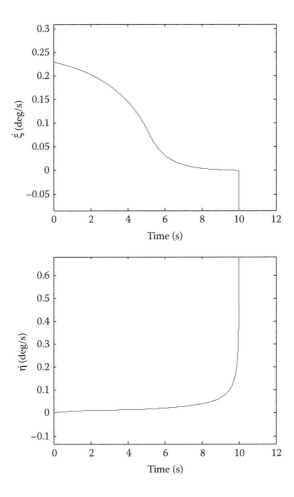

FIGURE 8.7 Simulation of closed-loop DACS vehicle engaging a target—LOS rate.

of the 5g limit imposed on the available acceleration, the demanded value may not be obtained. This is clearly seen in case of the body Y-acceleration in Figure 8.8. In the initial stages, the acceleration demanded by the guidance law is way too large and only the limited acceleration of 5g is available. Thus, the LOS rate is regulated, but not as rapidly as calculated by the guidance law.

The body Z-acceleration in Figure 8.8 naturally settles down to a value near 1g in response to the target dynamics in the presence of the gravity field. It must be noted that the 1g acceleration is not because it has been explicitly entered into the guidance law; it is based on the motion of the

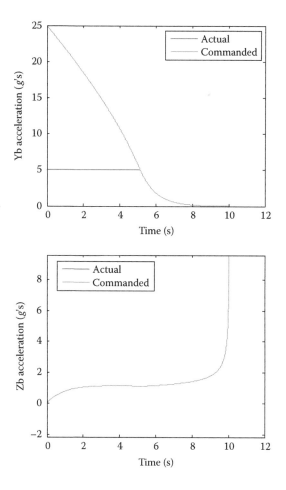

FIGURE 8.8 Simulation of closed-loop DACS vehicle engaging a target DACS body-axis acceleration.

target as estimated by the DACS seeker. Once again, the build-up in acceleration at $t = 10$ s is to be ignored.

This example of the flight dynamics, control, and guidance of the 10-thruster DACS vehicle is used as a capstone example of the various concepts and practices presented in the course of this text. The use of the dynamic inversion controller and a novel inversion-based guidance law is in keeping with our decision in Chapter 5 to use dynamic inversion as a convenient concept for control example in this text. In practice, any one of the several popular control design methods may be used and typically one would expect to see one of the many variants of the popular P-N guidance law to be used. However, since our emphasis in this text has been on the flight dynamics, and control and guidance are somewhat auxiliary topics in this context, rather than pick one or the other control design method, we have found it simpler and convenient to use the dynamic inversion framework. Readers are encouraged to apply their favorite control and guidance algorithms on this problem and indeed on any other that they may choose to explore. This book has hopefully provided the concepts and tools to explore a wide range of flight dynamics systems in operation and, no doubt, newer ones that will emerge from the imagination of the next generation of scientists and engineers.

REFERENCE

1. L. LaCroix and S. Kurzius, *Peeling the Onion: A Heuristic Overview of Hit-to-Kill Missile Defense in the Twenty-First Century*, edited by Manijeh Razeghi, Gail J. Brown, *Proceedings of SPIE*, 5732 Quantum Sensing and Nanophotonic Devices II, pp. 225–249, 2005. doi: 10.1117/12.583369.

Index

A